T0353554

The Sesiidae of Europe

Zdeněk Laštůvka & Aleš Laštůvka

The Sesiidae of Europe

Zdeněk Laštůvka & Aleš Laštůvka

Apollo Books
Stenstrup
2001

Text formatting, colour plates and artwork: Zdeněk Laštůvka.

Printed by: Litotryk Svendborg A/S.

Published by:

Apollo Books Aps.
Kirkeby Sand 19
DK-5774 Stenstrup
Denmark
apollobooks@vip.cybercity.dk
www.apollobooks.com

ISBN 87-88757-52-8

Front cover: *Synanthedon formicaeformis* (Esper).

Contents

Abstract

A comprehensive synthesis of information on 107 known European species of the Sesiidae. Short historical outline, brief information on the morphology, biology, phylogeny and distribution, collecting and rearing methods, list of European taxa and their hostplants in introductory chapters. Keys to European genera and species are provided. Diagnostic morphological characters, bionomics and distribution of each species are concisely described. Male and female genitalia, adult in colour, other diagnostic characters and distribution maps are illustrated.

The following new synonyms were established in the first edition: *Aegeria gaderensis* Králíček & Povolný, 1977 of *Synanthedon cephiformis* (Ochsenheimer, 1808); *Bembecia baumgartneri* Špatenka, 1992 of *B. himmighoffeni* (Staudinger, 1866); *Chamaesphecia rondouana* Le Cerf, 1922 of *C. mysiniformis* (Boisduval, 1840); *Sesia odyneriformis* Herrich-Schäffer, 1846 of *Chamaesphecia masariformis* (Ochsenheimer, 1808); *Sesia foeniformis* Herrich-Schäffer, 1846 and *Sesia oryssiformis* Herrich-Schäffer, 1846 of *Chamaesphecia anthraciformis* (Rambur, 1832) and *Chamaesphecia lastuvkai* Špatenka, 1987 of *C. empiformis* (Esper, 1783).

The following taxa (former species) are given as subspecies in this edition: *Synanthedon typhiaeformis* (Borkhausen, 1789) and *S. cruentata* (Mann, 1859) of *S. myopaeformis* (Borkhausen, 1789); *Synanthedon serica* (Alphéraky, 1882) of *S. formicaeformis* (Esper, 1783); *Bembecia psoraleae* Bartsch & Bettag, 1997 of *Bembecia albanensis* (Rebel, 1918); *Pyropteron sicula* Le Cerf, 1922 of *P. chrysidiformis* (Esper, 1782) and *Chamaesphecia colpiformis* (Staudinger, 1856) of *C. doleriformis* (Herrich-Schäffer, 1846).

The following new synonyms have been established in this edition: *Synansphecia* Căpuşe, 1973, syn. n. of *Pyropteron* Newman, 1832; *Synanthedon herzi* Špatenka & Gorbunov, 1992, syn. n. of *S. formicaeformis* (Esper, 1783) and *Bembecia daghestanica* Gorbunov, 1991, syn. n. of *B. puella* Laštůvka, 1989.

The following new combinations are introduced: *Pyropteron triannuliformis* (Freyer, 1842) comb. n.; *P. meriaeformis* (Boisduval, 1840) comb. n.; *P. hispanica* (Kallies, 1999) comb. n.; *P. koschwitzi* (Špatenka, 1992) comb. n.; *P. muscaeformis* (Esper, 1783) comb. n.; *P. kautzi* (Reisser, 1930) comb. n.; *P. aistleitneri* (Špatenka, 1992) comb. n.; *P. leucomelaena* (Zeller, 1847) comb. n.; *P. affinis* (Staudinger, 1856) comb. n.; *P. umbrifera* (Staudinger, 1870) comb. n.; *P. cirgisa* (Bartel, 1912) comb. n. and *P. mannii* (Lederer, 1853) comb. n.

Introduction

Preface

On a worldwide scale, the European clearwing moths are rather well-known. The family Sesiidae has recently become the object of extensive studies, and the amount of information obtained on this family is continuously increasing, largely through the use of synthetic sex pheromones, which facilitate their collection. The "Illustrated Key to European Sesiidae" was sold out within several months, since when ten new sesiid taxa (*Paranthrene insolita hispanica, Synanthedon stomoxiformis riefenstahli, S. perigordensis, S. geranii, Bembecia volgensis, B. abromeiti, B. blanka, B. psoraleae, B. pavicevici dobrovskyi* and *Pyropteron hispanica*) have been described and four species (*Tinthia hoplisiformis, Bembecia priesneri, B. flavida* and *Chamaesphecia maurusia*) have been recorded for the first time from Europe. Many additional faunistic and biological data have been obtained. Kallies (1999) proposed several changes in the taxonomy of *Synansphecia* as a result of his revision of this taxon. A comprehensive monograph of the Palaearctic Sesiidae has been published by Špatenka et al. (1999). Karsholt & Razowski (1996) compiled a checklist of European Lepidoptera, but incompetent corrections of faunistic data were done in the Sesiidae without the knowledge of the authors responsible (Laštůvka & Špatenka).

This, thoroughly revised second edition contains all updated information on this family, obtained both from the authors' own activities and from those of other persons, either published in part or based on personal communications. Some of the authors' continuous studies have led to taxonomic changes in a number of cases. The genera *Pyropteron* and *Synansphecia* have been found to be congeneric, and the species *Bembecia puella* and *B. daghestanica* conspecific. Further taxa previously considered to be separate species have been relegated as subspecies. Such cases include pairs of taxa which show no differences in the morphology of their genitalia, similar responses to sexual pheromones, identical bionomics, and absence of sympatry. The larger format of this edition has made it possible to enlarge some parts and to include important synonyms and other information.

Short historical outline

Descriptions of many European clearwing moth species can be found in the works of "ancient" authors, dating back to the late 18th and early 19th centuries. A separate family Sesiidae (or rather Sesiariae) was introduced for the first time by Boisduval (1828) in his "Index", and shortly thereafter also by Stephens (1828) under the name of Aegeriidae. The European species were comprehensively treated for the first time by Laspeyres (1801), later by e.g. Herrich-Schäffer (1846-1852) and Staudinger (1856). Variously extensive parts treating species of the family Sesiidae are found in the works of numerous other lepidopterists.

At the beginning of the 20th century and in the period between the two World Wars, important contributions to the increasing knowledge of the European species were made by M. Bartel, F. Le Cerf, Ch. Oberthür, R. Püngeler, and also by H. Kautz, H. Rebel, K. Predota, K. Schawerda, L. Schwingenschuss, B. Zukowsky, and others. The present classification of the family is based on Bartel´s Palaearctic Sesiidae in "Seitz" (Bartel, 1912). Important information, especially more or less complete synonymy, is contained in the catalogue by Dalla-Torre & Strand (1925). The second part of the 20th century brought monographs of the clearwing moths of certain countries or territories (Schwarz, 1953; Popescu-Gorj et al., 1958; Schnaider et al., 1961, etc.). In the 1950s to 1970s, the knowledge of European Sesiidae was enriched especially by I. Căpuşe, M. Fibiger, M. Králíček, C. M. Naumann, E. V. Niculescu, A. Popescu-Gorj, the members of the Schnaider family, R. Schwarz etc. The monograph of Scandinavian and Danish Sesiidae (Fibiger & Kristensen, 1974) was conceived in a modern way, based on the principal results of Naumann´s phylogenetic analysis of Holarctic taxa (Naumann, 1971).

During the past 20 years, numerous papers have been published pertaining to the taxonomy, bionomics and distribution of European clearwing moths, especially by K. Špatenka, O. Gorbunov, I. Toševski, C. Prola, F. Pühringer and a whole "team" of German lepidopterists. Laštůvka (1990f) presented an updated catalogue of the European species.

In studies of the individual species or faunae inhabiting certain territories correct determination is essential. Unfortunately, many (particularly earlier published) data are based on misidentifications, and also collection materials are often misidentified or remain unidentified. To a considerable extent, this situation is due to the fact that no comprehensive key for the determination of European species has been available since the beginning of the 20th century.

The present work is an attempt at filling this gap. It contains a key to species, accompanied by necessary illustrations, brief notes on bionomics, and maps showing the presently known distribution of European Sesiidae.

Determination

Depending on species variability and on availability of suitable distinguishing characters, individual taxa can be determined with the use of the keys with varying reliability, but the identification of some species requires considerable experience. Due to their mimicry, the external morphology of closely related species differs to varying degrees; on the other hand, distantly related species may be very similar to one another. In such similar species, the correct determination can be checked by studying their genitalia. In a few isolated cases, even the combination of external and genitalia characters may not lead to a satisfactory result, and the only unequivocal criterion may be, for example, knowledge of the hostplant of the species in question. In a large majority of species (more than 90 %), however, determination based solely on external characters is fairly easy and reliable.

The most important characters are given for each species. The genitalia are illustrated and therefore we have refrained from describing them, except for pointing out less conspicuous characters discriminating between closely related species.

Comments on the illustrations

In most species the male and female genitalia (usually the right valve, the tegumen-uncus complex, the saccus and, if necessary, the gnathos from below, the 8th female abdominal segment, the antrum, the ductus and the corpus bursae) and possibly other characters important for species identification (antenna, thorax, forewing, the external transparent area, discal spot, dorsal or ventral abdominal pattern, etc.) are figured.

The individual genitalia structures may show different degrees of variation and, moreover, various deformations may arise during dissection of genitalia. Each figure is based on an actual specimen and thus minor differences in the figures of closely related species do not reflect the true species-specific differences (if no special comments follow).

Acknowledgements

Gathering up all available data on the bionomics and especially the distribution of the individual species, and even compiling this manual would have been impossible without supplementary information and help provided by our numerous colleagues, to whom our warm thanks are due for cooperation.

We are particularly obliged to Mr. K. Špatenka (Prague) for numerous additions, comments and submission of necessary material. Valuable help was also provided by Messrs O. Gorbunov (Moscow), I. Toševski (Belgrade), M. Bąkowski (Poznań), R. Bläsius (Eppelheim), M. Fibiger (Sorø), T. & W. Garrevoet (Antwerpen), A. Kallies (Schwerin), F. Pühringer (Scharnstein) and C. Prola (Rome).

Additional information and assistance of various kinds were offered by the late B. Baker (Caversham, Berks), D. Bartsch (Stuttgart), D. Baumgarten (Hamburg), E. Bertaccini (Roncadello), E. Bettag (Dudenhofen), E. Blum (Neustadt/Wstr.), the late O. Brodský (Prague), J. Buszko (Toruń), F. Chládek (Brno), J. Cungs (Dudelange), E. Drouet (Mont Saint Aignan), C. Dutreix (Givry), G. Embacher (Salzburg), G. Fiumi (Forli), J. Gelbrecht (Berlin), the late D. Hamborg (Feldbach/Kassel), A. Hofmann (Linkenheim-Hochstetten), O. Karsholt (Copenhagen), Z. Kolev (Smolyan), U. Koschwitz (Eppenbrunn), M. Králíček (Kyjov), F. Krampl (Prague), the late Ph. Kristal (Bürstadt), M. Lödl (Vienna), G. Luquet (Paris), H. Malicky (Lunz), J. Marek (Brno), M. Meyer (Luxemburg), C. M. Naumann (Bonn), P. Parenzan (Bari), M. Petersen (Pfungstadt), the late E. Priesner (Munich), H. G. Riefenstahl (Hamburg), L. Ronkay (Budapest), N. Ryrholm (Uppsala), V. Sarto (Barcelona), H. Šefrová (Brno), T. Sobczyk (Hoyerswerda), D. Stadie (Eisleben), C. Taymans (Brussels), A. Vives (Madrid), H. Wegner (Lüneburg), S. Whitebread (Magden) and others.

We thank D. Povolný (Brno) for valuable comments on the manuscript of this work, and R. Obrtel (Brno) for translation of its essential parts into English. We are also grateful to Mr. M. Corley (Faringdon) for a linguistic check of the manuscript.

Abbreviations of museums, institutions and private collections

BMNH The Natural History Museum, London
IPPB Institute of Plant Protection, Beograd
LNK Staatliches Museum für Naturkunde, Karlsruhe
LNM Westfälisches Landesmuseum für Naturkunde, Münster
MGAB Muzeul National de Istorie Naturala „Grigore Antipa“, Bucuresti
MMB Moravské zemské muzeum, Brno
MNHP Muséum National d'Histoire Naturelle, Paris
NHMV Naturhistorisches Museum, Wien
NMP Národní muzeum, Praha
RMS Naturhistoriska Riksmuseet, Stockholm
SMNS Staatliches Museum für Naturkunde, Stuttgart
SMW Museum Wiesbaden, Wiesbaden
TLM Tiroler Landesmuseum Ferdinandeum
TMB Természettudományi Múzeum, Budapest
ZIMH Zoologisches Museum der Universität, Hamburg
ZISP Zoological Institute, Russian Acad. of Sciences, St. Petersburg
ZMHB Museum für Naturkunde, Humboldt-Universität, Berlin
ZMUK Zoological Museum, Kiev University, Kiev
ZSBS Zoologische Staatssammlung, München
AK Axel Kallies, Schwerin
IT Ivo Toševski, Beograd
KS Karel Špatenka, Praha
MK Milan Králíček, Kyjov
OG Oleg Gorbunov, Moskva
RK Radovan Kranjčev, Koprivnica
TG Theo Garrevoet, Antwerpen
TW Thomas Witt, Munich
ZL Zdeněk Laštůvka, Brno

Other abbreviations used in the text

A north, -ern
S south, -ern
W west, -ern
E east, -ern

The Morphology of the Sesiidae

Family characters

- characteristic "hymenopteriform" mimicry; variously marked wing transparency;
- forewing elongate, very narrow in the basal half;
- specific linking mechanism of wings (forewing and hindwing are linked by means of their recurved margins);
- median stem in the venation of forewing completely reduced; median stem of hindwing shifted markedly to fore edge of wing;
- abdomen often with yellow, red or white rings or margins;
- endophagous, non-pigmented larvae with slightly enlarged thoracic segments.

Adult (Text-fig. 1)

Wingspan of European species 8-48 mm.

HEAD.– Proboscis present or reduced; labial palpus smooth to long tufted; antenna filiform, setiform or clavate, in male usually ciliate or bipectinate.

THORAX.– Patagial scales collar-like, of various colours; tegula large, often of various colours; occasional scapular spot on base of forewing (Fig. 93a); often characteristic coloration or spots laterally below wings.

WINGS.– Forewing entirely transparent, or opaque, or with separate transparent areas (discal = anterior, longitudinal = posterior and external); usually 5 radial, 3 medial and 2 cubital veins present; some of them occasionally partly or completely fused; hindwing mostly transparent, with 3 median, 2 cubital and 4 anal veins.

LEGS.– Thin and relatively long, often with specific coloration particularly on hind tibia and/or on fore coxa.

ABDOMEN.– Often elongate with characteristic anal tuft and coloured margins or rings on some segments.

MALE GENITALIA (Tex-figs 2, 3).– Valve with simple, bi– or multifurcate hairs, occasionally with crista sacculi; tegumen and uncus separate, or fused to form simple complex with gnathos; gnathos with crista medialis and often with lateral wings; occasionally hair-covered scopula on uncus.

FEMALE GENITALIA (Text-fig. 3).– Ostium bursae situated on the 8th segment, or between the 7th and 8th segments; antrum tubular, usually sclerotised; lamella antevaginalis and lamella postvaginalis membranous, partly or entirely sclerotised, occasionally specialised.

Preadult stages (Text-fig. 4)

EGG.– Ellipsoidal, 0.4-1.1 mm long, light brown to black, with a structure of polygons and aeropyles of various size; the surface occasionally with small setae or spinules, particularly on the slightly flattened micropylar end; micropyle with a single or double rosette.

LARVA.– Nonpigmented, usually creamy white with a semiprognathous brown head; ocelli I-IV in a trapezium, V and VI separate, more ventral; prothoracic shield usually more sclerotised and darker; 3 lateral setae on prothorax, L2 and L1 on abdomen on a common pinaculum, L2 more dorsal than L1; crochets on prolegs in two transverse rows; the last abdominal segment sometimes with one or two small protuberances or spines dorsally.

PUPA.– Brown, frons with a pointed or chisel-shaped process; abdominal segments (2)3-6 in female and (2)3-7 in male with two rows of spines.

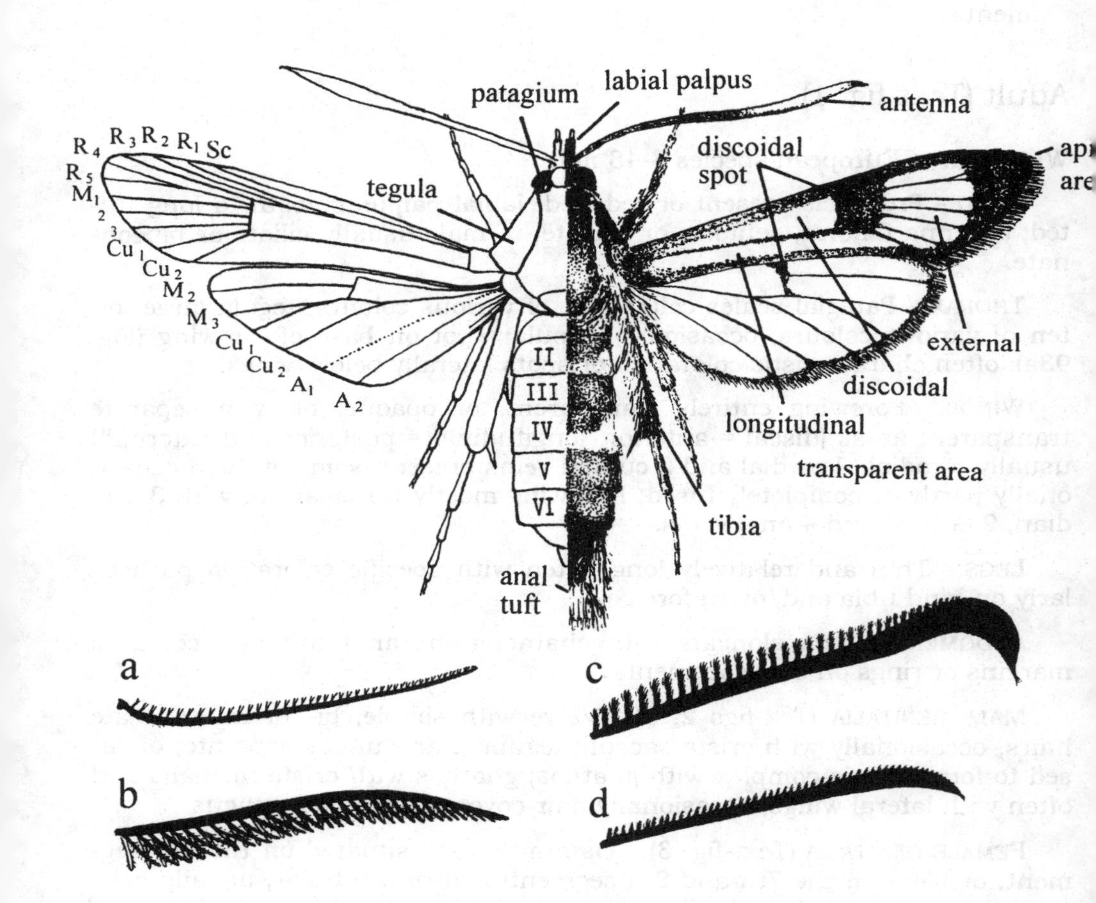

Text-fig. 1. Adult morphology; a-d: antenna, a, b: Tinthiinae, a: *Tinthia*, b: *Pennisetia*, c, d: Sesiinae, c: *Paranthrene*, d: *Bembecia*.

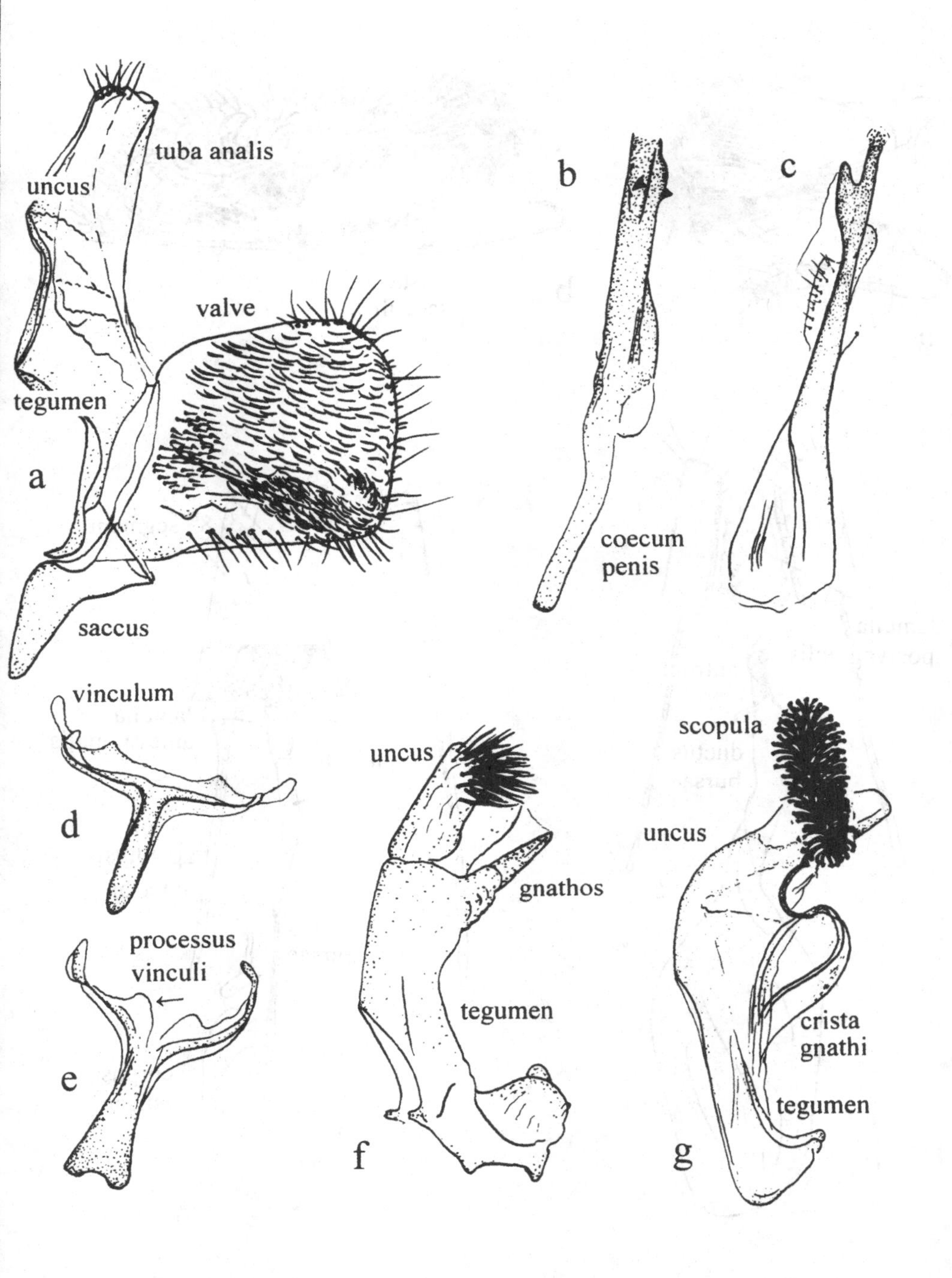

Text-fig. 2. Male genitalia; a: *Pennisetia hylaeiformis,* general view without left valve and aedeagus, b, c: aedeagus, b: *Tinthia myrmosaeformis*, c: *Synanthedon mesiaeformis*, d, e: saccus, d: *Sesia pimplaeformis*, e: *Synanthedon formicaeformis*, f, g: tegumen-uncus complex, f: *Sesia bembeciformis*, g: *Synanthedon mesiaeformis*.

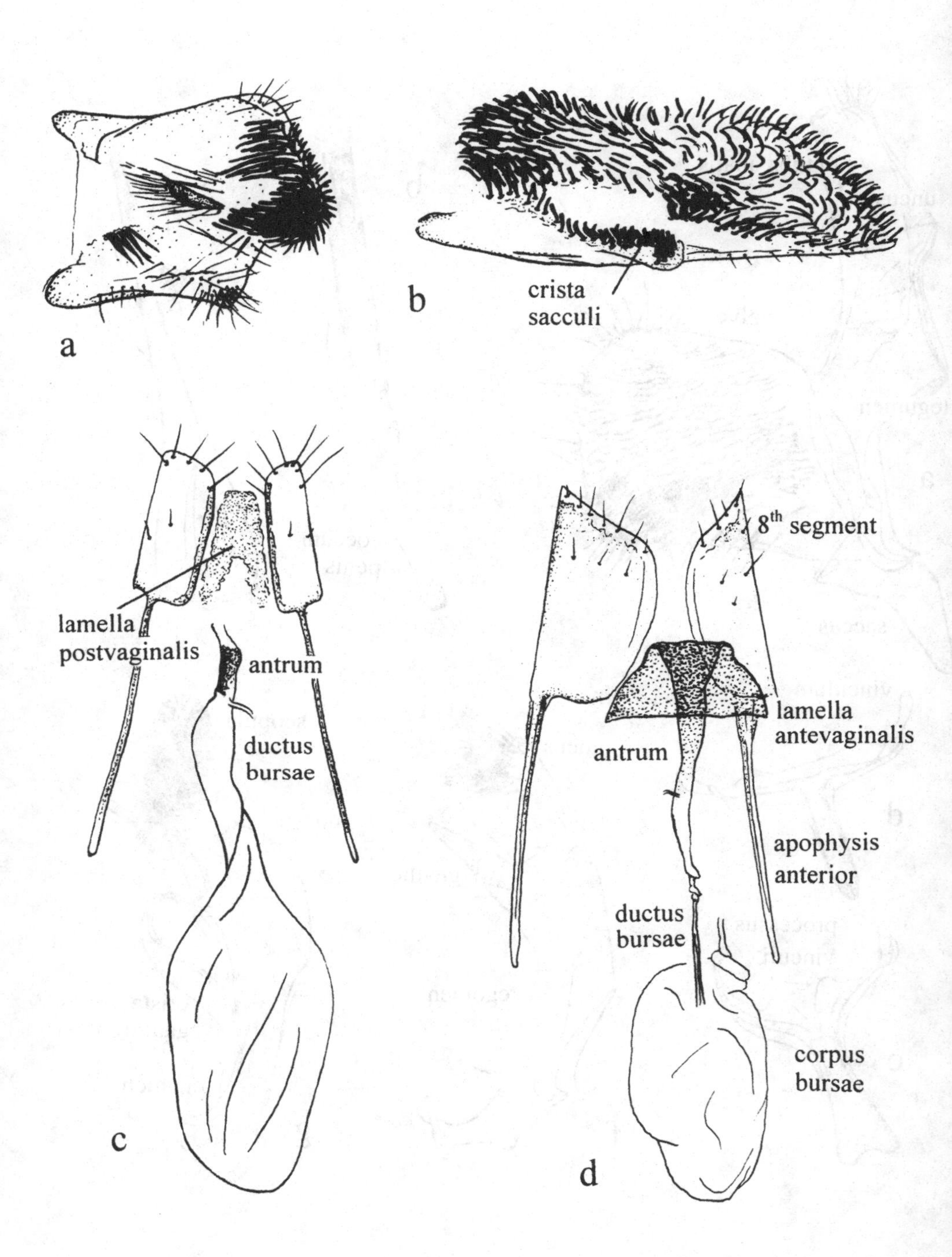

Text-fig. 3. Male and female genitalia; a, b: valve, a: *Sesia bembeciformis*, b: *Synanthedon mesiaeformis*, c, d: female genitalia, c: *Tinthia brosiformis*, d: *Synanthedon spheciformis*.

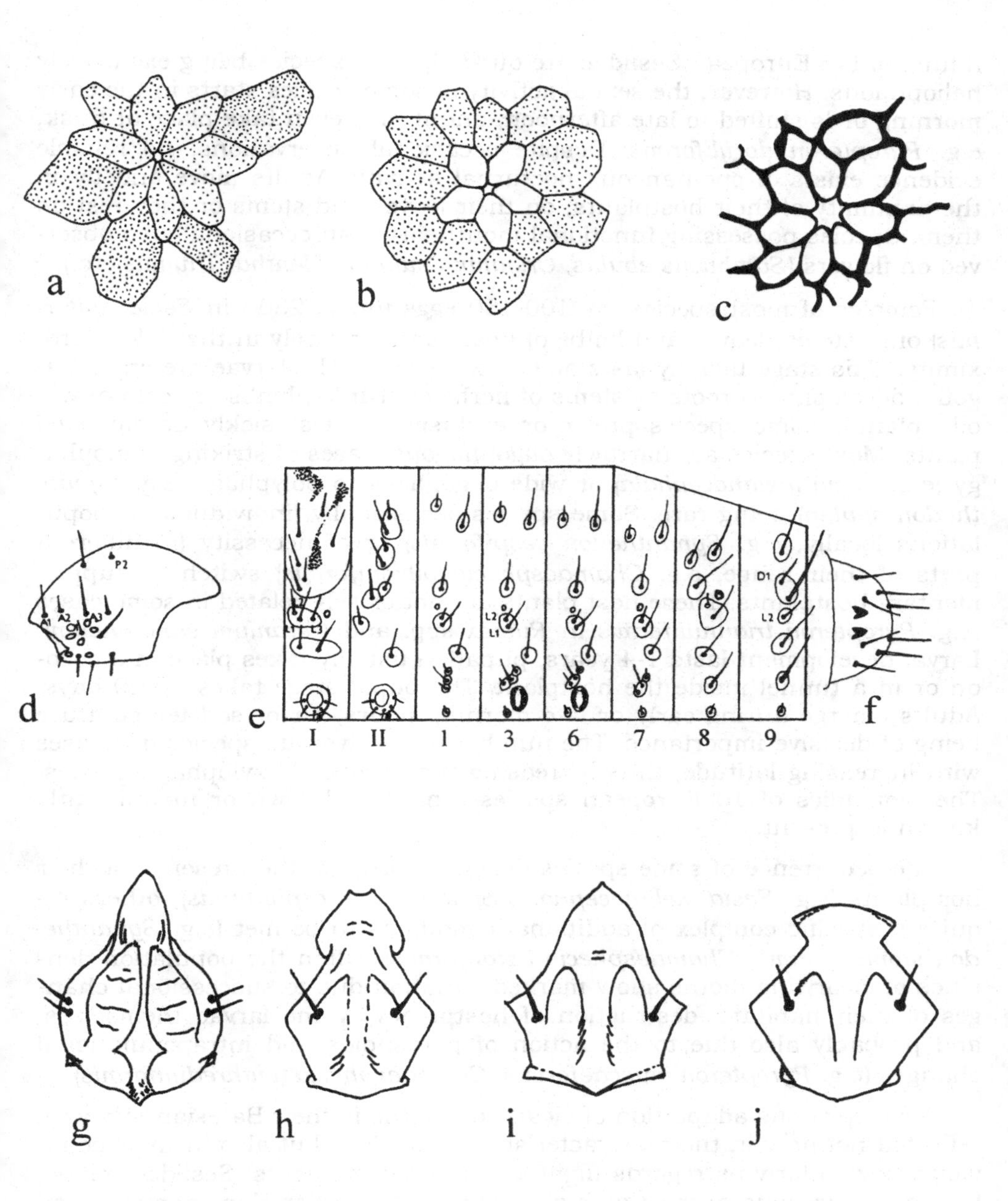

Text-fig. 4. Preadult stages; a-c: micropylar region of the egg, a: *Pennisetia hylaeiformis*, b: *Sesia apiformis*, c: *Paranthrene tabaniformis*, d-f: larval chaetotaxy, d: head of *Synanthedon myopaeformis* from lateral view, e: setal map of *Paranthrene tabaniformis*, f: last abdominal segment of *P. tabaniformis*, g-j: frontal process of pupa, g: *Pennisetia hylaeiformis*, h: *Synanthedon formicaeformis*, i: *S. culiciformis*, j: *S. andrenaeformis*.

Biology

Adults of the European Sesiidae are diurnal, most species being exclusively heliophilous. However, the sexual activity of some species starts in the early morning or is shifted to late afternoon, and it may even take place at dusk, e.g. *Pyropteron doryliformis.* Despite occasional observations, no reliable evidence exists of spontaneous nocturnal activity. Adults move and fly in the proximity of their hostplants, on their leaves and stems or they rest on them. Species possessing functional mouthparts can occasionally be observed on flowers (*Sambucus ebulus, Origanum vulgare, Mentha, Thymus* etc.).

Females of most species lay 100-150 eggs (up to 2500 in *Sesia apiformis*) onto stems, leaves and limbs of hostplants or rarely in their close proximity. This stage usually lasts one or two weeks. The larvae are endophagous, developing in roots or stems of herbs or trunks, limbs or roots of woody plants. Some species prefer or exclusively infest sickly or damaged plants. Most species are narrowly oligophagous, cases of striking monophagy (e.g. *Sesia melanocephala*) or wide oligophagy to polyphagy (e.g. *Synanthedon spuleri*) being rare. Some species occasionally (individuals or populations locally, e.g. *Synanthedon vespiformis*) or of necessity (in different parts of their range, e.g. *Chamaesphecia bibioniformis*) switch to supplementary hostplants. These host plants are not closely related in some cases (e.g. *Pyropteron triannuliformis* in *Rumex* spp. and *Geranium sanguineum*). Larval development lasts 1-4 years, pupation usually takes place in a cocoon or in a tunnel inside the hostplant. The pupal stage takes 10-20 days. Adults emerge during early or late morning hours, air or soil temperature being of decisive importance. The number of herbivorous species decreases with increasing latitude, thus increasing the number of xylophagous ones. The bionomics of 10 European species remain unknown or insufficiently known at present.

The occurrence of some species depends solely on the presence of their hostplants (e.g. *Sesia melanocephala, Synanthedon cephiformis*), others require a specific complex of additional conditions to be met (e.g. *Synanthedon stomoxiformis, Chamaesphecia astatiformis*). Often the population densities of clearwing moths show marked variation due to successional changes of their habitats, destruction of hostplants by the larvae themselves, and probably also due to the action of parasitoids and intrapopulational changes (e.g. *Pyropteron muscaeformis, Chamaesphecia tenthrediniformis*).

A conspicuous adaptation of clearwing moths is their Batesian mimicry, reflected not only in their characteristic morphology but also in their behaviour, particularly as regards flight and other movements. Sesiidae mimic, first of all, various species or higher taxa of the order Hymenoptera, and often the precise taxon mimicked can be unequivocally identified.

Several European species cause occasional or regular damage to their hosts – fruit, ornamental and forest trees. This is particularly true of *Synanthedon tipuliformis* on currants, *S. myopaeformis* on apple trees, *Pennisetia hylaeiformis* on raspberries, *Sesia apiformis* and *Paranthrene tabaniformis* on poplars and sometimes *Synanthedon formicaeformis* in osier beds.

List of Hostplants

Occasional trophic relationships are given in brackets. The hostplant names follow Flora Europaea (Tutin et al., 1964-1980) with minor changes.

Aceraceae

Acer spp. .... 34. *Synanthedon spuleri*

Asteraceae

Artemisia abrotanum .... 25. *Synanthedon uralensis*

Betulaceae

Alnus glutinosa .... 16. *Synanthedon mesiaeformis*
Alnus spp. .... 17. *Synanthedon spheciformis*
19. *S. culiciformis*
Betula pendula, B. pubescens .... 15. *Synanthedon scoliaeformis*
17. *S. spheciformis*
19. *S. culiciformis*
(34. *S. spuleri*)
Carpinus betulus .... 34. *Synanthedon spuleri*
Corylus avellana .... 30. *Synanthedon codeti*
34. *S. spuleri*
(18. *S. stomoxiformis*)

Caprifoliaceae

Lonicera nigra, L. xylosteum, L. tatarica, L. caerulea .... 24. *Synanthedon soffneri*
Viburnum lantana, V. opulus .... 23. *Synanthedon andrenaeformis*

Celastraceae

Euonymus europaea .... 33. *Synanthedon tipuliformis*

Cistaceae

Helianthemum spp., *Fumana* spp. .... 70. *Pyropteron affinis*

Convolvulaceae

Convolvulus arvensis, Convolvulus sp. .... 2. *Tinthia brosiformis*
Convolvulus spp. .... ? 1. *Tinthia tineiformis*

Cupressaceae

Juniperus communis .... 34. *Synanthedon spuleri*

Ebenaceae

Diospyros kaki .... ? 34. *Synanthedon spuleri*

Eleagnaceae

Hippophae rhamnoides .... 12. *Paranthrene tabaniformis*
(? 28. *Synanthedon myopaeformis*)

Euphorbiaceae

Euphorbia atlantis ....96. *Chamaesphecia anthraciformis*
Euphorbia austriaca ....99. *Chamaesphecia amygdaloidis*
Euphorbia ceratocarpa ....96. *Chamaesphecia anthraciformis*
Euphorbia characias .... 95. *Chamaesphecia bibioniformis*
96. *C. anthraciformis*
Euphorbia cyparissias .... 101. *Chamaesphecia leucopsiformis*
104. *C. empiformis*
Euphorbia epithymoides ....98. *Chamaesphecia euceraeformis*
Euphorbia esula ....105. *Chamaesphecia tenthrediniformis*
106. *C. astatiformis*
Euphorbia lucida .... 103. *Chamaesphecia hungarica*
Euphorbia macroclada .... 95. *Chamaesphecia bibioniformis*
Euphorbia myrsinites .... 95. *Chamaesphecia bibioniformis*
? 96. *C. anthraciformis*
Euphorbia nicaeensis ....96. *Chamaesphecia anthraciformis*
Euphorbia oblongifolia .... 102. *Chamaesphecia guriensis*
Euphorbia palustris ....97. *Chamaesphecia palustris*
Euphorbia rigida .... 95. *Chamaesphecia bibioniformis*
Euphorbia salicifolia ....105. *Chamaesphecia tenthrediniformis*
106. *C. astatiformis*
Euphorbia seguieriana .... 95. *Chamaesphecia bibioniformis*
Euphorbia serrata .... 95. *Chamaesphecia bibioniformis*
Euphorbia virgata .... 100. *Chamaesphecia crassicornis*

Fabaceae

Acanthyllis armata ....54. *Bembecia sirphiformis*
Anthyllis vulneraria .... 38. *Bembecia hymenopteriformis*
43. *B. ichneumoniformis*
49. *B. iberica*
Astragalus balearicus... 42. *B. abromeiti*
Astragalus angustifolius .... 39. *Bembecia lomatiaeformis*
Astragalus dipsaceus ....55. *Bembecia sanguinolenta*
Astragalus glycyphyllos .... 53. *Bembecia puella*
(52. *B. megillaeformis*)
Astragalus granatensis ....54. *Bembecia sirphiformis*
Astragalus monspessulanus ....54. *Bembecia sirphiformis*
? 55. *B. sanguinolenta*
Astragalus sigmoideus .... 53. *Bembecia puella*
Astragalus utriger .... ? 53. *Bembecia puella*
Calicotome spp. ....58. *Bembecia uroceriformis*
Chamaecytisus spp. ....52. *Bembecia megillaeformis*
58. *B. uroceriformis*
Colutea arborescens..52. *Bembecia megillaeformis*
54. *B. sirphiformis*
Coronilla emerus ....45. *Bembecia pavicevici*
58. *B. uroceriformis*
Coronilla minima .... 57. *Bembecia himmighoffeni*
Corothamnus procumbens................................. 58. *Bembecia uroceriformis*
(52. *B. megillaeformis*)

Dorycnium spp. ...... 43. *Bembecia ichneumoniformis*
Genista tinctoria ...... 52. *Bembecia megillaeformis*
Hedysarum candidum ...... 47. *Bembecia scopigera*
Hedysarum coronarium ...... 44. *Bembecia albanensis*
Hippocrepis spp. ...... 43. *Bembecia ichneumoniformis*
49. *B. iberica*
Lotus spp. ...... 38. *B. hymenopteriformis*
43. *B. ichneumoniformis*
49. *B. iberica*
57. *B. himmighoffeni*
Lotus cytisoides ...... 42. *Bembecia abromeiti*
Melilotus sp. ...... (49. *Bembecia iberica*)
Onobrychis spp. ...... 47. *Bembecia scopigera*
Ononis sp. ...... 48. *Bembecia priesneri*
Ononis rotundifolia, O. fruticosa ...... 46. *Bembecia fibigeri*
Ononis spinosa, O. repens, O. arvensis ...... 44. *Bembecia albanensis*
Psoralea bituminosa ...... 44. *Bembecia albanensis*
Spartium junceum ...... 58. *Bembecia uroceriformis*
Tetragonolobus spp. ...... 43. *Bembecia ichneumoniformis*
49. *B. iberica*
Trifolium fragiferum ...... 51. *Bembecia fokidensis*
Trifolium uniflorum ...... 50. *Bembecia blanka*
Ulex spp. ...... 58. *Bembecia uroceriformis*

Fagaceae

Castanea sativa ...... 29. *Synanthedon vespiformis*
Fagus sylvatica ...... 34. *Synanthedon spuleri*
(29. *S. vespiformis*)
Quercus spp. ...... 13. *Paranthrene insolita*
29. *Synanthedon vespiformis*
30. *S. codeti*
32. *S. conopiformis*
34. *S. spuleri*

Geraniaceae

Erodium cheilanthifolium ...... 67. *Pyropteron kautzi*
Geranium macrorrhizum ...... 35. *Synanthedon geranii*
Geranium rotundifolium ...... 73. *Pyropteron mannii*
Geranium sanguineum ...... 62. *Pyropteron triannuliformis*

Grossulariaceae

Ribes spp., *Grossularia* spp. ...... 33. *Synanthedon tipuliformis*

Hypericaceae

Hypericum perforatum ...... 94. *Chamaesphecia nigrifrons*

Juglandaceae

Carya pecan ...... 30. *Synanthedon codeti*

Lamiaceae

Ballota hirsuta ... 74. *Chamaesphecia mysiniformis*
Ballota nigra ... 90. *Chamaesphecia annellata*
Calamintha nepeta ... 82. *Chamaesphecia aerifrons*
Lavandula spp. ... 82. *Chamaesphecia aerifrons*
Marrubium spp. ... 75. *Chamaesphecia mysiniformis*
89. *C. oxybeliformis*
Mentha spp. ... 81. *Chamaesphecia alysoniformis*
82. *C. aerifrons*
Nepeta apuleii ... 79. *Chamaesphecia anthrax*
Nepeta nuda, N. parnassica, N. spruneri ... 76. *Chamaesphecia anatolica*
Nepeta tuberosa ... 80. *Chamaesphecia maurusia*
Origanum vulgare ... 77. *Chamaesphecia chalciformis*
82. *C. aerifrons*
Phlomis lychnitis, P. herba-venti ... 85. *Chamaesphecia ramburi*
Phlomis pungens ... 89. *Chamaesphecia oxybeliformis*
Salvia spp. ... 78. *Chamaesphecia schmidtiiformis*
84. *C. osmiaeformis*
86. *C. doleriformis*
Salvia sclarea ... 78. *Chamaesphecia schmidtiiformis*
82. *C. aerifrons*
86. *C. doleriformis*
92. *C. proximata*
Satureja spp. ... 82. *Chamaesphecia aerifrons*
Sideritis spp. ... 75. *Chamaesphecia mysiniformis*
Stachys recta, S. plumosa, S. thracica,
S. iberica, S. antherocalyx ... 88. *Chamaesphecia dumonti*
Stachys thirkei, S. germanica ... 87. *Chamaesphecia thracica*
Stachys spp. ... 75. *Chamaesphecia mysiniformis*
Thymus spp. ... 82. *Chamaesphecia aerifrons*

Loranthaceae

Loranthus europaeus ... 36. *Synanthedon loranthi*
(29. *S. vespiformis*)

Pinaceae

Abies alba, A. borisii-regis, A. cephalonica ... 37. *Synanthedon cephiformis*

Platanaceae

Platanus orientalis ... 30. *Synanthedon codeti*

Plumbaginaceae

Armeria spp. (*A. maritima*) ... 66. *Pyropteron muscaeformis*
Limonium gmelini ... 72. *Pyropteron cirgisa*
Limonium toletanum ... 65. *Pyropteron koschwitzi*
Limonium vulgare ... 71. *Pyropteron umbrifera*

Polygonaceae

Rumex spp. (s.l.) .... 59. *Pyropteron chrysidiformis*
60. *P. minianiformis*
61. *P. doryliformis*
62. *P. triannuliformis*
63. *P. meriaeformis*
64. *P. hispanica*

Rhamnaceae

Rhamnus spp., *Frangula* spp. .... 18. *Synanthedon stomoxiformis*

Rosaceae

Crataegus spp. ....28. *Synanthedon myopaeformis*
(18. *S. stomoxiformis*)
Eriobotrya japonica ....28. *Synanthedon myopaeformis*
Malus spp. ....28. *Synanthedon myopaeformis*
30. *S. codeti*
Potentilla recta, P. taurica ....3. *Tinthia myrmosaeformis*
Poterium minor ....4. *Tinthia hoplisiformis*
69. *Pyropteron leucomelaena*
Prunus spp. ....30. *Synanthedon codeti*
(29. *S. vespiformis*)
Pyrus spp. ....28. *Synanthedon myopaeformis*
Rosa canina (s.l.) .... 6. *Pennisetia bohemica*
Rubus idaeus ....5. *Pennisetia hylaeiformis*
Sorbus spp. ....28. *Synanthedon myopaeformis*
(18. *S. stomoxiformis*)

Salicaceae

Populus tremula ....10. *Sesia melanocephala*
Populus spp. ....7. *Sesia apiformis*
9. *S. pimplaeformis*
12. *Paranthrene tabaniformis*
14. *P. diaphana*
26. *Synanthedon melliniformis*
27. *S. martjanovi*
29. *S. vespiformis*
34. *S. spuleri*
Salix spp. ....8. *Sesia bembeciformis*
9. *S. pimplaeformis*
12. *Paranthrene tabaniformis*
14. *P. diaphana*
20. *Synanthedon formicaeformis*
21. *S. polaris*
22. *S. flaviventris*
26. *S. melliniformis*
29. *S. vespiformis*
34. *S. spuleri*

Scrophulariaceae

Scrophularia canina .... 93. *Chamaesphecia masariformis*
Verbascum spp. .... 93. *Chamaesphecia masariformis*

Ulmaceae

Ulmus spp. .... 34. *Synanthedon spuleri*
(29. *S. vespiformis*)

Tamaricaceae

Tamarix gallica, T. africana .... 31. *Synanthedon theryi*

Viscaceae

Viscum spp. .... 36. *Synanthedon loranthi*

Collecting and Rearing

Perfect knowledge of the bionomics is required for collecting adults as well as larvae. The adults can be collected by means of common entomological techniques, usually in the neighbourhood of their hostplants or in some genera on flowers. In recent years, attracting adults by means of sex pheromones has been the most frequent method. In such cases knowledge of the daily timing of sexual activity appears to be necessary which can be rather different in the individual species, as with knowledge of the ecological requirements and phenology.

The larvae are found in the respective parts of their hostplants. Infested herbaceous plants often wilt and dry off partly or wholly. The presence of larvae of many species is revealed by the production of tumescences, faeces, "sawdust", etc. In most cases, therefore, systematic destruction of hostplants is unnecessary. Xylophagous larvae can most suitably be collected during winter and early spring; the periods in which larvae developing in herbs can best be collected vary. In some cases the aerial part of the infested plant will dry off in autumn and fall off during the winter (*Pyropteron affinis, Chamaesphecia masariformis*). Thus, it is more convenient to collect such species in autumn. For other species it is more convenient to collect larvae shortly before pupation or in the pupal stage, the pupa being found in the basal parts of last year's stems or in a cocoon between stem bases (*Bembecia* spp.). Thus damage to the hostplants by the collector can be considerably limited. Rearing last instar larvae is mostly easy if proper humidity is maintained. It is possible to rear younger instar larvae of only a few species, for example in growing plants, in parts of tumours (*Synanthedon spuleri*) or on artificial diets (cf. e.g. Laštůvka, 1983b or Moraal, 1989).

Distribution

Species of the Sesiidae have populated most of the territory of Europe up to the southern edge of tundra in the north and up to the subalpine and partly also alpine zones in the mountains. Their present distribution developed as a result of climatic changes and opportunities for spreading from glacial refugia after the end of the Glacial Era and, to a considerable degree, also under the influence of man-made changes to habitats in Europe as a whole since the Atlantic epoch. Moreover, the distribution of some species was probably directly affected by introduction with their hostplants.

The distribution of clearwings is fairly satisfactorily known in northern, central and some regions of southern Europe, whereas many parts of eastern and southern Europe are still insufficiently known. Thus, the maps showing the distribution reflect the present state of knowledge obtained through the study of collection materials and reliable data in the literature; in some cases, precise boundaries of the ranges are unknown. Therefore, it is difficult to decide, in some cases, whether the illustrated "discontinuous" ranges are cases of actual disjunction or merely due to lack of faunal data. The ranges of many species more or less tally with those of their hostplants, and if the latter are disjunct their ranges are discontinuous as well (e.g. *Pyropteron muscaeformis* or *Chamaesphecia tenthrediniformis*). However, the range of several species is discontinuous, although the hostplant is distributed continuously and other ecological requirements of the species are apparently fulfilled (e.g. *Paranthrene insolita, Chamaesphecia nigrifrons*).

The widest ranges are occupied by species showing Eurasiatic (*Synanthedon scoliaeformis, S. spheciformis*) or Holarctic distribution (*Paranthrene tabaniformis, Synanthedon culiciformis*), making up some 10 % in Europe. The ranges of these species also reach furthest to the north, and one of them (*Synanthedon polaris*) shows an Arctic-Alpine disjunction of its range. Species showing European (*Sesia bembeciformis, Synanthedon cephiformis*) to Western Palaearctic (*Sesia apiformis, Paranthrene insolita*) distribution are represented by 24 %. As a rule, somewhat smaller ranges are occupied by Holomediterranean (4 %, *Tinthia tineiformis, Pyropteron leucomelaena*), West Mediterranean and North African (16 %, *Synanthedon codeti, Bembecia hymenopteriformis*), Atlantico-Mediterranean (3 %, *Pyropteron chrysidiformis, Chamaesphecia aerifrons*), Adriatico-Mediterranean (1 %, *Synanthedon melliniformis*), East Mediterranean in wide sense, i.e. incl. Pontic, Syrian and Iranian (23 %, *Tinthia myrmosaeformis, Chamaesphecia alysoniformis*), Caspian with variously extended ranges (5 %, *Pyropteron cirgisa, Chamaesphecia crassicornis*), West Siberian or Central Asiatic (3 %, *Synanthedon martjanovi, Bembecia sareptana*), and Caucasian species (1 %, *Chamaesphecia guriensis*). Three species are known only from the Balkan Peninsula (3 %, *Bembecia fokidensis, Pyropteron umbrifera*). 7 % fall into the category of so-called endemic species. This group includes true endemics, as well as species with insufficiently known distribution, such as taxa with unclear taxonomic status (*Pennisetia bohemica, Synanthedon geranii, Bembecia abromeiti, B. blanka, Pyropteron aistleitneri, P. kautzi, Chamaesphecia staudingeri, C. amygdaloidis*).

Phylogeny and Classification

Opinions concerning the systematic position of the Sesiidae have changed somewhat over the past decades and the family was placed in the Glyphipterigoidea, Tineoidea or Yponomeutoidea (cf. e.g. Meyrick, 1928; Popescu-Gorj et al., 1958; Bradley et al., 1972; Fibiger & Kristensen, 1974; Heath & Emmet, 1985). At present, the Sesiidae are classified in a separate superfamily Sesioidea of the suborder Ditrysia, either together with the families Choreutidae and Brachodidae (Heppner & Duckworth, 1981; cf. also Brock, 1971), only with Brachodidae (Minet, 1991; Karsholt & Razowski, 1996), or with the families Brachodidae and Castniidae (Kristensen, 1999).

Also, the internal division of the family has been developing and differing to varying extent (cf. Le Cerf, 1917; Niculescu, 1964; MacKay, 1968; Naumann, 1971; Fibiger & Kristensen, 1974; Bradley et al., 1972; Bradley & Fletcher, 1974; Duckworth & Eichlin, 1974, 1977; Heppner & Duckworth, 1981, etc.). Based on cladistic analysis, Naumann (1971) proposes the presence of two phylogenetical lines (subfamilies) within the family, viz. Tinthiinae (with the tribes Tinthiini and Pennisetiini) and Sesiinae (with the tribes Sesiini, Melittiini, Paranthrenini and Aegeriini). Bradley et al. (1972) and Bradley & Fletcher (1974) separate a third subfamily, Paranthreninae, comprising the most advanced tribes, Paranthrenini and Synanthedonini. Duckworth & Eichlin (1974, 1977) and Heppner & Duckworth (1981) have also set apart the subfamily Paranthreninae, albeit in a different sense, without Synanthedonini, so that the subfamily Sesiinae with the remaining tribes becomes a paraphyletic group. At any rate, the anagenetic changes in the tribe Paranthrenini are probably not so distinct as to require the establishment of a separate subfamily. For this reason, the division into two subfamilies in the sense of Naumann (1971) has been adopted in this work.

The family is represented worldwide by 9 tribes, about 120 genera and 1200 described species; 6 tribes, 11 genera and 107 species are treated here. However, the status of some taxa is still unclear. In some cases the present "species" (morphospecies) consists of a series of populations exhibiting geographical, bionomic or sexual chemical differences, and thus they could be considered to be several close biospecies (e.g. *Bembecia albanensis*, *Chamaesphecia bibioniformis*). On the other hand, some of the "separate" species may be mere subspecies, bionomic or morphological forms of another species (e.g. *Chamaesphecia anthrax*). In doubtful cases, we have usually accepted a general concept and the practicability of differentiating such taxa.

In classifying the Sesiidae, their mimicry must be considered, as it often results in systematically remote species developing similar characters (e.g. red-banded species or absence of the forewing transparency). On the other hand, groups of closely related species sometimes become markedly differentiated in habitus (e.g. *Bembecia sirphiformis* and *B. sanguinolenta*).

The type material has been revised in 72 European species (Špatenka & Laštůvka, 1988, 1990; Laštůvka, 1987; Špatenka, 1992b; Toševski, 1992, etc.); that of 35 species has not been found, had been lost or destroyed.

Check-list of the European Sesiidae

Synonyms are mentioned only exceptionally. Important synonyms are supplied in individual species, full synonymy given by Špatenka et al. (1999).

SESIIDAE Boisduval, 1828

TINTHIINAE Le Cerf, 1917

TINTHIINI Le Cerf, 1917

Tinthia Walker, [1865]

1. *tineiformis* (Esper, [1789])
2. *brosiformis* (Hübner, [1813])
3. *myrmosaeformis* (Herrich-Schäffer, 1846)
   3a. *myrmosaeformis myrmosaeformis* (Herrich-Schäffer, 1846)
   3b. *myrmosaeformis cingulata* (Staudinger, 1870)
4. *hoplisiformis* (Mann, 1864)

PENNISETIINI Naumann, 1971

Pennisetia Dehne, 1850

5. *hylaeiformis* (Laspeyres, 1801)
6. *bohemica* Králíček & Povolný, 1974

SESIINAE Boisduval, 1828

SESIINI Boisduval, 1828

Sesia Fabricius, 1775

7. *apiformis* (Clerck, 1759)
8. *bembeciformis* (Hübner, [1806])
9. *pimplaeformis* Oberthür, 1872
10. *melanocephala* Dalman, 1816

OSMINIINI Duckworth & Eichlin, 1977

Osminia Le Cerf, 1917

11. *fenusaeformis* (Herrich-Schäffer, 1852)

PARANTHRENINI Niculescu, 1964

Paranthrene Hübner, [1819]

12. *tabaniformis* (Rottemburg, 1775)
    12a. *tabaniformis tabaniformis* (Rottemburg, 1775)
    12b. *tabaniformis synagriformis* (Rambur, 1866)
13. *insolita* Le Cerf, 1914
    13a. *insolita polonica* Schnaider, [1939]
    13b. *insolita hispanica* Špatenka & Laštůvka, 1997
14. *diaphana* Dalla Torre & Strand, 1925

SYNANTHEDONINI Niculescu, 1964

Synanthedon Hübner, [1819]

15. *scoliaeformis* (Borkhausen, 1789)
16. *mesiaeformis* (Herrich-Schäffer, 1846)
17. *spheciformis* ([Denis & Schiffermüller], 1775)
18. *stomoxiformis* (Hübner, 1790)
    18a. *stomoxiformis stomoxiformis* (Hübner, 1790)
    18b. *stomoxiformis amasina* (Staudinger, 1856)
    18c. *stomoxiformis riefenstahli* Špatenka, 1997
19. *culiciformis* (Linnaeus, 1758)
20. *formicaeformis* (Esper, 1783)
    herzi Špatenka & Gorbunov, 1992, **syn.n.**
    20a. *formicaeformis formicaeformis* (Esper, 1783)
    20b. *formicaeformis serica* (Alphéraky, 1882)
21. *polaris* (Staudinger, 1877)
22. *flaviventris* (Staudinger, 1883)
23. *andrenaeformis* (Laspeyres, 1801)
    perigordensis Garrevoet & Vanholder, 1996
24. *soffneri* Špatenka, 1983
25. *uralensis* (Bartel, 1906)
26. *melliniformis* (Laspeyres, 1801)
27. *martjanovi* Sheljuzhko, 1918
28. *myopaeformis* (Borkhausen, 1789)
    28a. *myopaeformis myopaeformis* (Borkhausen, 1789)
    28b. *myopaeformis typhiaeformis* (Borkhausen, 1789)
    28c. *myopaeformis cruentata* (Mann, 1859)
    28d. *myopaeformis graeca* (Staudinger, 1870)
29. *vespiformis* (Linnaeus, 1761)
30. *codeti* (Oberthür, 1881)
31. *theryi* Le Cerf, 1916
32. *conopiformis* (Esper, 1782)
33. *tipuliformis* (Clerck, 1759)
34. *spuleri* (Fuchs, 1908)
35. *geranii* Kallies, 1997
36. *loranthi* (Králíček, 1966)
37. *cephiformis* (Ochsenheimer, 1808)
    gaderensis (Králíček & Povolný, 1977)

Bembecia Hübner, [1819]

38. *hymenopteriformis* (Bellier, 1860)
39. *lomatiaeformis* (Lederer, 1853)
40. *sareptana* (Bartel, 1912)
41. *volgensis* Gorbunov, 1994
42. *abromeiti* Kallies & Riefenstahl, 2000
43. *ichneumoniformis* ([Denis & Schiffermüller], 1775)
44. *albanensis* (Rebel, 1918)
    44a. *albanensis albanensis* (Rebel, 1918)
    44b. *albanensis tunetana* (Le Cerf, 1920)
    44c. *albanensis kalavrytana* (Sheljuzhko, 1924)
    44d. *albanensis psoraleae* Bartsch & Bettag, 1997

45. *pavicevici* Toševski, 1989
   45a. *pavicevici pavicevici* Toševski, 1989
   45b. *pavicevici dobrovskyi* Špatenka, 1997
46. *fibigeri* Laštůvka & Laštůvka, 1994
47. *scopigera* (Scopoli, 1763)
48. *priesneri* Kallies, Petersen & Riefenstahl, 1998
49. *iberica* Špatenka, 1992
50. *blanka* Špatenka, 2001
51. *fokidensis* Toševski, 1991
52. *megillaeformis* (Hübner, [1813])
53. *puella* Laštůvka, 1989
   daghestanica Gorbunov, 1991, **syn.n.**
54. *sirphiformis* (Lucas, 1849)
   astragali (Joannis, 1909)
55. *sanguinolenta* (Lederer, 1853)
56. *flavida* (Oberthür, 1890)
57. *himmighoffeni* (Staudinger, 1866)
   baumgartneri Špatenka, 1992
58. *uroceriformis* (Treitschke, 1834)

Pyropteron Newman, 1832
   Synansphecia Căpuşe, 1973, **syn. n.**

59. *chrysidiformis* (Esper, 1782)
   59a. *chrysidiformis chrysidiformis* (Esper, 1782)
   59b. *chrysidiformis sicula* Le Cerf, 1922
60. *minianiformis* (Freyer, 1842)
61. *doryliformis* (Ochsenheimer, 1808)
   61a. *doryliformis doryliformis* (Ochsenheimer, 1808)
   61b. *doryliformis icteropus* (Zeller, 1847)
62. *triannuliformis* (Freyer, 1842) **comb. n.**
63. *meriaeformis* (Boisduval, 1840) **comb. n.**
64. *hispanica* (Kallies, 1999) **comb. n.**
   atlantis auct., nec Schwingenschuss, 1935
65. *koschwitzi* (Špatenka, 1992) **comb. n.**
66. *muscaeformis* (Esper, 1783) **comb. n.**
67. *kautzi* (Reisser, 1930) **comb. n.**
68. *aistleitneri* (Špatenka, 1992) **comb. n.**
69. *leucomelaena* (Zeller, 1847) **comb. n.**
70. *affinis* (Staudinger, 1856) **comb. n.**
71. *umbrifera* (Staudinger, 1870) **comb. n.**
72. *cirgisa* (Bartel, 1912) **comb. n.**
73. *mannii* (Lederer, 1853) **comb. n.**

Dipchasphecia Căpuşe, 1973

74. *lanipes* (Lederer, 1863)

Chamaesphecia Spuler, 1910

75. *mysiniformis* (Boisduval, 1840)
rondouana Le Cerf, 1922
76. *anatolica* Schwingenschuss, 1938
77. *chalciformis* (Esper, [1804])
78. *schmidtiiformis* (Freyer, 1836)
79. *anthrax* Le Cerf, 1916
80. *maurusia* Püngeler, 1912
81. *alysoniformis* (Herrich-Schäffer, 1846)
82. *aerifrons* (Zeller, 1847)
82a. *aerifrons aerifrons* (Zeller, 1847)
82b. *aerifrons sardoa* (Staudinger, 1856)
83. *albiventris* (Lederer, 1853)
84. *osmiaeformis* (Herrich-Schäffer, 1848)
85. *ramburi* (Staudinger, 1866)
86. *doleriformis* (Herrich-Schäffer, 1846)
86a. *doleriformis doleriformis* (Herrich-Schäffer, 1846)
86b. *doleriformis colpiformis* (Staudinger, 1856)
87. *thracica* Laštůvka, 1983
88. *dumonti* Le Cerf, 1922
89. *oxybeliformis* (Herrich-Schäffer, 1846)
90. *annellata* (Zeller, 1847)
91. *staudingeri* (Failla-Tedaldi, 1890)
92. *proximata* (Staudinger, 1891)
93. *masariformis* (Ochsenheimer, 1808)
odyneriformis (Herrich-Schäffer, 1846)
94. *nigrifrons* (Le Cerf, 1911)
95. *bibioniformis* (Esper, 1800)
95a. *bibioniformis bibioniformis* (Esper, 1800)
95b. *bibioniformis tengyraeformis* (Boisduval, 1840)
96. *anthraciformis* (Rambur, 1832)
foeniformis (Herrich-Schäffer, 1846)
oryssiformis (Herrich-Schäffer, 1846)
97. *palustris* Kautz, 1927
98. *euceraeformis* (Ochsenheimer, 1816)
99. *amygdaloidis* Schleppnik, 1933
100. *crassicornis* Bartel, 1912
101. *leucopsiformis* (Esper, 1800)
102. *guriensis* (Emich, 1872)
103. *hungarica* (Tomala, 1901)
104. *empiformis* (Esper, 1783)
lastuvkai Špatenka, 1987
105. *tenthrediniformis* ([Denis & Schiffermüller], 1775)
106. *astatiformis* (Herrich-Schäffer, 1846)

Weismanniola Naumann, 1971

107. *agdistiformis* (Staudinger, 1866)

Keys to European Taxa

Key to the subfamilies, tribes and genera

1. Antenna without terminal scale-pencil (filiform or setiform)(Text-fig. 1, Figs 1a, 5a)..Tinthiinae 2
– Antenna with terminal scale-pencil (clavate)(Text-fig. 1, Fig. 9a) ..Sesiinae 3

2. Forewing opaque, without transparent areas (Fig. 1b) . Tinthiini (*Tinthia*, p. 44)
– Forewing with transparent areas (Fig. 5b) .... Pennisetiini (*Pennisetia*, p. 46)

3. Entire forewing or one area only transparent (Fig. 7a), or nearly completely opaque with small transparency near base (Fig. 13b) or in distal part; discal spot of forewing not sharply delimited towards discal transparent area (Figs 13a, 14a); forewing without red; male antenna bipectinate (Figs 7a, 13) 4
– Forewing with 2-3 separate transparent areas (Text-fig. 1)(exceptionally with external transparent area only); transparent areas occasionally covered with whitish or yellowish scales; discal spot of forewing sharply delimited from the discal transparent area and/or discal transparent area is reduced or absent; at the same time forewing without transparency near base; male antenna ciliate or exceptionally without cilia; forewing exceptionally entirely opaque, at the same time partly or entirely red or with a red spot 5

4. Forewing entirely or mostly transparent (Fig. 7a), or partly covered with translucent scales (Fig. 10a); veins M3 and Cu1 of hindwing stalked from the crossvein; crossvein of hindwing very oblique without distinct discal spot (Fig. 7a); a robust species .... Sesiini (*Sesia*, p. 47)
– Forewing nearly entire opaque with small transparency near base (Fig. 13b) or exceptionally with some transparent cells distally (Fig. 13a); veins M3 and Cu1 not stalked; crossvein of hindwing slightly oblique, with discal spot (Fig. 13).... .... Paranthrenini (*Paranthrene*, p. 50)

5. Small species (wingspan up to 16 mm); the crossvein of hindwing shifted distally, without discal spot; roots of veins M3, Cu1 and Cu2 very close to one another (Fig. 11a); antenna without cilia in both sexes; (transparent areas very small; segments 2, 4, 6 with narrow yellow rings) .... Osminiini (*Osminia*, p. 49)
– The crossvein of hindwing in normal position, usually with a discal spot; veins M3 and Cu1 stalked, exceptionally their stalk very short and both apparently arising from the crossvein .... Synanthedonini 6

6. Longitudinal transparent area reaches discal spot (often Cu2) (Text-fig. 1); proboscis present .... *Synanthedon* (p. 52)
– Longitudinal transparent area does not reach discal spot; if so exceptionally, then proboscis reduced 7

7. Proboscis distinctly reduced (very short or absent, light) 8
– Proboscis present, dark .... *Pyropteron, Dipchasphecia,. Chamaesphecia*

8. Discal spot of forewing usually partly or entirely orange or yellow; hindwing transparent; abdomen with yellow or white rings; forewing exceptionally without transparency, partly reddish, abdomen without rings ... *Bembecia* (p. 63)
– Discal spot of forewing without yellow or red spot; transparent areas covered with whitish scales, not transparent; abdomen slender ... *Weismanniola* (p. 97)

Key to the genus *Tinthia*

1. Abdomen with distinct yellow bands; anal tuft entirely or partly yellow............ .... 4. *T. hoplisiformis*
– Abdomen without distinct yellow bands or uniformly coloured; anal tuft without yellow coloration 2
2. Forewing and abdomen brown; hind tibia not orange 3
– Forewing and abdomen brown-black or black, with bluish metallic sheen; hind tibia orange in middle 4
3. Proboscis present; distal part of forewing without whitish yellow or yellow spot; 1st abdominal segment not white or whitish yellow dorsally, occasionally with ochreous scales laterally (Figs 1b, c) .... 1. *T. tineiformis*
– Proboscis reduced; a yellowish spot in distal part of forewing; 1st abdominal segment white or yellowish (Figs 2a, b) .... 2. *T. brosiformis*
4. Abdomen without white rings .... 3. *T. myrmosaeformis myrmosaeformis*
– Abdomen with some narrow indistinct white or yellowish white rings (Fig. 3a).... .... 3. *T. myrmosaeformis cingulata*

Key to the genus *Pennisetia*

1. Yellow rings on 4th-6th (7th in male) segments of the same width dorsally (Fig. 5c); labial palpus with black scales; thorax dark laterally below forewing......... .... 5. *P. hylaeiformis*
– Yellow ring on 6th (and 7th in male) segment very broad, distinctly broader than other rings on the foregoing segments (Fig. 6b); labial palpus deep yellow without black scales; thorax with a yellow spot laterally below forewing........... .... 6. *P. bohemica*

Key to the genus *Sesia*

1. Yellow rings on the abdomen narrow, distinctly narrower than dark parts of segments (Fig. 10a) .... 10. *S. melanocephala*
– Yellow rings on the abdomen broad, some segments wholly or nearly wholly yellow (exceptionally abdomen without rings, black) 2
2. Patagial collar yellow; tegula black (Fig. 8a) .... 8. *S. bembeciformis*
– Patagial collar wholly or partly black; tegula with large yellow anterior spot (Fig. 7a) 3
3. 4th abdominal segment without yellow coloration dorsally (Fig. 7a); labial palpus yellow (quite exceptionally palpus and abdomen black); thorax without yellow spot laterally below forewing .... 7. *S. apiformis*
– 4th abdominal segment yellow dorsally with small round or triangular black spot (Fig. 9a); basal segment of labial palpus black ventrally; thorax with yellow spot laterally below forewing .... 9. *S. pimplaeformis*

Key to the genus *Paranthrene*

1. Tegula with yellow border up to patagium; metathorax with broad V-shaped yellow border; antenna distinctly longer than one half of forewing; forewing with 1-7 transparent cells distally; proboscis short, yellowish (Figs 13a,b) 2
– Tegula without yellow border up to patagium, black or with yellow spot or with short yellow posterior border (Fig. 12b); metathorax without V-shaped yellow coloration; antenna nearly as long as one half of forewing; proboscis normal, dark brown or black 3

2. Transparent areas of forewing distinct; abdominal rings of various width, on 3rd segment narrow or indistinct .... 13. *P. insolita polonica*
– Transparent areas of forewing strongly reduced; abdominal rings of equal width .... 13. *P. insolita hispanica*

3. 1-3 transparent cells between the veins M1-Cu1 of forewing; male antenna finely pectinate; all abdominal segments with yellow rings dorsally, on 2nd, 6th (and 7th in male) segments slightly broader; patagial collar continuously yellow dorsally (Fig. 14a).. 14. *P. diaphana*
– Forewing without transparent cells distally; male antenna distinctly pectinate; abdomen with 3(4)–6(7) yellow rings (Figs 12a,b); patagial collar black or yellow only laterally 4

4. Forewing dark brown; thorax wholly black dorsally or metathorax with 2 yellow spots; abdomen usually with 3 (4) rings, exceptionally rings on additional or on all segments (Fig. 12a) .... 12. *P. tabaniformis tabaniformis*
– Forewing light brown; tegula yellow at the back and 2 yellow spots on metathorax (i.e. 4 yellow spots on thorax); all abdominal segments with yellow rings (Fig. 12b) .... 12. *P. tabaniformis synagriformis*

Key to the genus *Synanthedon*

1. Eye with white border (Fig. 33b); antenna without white part; forewing brown-black or black without reddish coloration; apical area dark or with light spots between veins; abdomen with narrow yellow rings on 2nd, 4th, (6th) segments in female and on 2nd, 4th, (6th) and 7th segments in male; a small species 19
– Different combination of characters (abdomen with different number or coloration of rings, or forewing with red or reddish parts, or eye without white border, or antenna with white part) 2

2. Whole 4th abdominal segment red dorsally 3
– 4th abdominal segment not red dorsally 11

3. Forewing apical area red .... 20. *S. formicaeformis*
– Forewing apical area not red 4

4. Thorax black laterally below forewing; 4-6th abdominal segments red ventrally and anal flaps black .... 18. *S. stomoxiformis*
– Thorax with reddish or orange spot laterally below forewing (Fig. 19a); abdomen of another coloration ventrally 5

5. Antenna black 6
– Antenna distally white or with white scales 10

6. Labial palpus ventrally white in male and brown in female; white border before eye; male abdominal segments 4-6 partly or entirely white ventrally, female segment 4 reddish ventrolaterally and dark medially 7
– Labial palpus dark brown ventrally and eye without white border, or labial palpus orange reddish; 4th segment entirely or nearly entirely red, orange or deep yellow ventrally 8

7. Only 4th abdominal segment red dorsally .. 28. *S. myopaeformis myopaeformis*
– Segments 2-4 partly or entirely red dorsally .... 28. *S. myopaeformis graeca*

8. Forewing covered with reddish or orange scales basally or at the anal margin; transparent areas usually large, external transparent area broader than apical area; abdominal ring usually orange-red or red 9
– Forewing without red scales basally or at the anal margin; transparent areas small, external transparent area narrower than apical area; abdominal ring deep yellow to orange .... 24. *S. soffneri*

9. Labial palpus orange ventrally; eye white-bordered .... 19. *S. culiciformis*
– Labial palpus dark brown to black, long tufted in male; eye without white border; male abdomen reddish orange ventrally from 4th segment to anal flaps... .... 25. *S. uralensis*

10. Narrow red ring dorsally on 2nd segment ... 28. *S. myopaeformis typhiaeformis*
– Abdominal segments 2-4 continuously red .... 28. *S. myopaeformis cruentata*

11. Antenna white or yellow or at least with yellow or white scales distally, or discal spot of forewing conspicuously broad, nearly square (Text-fig. 1)........... 12
– Antenna entirely dark; discal spot of forewing normally broad 14

12. 2nd abdominal segment dorsally and 4th segment ventrally with narrow yellow-white ring; labial palpus yellow-white ventrally; eye without white border (frons black) .... 17. *S. spheciformis*
– 2nd and 4th segment with yellow or orange ring dorsally; labial palpus orange ventrally; eye narrowly white-bordered 13

13. Abdominal rings narrow; 4th segment yellow-white ventrally; anal tuft usually rusty; hind tibia yellowish with dark end or dark dorsally; discal spot of forewing conspicuously broad ....15. *S. scoliaeformis*
– Ring on 4th segment broad; abdominal segments 4 and 5 (in male) deep yellow ventrally; anal tuft black; hind tibia orange-yellow with black band distally; forewing discal spot normally broad .... 16. *S. mesiaeformis*

14. Abdomen with broad orange-yellow ring on 4th segment (usually entire segment); other segments without rings 6
– Abdomen with 2-5 yellow rings (exceptionally only 1 very narrow ring) 15

15. Discal spot of forewing with red spot or almost entirely red; reddish coloration often in other parts of wing (anal margin) 16
– Forewing without red or orange-red coloration; abdominal segments 4-6 whitish or yellow ventrally; eye without white border; 2nd and 4th abdominal segments with yellow ring .... 23. *S. andrenaeformis*

16. Labial palpus orange ventrally; eye without white border; thorax dark dorsally; 2nd, 4th and 6th segments with narrow yellow rings (rings on 2nd and 6th segments occasionally indistinct); external transparent area usually relatively narrow (Fig. 21a) .... 21. *S. polaris*

– Labial palpus yellow ventrally; eye with white border; metathorax or tegula with yellow coloration; 2nd, 4th, (6th) and 7th segments with variously broad yellow rings; external transparent area usually broad 17

17. Abdominal segments 2, 4, (5), 6 and 7 (in male) with yellow rings; anal flaps yellow in male (Fig. 29b); discal spot of forewing nearly entirely red 18
– Abdominal segments 2, 4, (6), 7 in male and 2, 4, 6 in female with yellow rings dorsally; anal flaps black in male (Fig. 30a); discal spot of forewing broadly black towards base, particularly in male .... 30. *S. codeti*

18. Tegula with narrow yellow border; 4th and 6th abdominal segments with yellow ring ventrally; transverse yellow spot on metathorax, metathorax black laterally (Fig. 29a) .... 29. *S. vespiformis*
– Tegula with yellow anterior spot; all segments with yellow rings ventrally; metathorax yellow dorsally and laterally (Fig. 31a) .... 31. *S. theryi*

19. 4th segment entirely yellow ventrally, and/or other segments (5, 6) yellow or whitish yellow ventrally (Fig. 26a); discal spot of forewing occasionally with yellow scales 20
– 4th segment with only distal yellow border ventrally, other segments without continuous yellow coloration; discal spot of forewing without yellow scales .. 22

20. Anal flaps in male black; anal tuft in female black with ochreous scales laterally; external transparent area usually very narrow (Fig. 22a); abdominal segments 2, 4, 6 with yellow rings .... 22. *S. flaviventris*
– Anal flaps in male yellow distally; anal tuft in female almost entirely yellow; external transparent area usually broader or very broad; abdominal segments 2, 4 or 2, 4, 7 in male and 2, 4 or 2, 4, 6 in female with yellow rings 21

21. External transparent area nearly square or slightly higher than broad; abdominal segments 2, 4 with narrow yellow rings .... 27. *S. martjanovi*
– External transparent area broad (apical area nearly absent); segments 2, 4, 7 in male and 2, 4, 6 in female with yellow rings .... 26. *S. melliniformis*

22. Apical area of forewing with distinctly orange spots between veins; transverse yellow spot on metathorax (Fig. 32a) .... 32. *S. conopiformis*
– Apical area of forewing dark, or with ochreous or brownish spots; metathorax without transverse yellow spot 23

23. Anal tuft in male with yellow scales in middle, in female almost entirely yellow distally; more or less distinct yellow spot on metathorax 24
– Anal tuft in male black, in female entirely black or with individual yellow scales; yellow spot on metathorax absent 26

24. External transparent area square or broader than high, if narrow, then ochreous spots in apical area indistinct and forewing dark 25
– External transparent area distinctly higher than broad; apical area with distinct ochreous spots; margins of forewing more or less ochreous . 35. *S. geranii*

25. Fore coxa entirely yellow laterally (Fig. 37b); external transparent area nearly square or higher than broad; discal spot of hindwing broad up to base of M2 (Fig. 37a) .... 37. *S. cephiformis*
– Fore coxa yellow up to 2/3 laterally (Fig. 36b); external transparent area broader than high; discal spot of hindwing small, scarcely reaches to base of M2 (Fig. 36a) .... 36. *S. loranthi*

26. Apical area of forewing with ochreous or brownish spots between veins; external transparent area narrow with straight distal border (Fig. 33a) .... 33. *S. tipuliformis*
– Apical area dark or with only indistinct spots; external transparent area slightly convex towards apex (Fig. 34a) .... 34. *S. spuleri*

Key to the genus *Bembecia*

1. Forewing without transparent areas, largely rusty-red; abdomen without rings ....55. *B. sanguinolenta*
– Transparent areas present, or at least external transparent area indicated; if exceptionally transparent areas entirely covered with dark scales, abdomen with yellow or white rings 2
2. External transparent area small, narrower than apical area, round or distinctly broader than high, with 3 slightly elongate cells and/or with one further small cell below (Fig. 44a), or external transparent area narrow (Figs 57a,b); quite exceptionally external transparent area nearly or entirely absent 16
– External transparent area large, round or oval in shape (broader than high), with 4 or more cells and broader than apical area (Fig. 43a) 3
3. Abdominal segments 2-6 (7) with orange-reddish rings; tegula with orange-reddish margin; external transparent area large, apical area almost absent; forewing margins black .... 42. *B. abromeiti*
– Abdominal segments with whitish yellow to orange-yellow rings; apical area present; if exceptionally nearly absent, forewing margins not entirely black .. 4
4. Male antenna with distinct yellowish ochreous part distally; rings on 2nd, 4th, 6th and 7th segments broader, 3, 5 narrow .... 45. *B. pavicevici*
– Male antenna without distinct yellowish ochreous part distally, or female 5
5. Yellow rings on 2-6th (7th) segments nearly of the same width, ring on 5th segment occasionally narrower or absent; ring on 3rd segment occasionally slightly narrower, at least distinct dorsally (occasionally interrupted laterally); abdominal segments 2-6(7) with narrow rings ventrally, rings on 2nd and 3rd segments occasionally indistinct 6
– Abdominal rings of dissimilar width or some of them absent; yellow ring on 3rd segment indistinct or absent; occasionally present, then apical area of forewing absent 8
6. Forewing with only yellow coloration (without orange); small species .... 50. *Bembecia blanka*
– Forewing with yellow and orange coloration; usually larger species 7
7. Yellow ring on 3rd segment of the same width as that on 2nd segment, dorsally and laterally .... 43. *B. ichneumoniformis*
– Yellow ring on 3rd segment slightly narrower than that on 2nd segment, interrupted or thinned laterally .... 47. *B. scopigera*
8. Base of hindwing and fringes near base red (anal margin of forewing and apical area orange-red; abdominal segments 2, 4, 6 dorsally and 4 ventrally with yellow ring .... 40. *B. sareptana*
– Hindwing base and fringes near base not red 9

9. Apical area of forewing narrow, yellow; 4th abdominal segment almost entirely yellow dorsally, occasionally other segments (2, 6, 7 and 3, 5) with yellow rings of varying width; metathorax with long whitish grey hairs laterally; a robust species (wingspan more than 25 mm) .... 39. *B. lomatiaeformis*
– Apical area of forewing yellow, orange, dark with orange spots or absent; abdominal rings yellow; hairs on metathorax laterally not whitish grey and long; wingspan usually less than 25 mm 10

10. Apical area of forewing quite absent (Fig. 53a), or (in female) very narrow with orange spots indistinctly limited towards external transparent area; (abdominal rings 2, 4, 6, 7 broader and 3, 5 narrow or absent in male; in female rings on 2nd, 4th and 6th segments dorsally, 4th ventrally (Fig. 53b), anal tuft black) .... 53. *B. puella*
– Apical area distinct; anal tuft in female not entirely black 11

11. Abdominal segment 3 without yellow ring; light hairs in dorsal part of anal tuft darker than yellow abdominal rings, ochreous-yellow or rusty 12
– Dorsal ring of 3rd segment present or at least indicated; anal tuft almost entirely yellow dorsally, of the same coloration as abdominal rings 15

12. Forewing dark brown to black with orange anal margin, orange discal spot and orange spots between veins in apical area 13
– Forewing brown with yellow anal margin, partly yellow discal spot and yellow apical area 14

13. Hind tibia in proximal and distal part dark dorsally, only in middle narrowly yellow or at least with distinct black ring distally; female antenna ochreous-yellow with black apex .... 51. *B. fokidensis*
– Hind tibia nearly entirely yellow or orange, or with undefined blackish ring distally; female antenna black or black-brown basally 21

14. Discal spot of forewing almost entirely yellow; apical area yellow without dark veins .... 56. *B. flavida*
– Discal spot almost entirely brown; apical area with dark veins.. 41. *B. volgensis*

15. Hind tibia and tarsus orange, distinctly darker than abdominal rings; dark coloration on the inner side of hind tibia distally indistinct or absent; abdominal rings ventrally nearly of the same width .... 47. *B. scopigera*
– Hind tibia and tarsus of the same coloration as abdominal rings; dark coloration (ring) on hind tibia distally usually distinct and continuing on inner side; abdominal rings ventrally of varying width .... 54. *B. sirphiformis*

16. Segments 2, 4, 6 with narrow white rings; male forewing grey-brown or black with small orange spot on discal spot and with paler spots in apical area; abdomen covered with greyish yellow scales; female black, almost without transparent areas on forewing; a small species .... 38. *B. hymenopteriformis*
– Abdominal rings yellow; forewing and legs usually with yellow or orange coloration 17

17. Forewing covered nearly all over with black scales; transparent areas small or absent; reddish spot on discal spot; 2nd and 4th (often entire) segments with whitish yellow rings; a robust species (Fig. 39a) .... 39. *B. lomatiaeformis*
– Forewing with transparent areas; more abdominal rings; if quite exceptionally forewing without transparent areas, nearly all segments with narrow yellow rings or very small species (wingspan to 16 mm) 18

18. Abdominal rings on segments 2, 4, 6 or additional rings only indicated 19
– Abdominal rings or margins also on other segments 24

19. Antenna black with white scales or spot before apex .... 58. *B. uroceriformis*
– Antenna without white spot or scales before apex 20

20. Antenna black 21
– Antenna ochreous yellow with black apex 23

21. Longitudinal transparent area present; anal tuft more or less dark; hind tibia orange; apical area with dark veins 22
– Longitudinal transparent area absent; anal tuft almost entirely yellow; hind tibia yellow; apical area without dark veins .... 56. *B. flavida*

22. Discal spot of hindwing reaches to M2 .... 52. *B. megillaeformis*
– Discal spot of hindwing reaches to M3-Cu1 .... 41. *B. volgensis*

23. Longitudinal transparent area absent; hind tibia proximally and distally black, yellow in middle, not strongly hairy .... 51. *B. fokidensis*
– Longitudinal transparent area present; hind tibia yellow with dark ring distally, strongly hairy (Fig. 54b) .... 54. *B. sirphiformis*

24. Abdominal segments 2, 3, 4, 5, 6, (7) with yellow or whitish yellow rings of the same width; ring on 5th segment occasionally narrower or exceptionally absent; external transparent area broader than high, consisting of 3 cells between R5-M3, occasionally with very small cell between M3 and Cu1 (Fig. 44a), quite exceptionally external transparent area absent .... 44. *B. albanensis*
– Abdominal rings of unequal width, usually broader on segments 2, 4, 6, (7) and distinctly narrower on 3rd (and 5th) segment 25

25. External transparent area very small, round (3 cells) or slightly or clearly broader than high (not higher than broad), consisting of 3 cells, occasionally with a further small cell below them, more or less distinctly delimited from apical area; discal spot of forewing almost entirely yellow or with conspicuous orange-red spot 26
– External transparent area larger, round or higher than broad, consisting of at least 4 cells 28

26. Discal spot of forewing with orange to red spot; if almost entirely orange-yellow in female, then antenna not black and without white subapical spot; male antenna dark or ochreous brown basally, in female usually brown or orange-yellow in basal 2/3 27
– Discal spot almost entirely yellow; apical area in male with yellow spots between veins to almost completely dark, in female conspicuously yellow; male antenna black, female with white subapical spot .... 58. *B. uroceriformis*

27. Apical area of forewing with distinct orange spots between veins to almost completely dark; 2nd abdominal segment with conspicuous broad yellow border ventrally; yellow ring of 5th segment narrower than those on 4th and 6th segments .... 49. *B. iberica*
– Apical area entirely yellow, with conspicuously dark veins only distally; 2 nd abdominal segment with interrupted yellow margin; 5 th segment with broad yellow ring .... 48. *B. priesneri*

28. Antenna in distal half with ochreous-yellow part (male) or basal 3/4 entirely ochreous–yellow, apex dark (female); 2nd, 4th, 6th (7th) segments with slightly broader and 3rd and 5th with narrow yellow rings; hind tibia in distal 1/3 dark, not conspicuously hairy; segments 4-6 (7) narrowly bordered ventrally ... . 45. *B. pavicevici*
– Antenna entirely black or with white spot (scales) before apex; if yellow or ochreous basally, hind tibia and 1st tarsal segment conspicuously hairy; hind tibia yellow, with dark ring distally; segments 4-6 (7) almost entirely yellow ventrally 29

29. External transparent area small, narrow, indistinctly delimited from yellow or orange apical area with brown veins, or transparent area clearly delimited, apical area dark and discal spot of forewing almost entirely dark, only distally partly yellow (Figs 57a,b)(male antenna black, in female covered with white scales subapically; discal and external transparent area exceptionally absent; in male segments 2, 4, 6 with broader and 3, 5 with narrow yellow rings, in female 2, 4, 5, 6, occasionally 3 with broad rings; longitudinal transparent area in female absent) .... 57. *B. himmighoffeni*
– External transparent area distinctly delimited from yellow or yellow-orange apical area; longitudinal transparent area present in female 30

30. Discal spot of forewing almost entirely yellow or orange; external transparent area usually square, or slightly higher than broad (Fig. 46a); male antenna black, in female black or with white scales before apex; male abdominal segments 2-7 with yellow rings ventrally, rings on 2nd and 3rd segments slightly narrower or discontinuous .... 46. *B. fibigeri*
– Discal spot of forewing almost entirely brown; external transparent area usually broader than high (Fig. 54a); male antenna black or ochreous basally, in female basal 2/3 orange, apex dark; male abdominal segments 4-7 almost entirely yellow ventrally, rings on 2nd and 3rd segments absent or only indicated .... 54. *B. sirphiformis*

Key to the genera *Pyropteron*, *Dipchasphecia* and *Chamaesphecia*

1. Forewing partly orange or red, at least anal margin red basally 2
– Forewing without orange or red (occasionally orange spots in apical area) 8

2. Antenna entirely black, or with whitish (ochreous) spot before apex 3
– Antenna yellow (continuous or individual scales), (abdominal segments 2, 4, 6 with narrow silvery white rings; forewing and abdomen in male brown with varying quantity of yellow scales, and/or rings; tegula in female with red border; abdomen distinctly covered with red scales) .... 61. *P. doryliformis*

3. Abdomen with 2-4 narrow yellow or white rings; if only 1 ring present or rings quite absent, 1st tarsal segment of hind legs conspicuously hairy (distinctly more than other segments) 4
– Abdomen with 1 narrow white ring, without rings or with broad red rings; 1st tarsal segment of hind legs not conspicuously hairy 6

4. Tegula with yellow border (distinctly so in female); hairs on vertex rusty yellow; abdomen with yellow rings on 2nd, 4th, 6th (7th) segments (Fig. 60a); yellow scapular spot on forewing .... 60. *P. minianiformis*

– Tegula without yellow border; hairs on vertex in male dark; abdominal rings narrow, yellowish white or white; if white ring present on 2nd segment, scapular spot on forewing base indistinct (white) or absent 5

5. 2nd abdominal segment with narrow white or yellowish white ring .... 59. *P. chrysidiformis sicula*
– 2nd abdominal segment without ring .... 59. *P. chrysidiformis chrysidiformis*

6. Tegula with red border; abdomen entirely black, or with some red rings dorsally or ventrally; labial palpus entirely black; occasionally narrow white ring on 4th segment .... 96. *C. anthraciformis*
– Tegula and abdomen without red; labial palpus partly white 7

7. Abdomen black (without rings) .... 77. *C. chalciformis*
– Abdomen with narrow white ring on 4th segment .... 78. *C. schmidtiiformis*

8. Hindwing entirely transparent 9
– Large discal spot of hindwing connected by opaque band with broad wing border; (a robust, brown species; external transparent area narrow, high; longitudinal transparent area absent; whitish ring on 4th segment) .. 71. *P. umbrifera*

9. Abdomen entirely white ventrally (not only individual white scales), very slender (external transparent area of 4-5 cells; segments 2, 4, 6 with white rings)... .... 83. *C. albiventris*
– Abdomen not entirely white ventrally (occasionally with white scales) 10

10. Antenna dark dorsally and ventrally, occasionally with white or yellowish spot and with yellow basal segment; if exceptionally brown dorsally (with pale subapical spot) and ochreous externally, at the same time a robust species with narrow external transparent area and with white rings on 2nd and 4th segments only 11
– Antenna externally (or ventrally) entirely or partly whitish, yellow, ochreous (at least discontinuous scales), occasionally with yellow subapical spot 31

11. Body black without other colour markings .... 96. *C. anthraciformis*
– Body not entirely black 12

12. Frons white, yellow, clearly orange, or at least eye with white or clear yellow border (Fig. 106a) 13
– Frons brown to black, occasionally with metallic sheen 27

13. External transparent area narrow and high (higher than broad)(Fig. 73a); abdominal ring on 6th segment absent, or labial palpus in male with long black hairs, or hind tibia in female partly orange 14
– External transparent area nearly round or broader than high; if exceptionally slightly higher than broad, abdominal ring on 6th segment present, labial palpus in male not long hairy and hind tibia in female not orange 15

14. Labial palpus in male with long black scales; frons yellow; labial palpus in female orange, slightly hairy; frons orange; hind tibia in female orange in basal half, black distally; abdominal segments 2, 4, 6 with white rings; wingspan up to 18 mm .... 73. *P. mannii*
– Labial palpus shortly hairy; frons white; hind tibia in female not orange; segments 2 and 4 with white rings; wingspan over 20 mm .... 72. *P. cirgisa*

15. External transparent area small, broader than high, exceptionally round, of 3 cells, occasionally with additional very small cells above R4 and below M3 . 16

– External transparent area higher and larger, consisting of 4-5 cells (occasionally relatively narrow) 22

16. Abdominal segments 2, 4, 6 with more or less distinct white or yellow rings 17
– 2nd segment without white ring 20

17. Distinct dorsal spot line on abdomen; antenna without white subapical spot .... 75. *C. mysiniformis*
– Dorsal spot line on abdomen indistinct or absent 18

18. Abdominal rings white; body and wings greyish brown to black, more or less covered with yellow scales; labial palpus not orange 19
– Abdominal rings yellow; body and wings dark brown-black, with metallic sheen; labial palpus and hind tibia partly orange .... 67. *P. kautzi*

19. Longitudinal transparent area absent; abdominal ring on 2nd segment often indistinct; body and wings with varying quantity of yellow-ochreous scales; antenna in male without distinct white subapical spot .... 63. *P. meriaeformis*
– Longitudinal transparent area in male present, in female mostly indicated basally; abdominal ring on 2nd segment distinct; male with white subapical spot on antenna; body and wings without yellow scales .... 64. *P. hispanica*

20. Tegula bordered yellowish white; antenna in female without white subapical spot; longitudinal transparent area in male at least basally present; thorax whitish laterally below forewing; ground coloration greyish silver . 70. *P. affinis*
– Tegula with yellow border; antenna in female with white subapical spot; thorax yellow laterally below forewing; ground coloration brown to brown-black 21

21. Longitudinal transparent area in male present; thorax below forewing with distinct yellow spot; frons more or less white or yellowish .. 69. *P. leucomelaena*
– Longitudinal transparent area in male absent; yellow lateral spot on thorax usually indistinct; frons dark in middle with white borders before eyes; abdominal segments (2), 4, 6 with white rings; coloration of body and wings greyish brown to black with varying quantity of yellow scales .... 63. *P. meriaeformis*

22. Abdominal segments 4 and 6 with white rings .... 69. *P. leucomelaena*
– Abdominal segments 2, 4, 6 with white rings 23

23. Discal spot of hindwing more or less of same width to M2, suddenly ending or narrowing between M2 and M3 (Fig. 62a); longitudinal transparent area in female usually present; labial palpus in male more or less smooth 24
– Discal spot of hindwing narrowed to M2 (triangular)(Fig. 66a); longitudinal transparent area in female usually absent; labial palpus in male slightly hairy 26

24. Ground coloration yellowish brown; anal tuft in male usually trifid; antenna in male without white subapical spot .... 62. *P. triannuliformis*
– Ground coloration silver (whitish)-grey (brown-black); anal tuft in male simple; antenna with white subapical spot also in male (in European specimens) ... 25

25. Ground coloration extremely dark, black, with metallic sheen; abdomen without whitish coloration (scales) laterally (at the pleura); hind tibia black, white only in fore part laterally .... 68. *P. aistleitneri*
– Ground coloration grey to brown-black; abdomen usually white laterally on most segments; hind tibia white in fore part and with some white scales at the end .... 64. *P. hispanica*

26. Spots in apical area of forewing yellowish or ochreous; tegula with yellow border; ground coloration more yellowish brown .... 66. *P. muscaeformis*
– Spots in apical area of forewing pale ochreous; tegula with whitish or whitish yellow border; ground coloration whitish grey .... 65. *P. koschwitzi*

27. Distinct continuous dorsal spot line on abdomen; 4th segment with white ring (Fig. 101b); tegula with whitish border .... 101. *C. leucopsiformis*
– Abdomen without distinct dorsal spot line (occasionally individual yellow spots on fore segments); abdominal segments (2), 4, 6 (7) with white or yellow rings; tegula with yellow border or entirely black 28

28. Abdominal segments 4, 6 (7) with yellow or yellowish white rings; hind tibia black with distinct white or whitish yellow distal end (Fig. 94a); anal tuft in female often with ochreous scales .... 94. *C. nigrifrons*
– Abdominal segments 4, 6 or 2, 4, 6 with white rings (white ring on 6th segment occasionally indistinct); hind tibia without distinct white or yellow distal end; anal tuft in female dark 29

29. External transparent area round or high, usually of 5 cells (Fig. 79b); labial palpus in male long black hairy (Fig. 79a); thorax dark below forewing 30
– External transparent area very small, usually slightly broader than high, of 3 cells (Fig. 82b); labial palpus in male nearly smooth, white (Fig. 82a); thorax laterally below forewing with yellow spot .... 82. *C. aerifrons aerifrons*

30. Forewing and body greyish black; external transparent area round; abdominal segments 4, 6 with white rings .... 79. *C. anthrax*
– Forewing and body black; external transparent area high, narrow; abdominal segments 4, 6 in male and 2, 4, 6 in female with white rings .. 80. *C. maurusia*

31. Frons light, white, yellow, orange or at least eyes with white borders 32
– Frons dark (brown to black), occasionally with metallic sheen, at least with individual yellowish or orange scales before eyes 49

32. Abdomen with clear yellow rings of varying width; segments 4, 6 (7) almost entirely yellow; narrow white rings absent; (labial palpus in male yellow with long black scales ventrally; in female yellow; hind tibia strongly hairy, deep yellow, with black ring distally; distinct yellow scapular spot) . 92. *C. proximata*
– Abdomen with only narrow white or yellowish white rings, or in combination with broader variously sharp yellow or orange rings of varying width, i.e. rings yellow or orange in proximal parts and with white border distally 33

33. Abdomen with only narrow white or yellowish white rings without broader clear yellow or orange rings and without groups of clear yellow, greenish yellow or orange scales on abdomen and on wings; if exceptionally abdomen with groups of ochreous-orange scales, hind tibia strongly hairy, ochreous 34
– Abdomen with narrow white to yellowish white rings on distal border of segments (usually segments 2, 4, 6, 7) and with broader yellow to orange rings before them, or with more extensive scattered clear yellow coloration 42

34. Antenna whitish externally; transparent areas small, covered with whitish scales; cells of external transparent area get shorter from R5 to Cu2 distally (Fig. 74a) .... 74. *D. lanipes*
– Antenna yellow to rusty externally; transparent areas usually not covered with white scales 35

35. External transparent area distinctly broader than high, small, of 3 cells (exceptionally another 2 very small cells)(Fig. 75a); hind tibia not strongly ochreous hairy; (segments 2, 4, 6 with narrow white rings) 41
– External transparent area not distinctly broader than high, usually round or higher than broad, of 4-5 cells; if exceptionally small, of 3 cells, broader than high, hind tibia strongly ochreous hairy (Fig. 86d) 36

36. Discal spot of hindwing broadly reaching M2 (Fig. 86b) 37
– Discal spot of hindwing narrow, usually reaching M3, only slightly narrower between M2 and M3 than between M1 and M2 (Fig. 95b) .. 95. *C. bibioniformis*

37. Labial palpus yellow, with long black scales ventrally, without white coloration; apical area of forewing brownish, without pale spots .... 61. *P. doryliformis*
– Labial palpus at least partly white; apical area at least with indicated or indistinct spots between veins 38

38. Whitish yellow ring on 4th segment, distinctly broadened dorsally and laterally; antenna clear brown (above); external transparent area conspicuously narrow (Fig. 84a) .... 84. *C. osmiaeformis*
– Abdomen with 2-3 rings, not broadened dorsally; antenna dark; external transparent area not conspicuously narrow 39

39. Ground coloration brown to ochreous-brown; discal spot of forewing slightly concave towards external transparent area; apical area with ochreous spots between veins (Fig. 86a) 40
– Ground coloration grey with varying quantity of whitish scales; discal spot of forewing more or less straight towards external transparent area; light greyish spots in apical area distinct in its distal half (Fig. 85a) .... 85. *C. ramburi*

40. 2nd, 4th and 6th abdominal segments with whitish ring; hind tibia slightly light ochreous hairy (Fig. 86c) .... 86. *C. doleriformis colpiformis*
– 2nd segment without ring, 4th and 6th one with ring; hind tibia distinctly deep ochreous hairy (Fig. 86d) .... 86. *C. doleriformis doleriformis*

41. Forewing borders, apical area, abdomen and anal tuft unicolorous dark brown; segments 2, 4, 6 with white rings dorsally; abdomen without dorsal spot line and without any rings ventrally .... 83. *C. albiventris*
– Forewing borders and abdomen more or less covered with whitish or ochreous scales; apical area with more light spots (at least indicated); anal tuft with greyish hairs in female; usually white or yellowish white dorsal spot line on abdomen .... 75. *C. mysiniformis*

42. Labial palpus yellow, without white colour, with long black scales ventrally; antenna with whitish scales subapically .... 61. *P. doryliformis*
– Labial palpus smooth, partly white, exceptionally without white coloration, with black scales ventrolaterally; antenna without white scales before apex . 43

43. Scapular spot on forewing base (Fig. 93a) 44
– Scapular spot absent 48

44. Apical area of forewing with distinct yellowish or ochreous spots between veins; usually without sharply defined yellow rings on abdomen (yellow or greenish yellow coloration present) .... 95. *C. bibioniformis*
– Apical area of forewing without distinct light spots, unicolorous or with indistinct spots; more or less sharply defined orange rings on abdomen 45

45. Forewing narrow; external transparent area slightly broader than high, of 3-5 cells in male and 3 in female (Fig. 90a) .... 90. *C. annellata*, (91. *C. staudingeri*)
– Forewing slightly broader distally; external transparent area round 46

46. Transparent areas relatively small, as broad as apical area or narrower, round, of 5 cells in male and 3-(5) in female (Fig. 87a) .... 87. *C. thracica*
– Transparent areas large, usually broader than apical area, of 5 cells in male and 3-5 in female (Figs 88a, 89a) 47

47. Distinct orange rings on abdomen .... 89. *C. oxybeliformis*
– Orange rings usually less defined .... 88. *C. dumonti*

48. Forewing and abdomen in male more or less covered with yellow scales; male forewing conspicuously broad with yellow apical area with dark veins; abdomen slender; in female abdomen and forewing only partly covered with yellow scales or almost without them; spots in apical area indistinct or absent (Fig. 106b); discal spot of hindwing tapers from M2 to M3 .... 106. *C. astatiformis*
– Forewing and abdomen of both sexes more or less covered with ochreous yellow scales; male forewing not conspicuously broad and abdomen not distinctly slender; discal spot of hindwing reaches M2 (Fig. 76) .... 76. *C. anatolica*

49. Abdomen with conspicuous orange or deep yellow rings, without narrow white rings; scapular spot present (Fig. 93a) .... 93. *C. masariformis*
– Abdomen with several narrow white or yellow rings; scapular spot of forewing absent; if present, ground coloration inconspicuous, light brown or greyish .. 50

50. Longitudinal transparent area absent; other transparent areas very small, covered with yellow scales; forewing narrow, dark; abdomen slender, with narrow yellow rings on 2nd, 4th and 6th segments; transverse yellow spot on metathorax (Fig. 81a) .... 81. *C. alysoniformis*
– Transparent areas not covered with yellow scales; longitudinal transparent area present or absent; 1-3 white or yellowish white rings on abdomen; metathorax without distinct transverse yellow spot 51

51. External transparent area distinctly broader than high, small, exceptionally round, usually of 3 cells (exceptionally another 2 very small cells); 1 or 3 white or yellowish white rings on abdomen 52
– External transparent area more or less round or higher than broad, often large, usually of 4-5 cells; abdomen with 1-3 white or yellowish white rings 54

52. Apical area of forewing unicolorous or with only indistinct spots; longitudinal transparent area in female absent, in male absent or indicated basally; dorsal spot line of abdomen absent or indistinct (individual spots on fore segments); abdomen with narrow rings on 2nd, 4th and 6th segments (Sardinia and Corsica only) .... 82. *C. aerifrons sardoa*
– Apical area of forewing with distinct spots between veins; abdomen with 1 or 3 rings; if 3 rings present, longitudinal transparent area in male present, in female at least indicated; distinct dorsal spot line on abdomen 53

53. Dorsal spot line on abdomen continuous; 4th segment with white ring; apical area of forewing with light spots between R5 and M3 (Figs 101a, b)............ .... 101. *C. leucopsiformis*
– Dorsal spot line of abdomen discontinuous; 2nd, 4th and 6th segments with white ring; apical area of forewing with light spots between R3 and Cu1 (Figs 100a, b) .... 100. *C. crassicornis*

54. Labial palpus, and/or fore coxa at least partly white 55
– Labial palpus and fore coxa not white, usually yellow or dark 56

55. Labial palpus in male with long black scales ventrally, white dorsolaterally; in female white ventrally, nearly smooth; ground coloration brownish black to black; segments 4, 6 with narrow white rings .... 79. *C. anthrax*
– Labial palpus smooth or only slightly hairy, with different combination of characters; if labial palpus long hairy, body brown 38

56. One white or yellowish white ring on 4th segment; hind tibia not clearly yellow with dark ring distally 57
– Narrow white rings on (2nd), 4th and 6th segments and varying quantity of yellow scales (rings); hind tibia yellow, usually with dark ring distally 60

57. Antenna of male dark dorsally, of female with whitish spot subapically 58
– Antenna brown to yellowish brown, occasionally with dark apex 59

58. External transparent area nearly as broad as apical area; dark margins of forewing of normal width (Fig. 98a) .... 98. *C. euceraeformis*
– External transparent area narrow, narrower than apical area; dark margins of forewing broad, transparent areas small (Fig. 99a); ground coloration dark .... 99. *C. amygdaloidis*

59. Wingspan less than 20 (22) mm; external transparent area conspicuously narrow, high (Fig. 84a); hind tibia proximally ochreous, distally brown to dark brown .... 84. *C. osmiaeformis*
– Wingspan over 22 (20) mm; external transparent area large, more or less round; hind tibia without distinct contrasts; antenna pale brown (yellowish brown), dark apically; body brown to yellowish brown .... 97. *C. palustris*

60. Apical area of forewing unicolorous, without yellow spots between veins; antenna with white scales subapically, or largely brown; hind tibia ochreous yellow, with darker shade distally .... 61. *P. doryliformis*
– Apical area of forewing with yellow or ochreous spots between veins (at least indicated)(Fig. 104a); antenna entirely black dorsally; hind tibia yellow, darker proximally and with black ring distally 61

61. Forewing and abdomen brownish or ashy greyish black; segments 4, 6 (7) with narrow white rings; entire 4th segment covered with whitish scales, whitish scales partly on segments 6, 7 (Caucasus only) .... 102. *C. guriensis*
– Ground coloration less dark, clearer, with more yellow or greenish yellow scales on wings, on abdomen and anal tuft; often also 2nd segment with white ring and yellow coloration 62

62. Yellow scales on abdomen more greenish yellow; ground coloration dark, indistinct; 2nd abdominal segment often without white ring ... 103. *C. hungarica*
– Clear yellow scales on abdomen; 2nd segment with white ring 63

63. External transparent area round or broader than high; less conspicuous yellow spot between R3 and R4 above transparent area (Fig. 104a); forewing narrow distally; yellow coloration on abdomen usually scattered, without sharply defined yellow rings .... 104. *C. empiformis*
– External transparent area higher than broad; a conspicuous yellow spot between R3 and R4 above transparent area (Fig. 105b); forewing relatively broad distally; yellow rings on abdomen more or less sharply defined.............
.... 105. *C. tenthrediniformis*

Systematic Treatment of the European Sesiidae

Sesiidae Boisduval, 1828

Sesiariae Boisduval, 1828: 29. Type genus: *Sesia* Fabricius, 1775.
Aegeriidae Stephens, 1828: 136. Type genus: *Aegeria* Fabricius, 1807.

Tinthiinae Le Cerf, 1917

Tinthiinae Le Cerf, 1917: 148. Type genus: *Tinthia* Walker, [1865].
Zenodoxinae MacKay, 1968: 5. Type genus: *Zenodoxus* Grote & Robinson, 1868.

SUBFAMILY CHARACTERS.– Antenna filiform or setiform without scale tuft apically; A1 of hindwing developed, A4 absent; tegumen and uncus separate, usually without gnathos; vinculum without processes; distinct coecum penis.

The subfamily is represented by 2 tribes: Tinthiini and Pennisetiini, 2 genera and 6 known species in Europe.

Tinthiini Le Cerf, 1917

Tinthiinae Le Cerf, 1917: 148. Type genus: *Tinthia* Walker, [1865].

TRIBE CHARACTERS.– Antenna filiform, ciliate in male (in European species); forewing opaque; Cu1 of hindwing arising before crossvein; the base of M2 approximated to M1; tegumen broad without gnathos; uncus slender, elongate. 1 genus with 4 species in the region.

Tinthia Walker, [1865]

Tinthia Walker, [1865]: 23.
Type species: Tinthia varipes Walker, [1865]; by monotypy.
Microsphecia Bartel, 1912: 414.
Type species: *Sphinx tineiformis* Esper, [1789] sensu Bartel, 1912 [= *Sphinx brosiformis* Hübner, [1813]]; by original designation.

GENUS CHARACTERS.– Veins R1-R5 of forewing separate, Cu2 reduced or occasionally absent. Larva in the root of herbaceous plants of Rosaceae and Convolvulaceae. 4 species in the region.

1. *Tinthia tineiformis* (Esper, [1789])

(Pl. 2: 1)

Sphinx tineiformis Esper, [1789]: 9, pl. 38, fig. 4. Type locality: Italy, Florence. Type material: not traced.

DIAGNOSIS.– Wingspan 12-18 mm; proboscis developed; forewing without conspicuous yellowish white spot near outer edge (Fig. 1b); 1st abdominal segment not distinctly whitish (Fig. 1c).

GENITALIA.– Fig. 1; uncus thin, tegumen hooked basally; valve broadest in the middle; saccus and coecum penis of aedeagus short; 8th segment of female and antrum elongate; signum in corpus bursae.

BIOLOGY AND HABITAT.– Xerothermic seminatural to ruderal treeless places; hostplant: probably *Convolvulus* sp. (Špatenka, pers. comm.); adult V-VII.

DISTRIBUTION.– Holomediterranean: S Europe.

2. *Tinthia brosiformis* (Hübner, [1813])

(Pl. 2: 5)

Sphinx brosiformis Hübner, [1813]: pl. 25, fig. 116. Type locality: Slovakia, Štúrovo. Type material: neotype ♂, NHMV (des. by Laštůvka, [1987]).

DIAGNOSIS.– Wingspan 11-16 mm; proboscis reduced; forewing with conspicuous yellowish white spot near outer edge (Fig. 2a); 1st abdominal segment white or yellowish (Fig. 2b, cf. Laštůvka, [1987]).

GENITALIA.– Fig. 2; uncus more robust than in *T. tineiformis*; tegumen without hooks basally; valve broadest basally; saccus and coecum penis longer; in female 8th segment and antrum short; corpus bursae without signum.

BIOLOGY AND HABITAT.– Xerothermic seminatural to ruderal grassy places; hostplants: *Convolvulus arvensis* L., *Convolvulus* sp.; larva 1 year in root (Laštůvka, 1985); adult VI-IX.

DISTRIBUTION.– E Mediterranean-Asiatic: SE central, SE and E Europe.

3. *Tinthia myrmosaeformis* (Herrich-Schäffer, 1846)

(Pl. 2: 2; 9: 1)

Paranthrena myrmosaeformis Herrich-Schäffer, 1846: 59, pl. 6, figs 30, 31. Type locality: Turkey. Type material: not traced.

Paranthrene myrmosiformis var. *cingulata* Staudinger, 1870: 100. Type locality: Greece, Mt. Parnass, Kastri. Type material: lectotype ♂, ZMHB (des. by Căpuşe, 1971).

DIAGNOSIS.– Wingspan 14-23 mm; forewing dark brownish black or black with bluish metallic sheen; hind tibia orange in middle; abdomen in nominotypical form usually without rings, in ssp. *cingulata* with narrow yellowish white rings on 3, 5, 6, 7th segments in male (Fig. 3a), 4, 5, 6th in female and with yellow spots on metathorax laterally.

GENITALIA.– Fig. 3.

BIOLOGY AND HABITAT.– Natural or seminatural grassy or shrubby places; hostplants: *Potentilla recta* L. (Laštůvka & Laštůvka, 1980); *Potentilla taurica* Willd. (Toševski, pers. comm.); larva 1 year in root; adult V-VII.

DISTRIBUTION.– E Mediterranean: ssp. *cingulata* in Albania, Macedonia, Greece and western Bulgaria; nominotypical ssp. in Bulgaria, European Turkey, SE Romania, Crimea and southern Russia.

REMARKS.– The form (ssp.) *cingulata* is considered by some authors as a separate species (cf. Špatenka et al., 1993, 1999).

4. ***Tinthia hoplisiformis*** (Mann, 1864)

(Pl. 9: 2)

Paranthrene hoplisiformis Mann, 1864: 176. Type locality: Turkey, Brussa [Bursa]. Type material: lectotype ♀, ZMHB (des. by Špatenka & Laštůvka, 1988).

DIAGNOSIS.– Wingspan 15-24 mm; metathorax with four yellow spots; forewing yellowish brown to brown; abdomen brown or black with broad yellow rings on the 3rd, 5th, 6th (7th) segments; hind tibia yellow. European specimens are slightly larger with darker forewings.

GENITALIA.– Fig. 4.

BIONOMICS AND HABITAT.– Grassy or shrubby places, in Europe mesophilous habitats from 400 to 800 m above sea level; hostplant: *Poterium minor* Scop. (Gorbunov, pers. comm.); larva 1 year in root; adult VI.

DISTRIBUTION.– E Mediterranean: only several localities in northern Greece (Lingenhöle, Bartsch, pers. comm., Petersen & Bartsch, 1998).

Pennisetiini Naumann, 1971

Pennisetiini Naumann, 1971: 55. Type genus: *Pennisetia* Dehne, 1850.

TRIBE CHARACTERS.– Antenna setiform, bipectinate or ciliate in male; M3 and Cu1 of hindwing long stalked; uncus large; tegumen small; ductus bursae long. 1 genus with 2 species in the region.

Pennisetia Dehne, 1850

Pennisetia Dehne, 1850: 28.
Type species: *Pennisetia anomala* Dehne, 1850 [= *Sesia hylaeiformis* Laspeyres, 1801], by monotypy.

GENUS CHARACTERS.– Antenna setiform, bipectinate in male; reduced proboscis; transparent areas present; juxta large, tubular; lamella postvaginalis in female broad, sclerotised; ductus bursae spirally twisted. Larva in the root and basal part of the stem of *Rubus* and *Rosa* (Rosaceae). 2 species in the region.

***5. Pennisetia hylaeiformis* (Laspeyres, 1801)**

(Pl. 1: 1)

Sesia hylaeiformis Laspeyres, 1801: 14. Type locality: Germany. Type material: lost.

DIAGNOSIS.– Wingspan 18-31 mm; labial palpus with black scales; thorax black below forewing; all yellow rings on abdomen of the same width (Fig. 5c).

GENITALIA.– Fig. 5.

BIOLOGY AND HABITAT.– Forest edges and clearings, river banks, raspberry crops; hostplant: *Rubus idaeus* L.; larva 2 years in root; adult VII-VIII(IX).

DISTRIBUTION.– Eurasiatic: central, N, parts of S Europe, not in Britain.

REMARKS.– The relationship to the North American taxon, *Pennisetia marginata* (Harris, 1839), is not known.

6. ***Pennisetia bohemica*** Králíček & Povolný, 1974

(Pl. 1: 2,3)

Pennisetia bohemica Králíček & Povolný, 1974: 165. Type locality: Czech Republic, Prague. Type material: holotype ♂, MK.

DIAGNOSIS.– Wingspan 23-32 mm; labial palpus deep yellow without black scales; thorax with yellow spot laterally below forewing; broad yellow rings on abdominal segments 6 and 7 (in male), narrow on other segments (Fig. 6b).

GENITALIA.– Fig. 6.

BIOLOGY AND HABITAT.– Shrubby, rocky places and slopes, roadsides, embankments; hostplant: *Rosa canina* L. s.l.; larva 2 years in root; pupation in gallery without cocoon (Schwarz & Tolman, 1961); adult VII-VIII(IX).

DISTRIBUTION.– Central Bohemia (Silbernagel, 1943; Priesner & Špatenka, 1990), Caucasus: Teberda (Gorbunov, pers. comm.).

REMARKS.– The origin of this taxon and the relationship to the East Palaearctic *Pennisetia pectinata* (Staudinger, 1883) requires clarification.

Sesiinae Boisduval, 1828

Sesiariae Boisduval, 1828: 29. Type genus: *Sesia* Fabricius, 1775.

SUBFAMILY CHARACTERS.– Antenna clavate with scale tuft apically; A1 of hindwing indicated, covered with line of scales; distinct gnathos in male genitalia; vinculum with processes; valve and/or uncus often with specialised hairs.

The subfamily is represented by 4 tribes: Sesiini, Osminiini, Paranthrenini and Synanthedonini, 9 genera and 101 known species in Europe.

Sesiini Boisduval, 1828

Sesiariae Boisduval, 1828: 29. Type genus: *Sesia* Fabricius, 1775.

TRIBE CHARACTERS.– Antenna bipectinate in male; R4-R5 of forewing and M3-Cu1 of hindwing stalked; valve with simple large hairs; tegumen and uncus separate; gnathos distinct; antrum usually membranous. 1 genus with 4 species in the region.

Sesia Fabricius, 1775

Sesia Fabricius, 1775: 547.
Type species: *Sphinx apiformis* Clerck, 1759; des. by Latreille, 1810.
Aegeria Fabricius, 1807: 288.
Type species: *Sphinx apiformis* Clerck, 1759; des. by Westwood, 1840.

GENUS CHARACTERS.– Large species with distinct "vespoid" mimicry; proboscis reduced; forewing entirely transparent or covered partly with translucent scales; aedeagus with coecum penis and with small sclerites apically; uncus bilobed. Larva xylophagous in Salicaceae (in European species). 4 species in the region.

7. *Sesia apiformis* (Clerck, 1759)

(Pl. 1: 13,14)

Sphinx apiformis Clerck, 1759: pl. 9, fig. 2. Type locality: not mentioned. Type material: lost.

DIAGNOSIS.– Wingspan 31-48 mm; labial palpus yellow; patagial collar black; tegula in fore part with extensive yellow spot; broad yellow rings on abdomen; 4th abdominal segment without ring (Fig. 7a).

GENITALIA.– Fig. 7.

BIOLOGY AND HABITAT.– Light forests, river banks, windbreaks, parks; hostplants: *Populus* spp. (*Salix* spp.); larva 3-4 years in basal part of trunk and in roots (Laštůvka, 1983b); pupa in tough cocoon; adult V-VIII.

DISTRIBUTION.– W Palaearctic: S, central and parts of N Europe.

8. *Sesia bembeciformis* (Hübner, [1806])

(Pl. 1: 15,16)

Sphinx bembeciformis Hübner, [1806]: 92, pl. 20, fig. 98. Type locality: Belgium, Brussels. Type material: lectotype ♀, NHMV (des. by Naumann, 1971).

Sphinx crabroniformis Lewin, 1797: 2 [nec *Sphinx crabroniformis* Denis & Schiffermüller, 1775]. Type locality: Great Britain. Type material: not traced.

Aegeria montelli Löfqvist, 1922: 82. Type locality: Finland, Kuopio (here restricted). Type material: lost.

DIAGNOSIS.– Wingspan 28-43 mm; patagial collar yellow; tegula black (Fig. 8a).

GENITALIA.– Fig. 8.

BIOLOGY AND HABITAT.– Scrub, meadows, forest edges, river bank vegetation; hostplants: *Salix* spp. (*Salix cinerea* L., *S. aurita* L., *S. caprea* L., etc.); larva 3-4 years in trunk; pupation in fine cocoon at the end of gallery; adult VI-VIII.

DISTRIBUTION.– European: central, W and N Europe.

9. *Sesia pimplaeformis* Oberthür, 1872

(Pl. 1: 11)

Sesia pimplaeformis Oberthür, 1872: 486, pl. 21, fig. 3. Type locality: Turkey, Van. Type material: not traced.

Trochilium maculiferum Staudinger, 1895: 290. Type locality: Syria, Akbés [Eibes]. Type material: holotype ♀, ZMHB.

DIAGNOSIS.– Wingspan 27-35 mm; labial palpus black ventrally; thorax with yellow spot below forewing; tegula with yellow spot in fore part; 4th abdominal segment yellow with round or triangular black spot dorsally (Fig. 9a).

GENITALIA.– Fig. 9.

BIOLOGY AND HABITAT.– Windbreaks, solitary poplar or willow trees, river banks; hostplants: *Populus* spp., *Salix* spp.; larva 2 years in trunk; pupation without cocoon in gallery (Zukowsky, 1935; Gorbunov, pers. comm.); adult V-VII.

DISTRIBUTION.– E Mediterranean: S Macedonia (Toševski, pers. comm.), SW Bulgaria (captured by Chládek), Greece (one specimen in ZMHB), Turkey, S Russia (Gorbunov, pers. comm.).

10. Sesia melanocephala Dalman, 1816

(Pl. 1: 12)

Sesia melanocephala Dalman, 1816: 217. Type locality: S Sweden. Type material: holotype ♀, RMS.

DIAGNOSIS.– Wingspan 27-38 mm; tegula bordered yellow anteriorly and posteriorly; forewing almost entirely transparent (Fig. 10a), partly covered with translucent brownish scales; abdomen with narrow yellow rings on segments 1, 2, 3, 5, 6 (7)(Fig. 10a).

GENITALIA.– Fig. 10.

BIOLOGY AND HABITAT.– Open forests, forest edges, groups of trees, roadsides; hostplant: *Populus tremula* L.; larva 3 years in trunk; pupation in dry twig, without cocoon; adult VI-VII.

DISTRIBUTION.– Eurasiatic: central, E, parts of W and N Europe.

Osminiini Duckworth & Eichlin, 1977

Osminiini Duckworth & Eichlin, 1977: 26. Type genus: *Osminia* Le Cerf, 1917.

TRIBE CHARACTERS.– Antenna without cilia in both sexes; R4 and R5 of forewing stalked; Cu1 of hindwing arising from the crossvein; crossvein shifted distally (Fig. 11a); tegumen, uncus and gnathos separate; uncus large; gnathos without crista medialis; corpus bursae with large signum. 1 genus with a single species in the region.

Osminia Le Cerf, 1917

Osminia Le Cerf, 1917: 327.
Type species: *Osminia ferruginea* Le Cerf, 1917; by original designation.

GENUS CHARACTERS.– See tribe characters; antenna clavate without cilia in both sexes; proboscis present. 1 species in the region (see Laštůvka, 1984).

11. Osminia fenusaeformis (Herrich-Schäffer, 1852)

(Pl. 2: 9)

Sesia fenusaeformis Herrich-Schäffer, 1852: 48. Type locality: Turkey. Type material: lectotype ♂, ZMHB (des. by Špatenka & Laštůvka, 1988).

DIAGNOSIS.– Wingspan 11-15 mm; forewing with 2 very small transparent areas; abdomen with yellow rings on segments 2, 4, 6.

GENITALIA.– Fig. 11.

BIOLOGY AND HABITAT.– Grassy places; bionomics unknown; adult IV-VI.

DISTRIBUTION.– E Mediterranean: in Europe known only from Crete.

Paranthrenini Niculescu, 1964

Paranthreniinae [sic] Niculescu, 1964: 34. Type genus: *Paranthrene* Hübner, [1819].

TRIBE CHARACTERS.– Antenna bipectinate or ciliate in male; Cu1 of hindwing arising before crossvein; tegumen and gnathos very small; uncus large; antrum short, ring-shaped; ductus bursae long, slender; corpus bursae often with transverse folds. 1 genus with 3 species in the region.

Paranthrene Hübner, [1819]

Paranthrene Hübner, [1819]: 128.
Type species: *Sphinx vespiformis* sensu Newman, 1832 [= *Sphinx tabaniformis* Rottemburg, 1775]; des. by Newman in Westwood, 1840.
Sciapteron Staudinger, 1854: 43.
Type species: *Sphinx asiliformis* [Denis & Schiffermüller], 1775 [= *Sphinx tabaniformis* Rottemburg, 1775]; by monotypy.

GENUS CHARACTERS.– Antenna bipectinate in male; forewing opaque, transparent basally and/or apically; valve with specialised furcate hairs dorsally. Larva xylophagous. 3 species in the region.

12. Paranthrene tabaniformis (Rottemburg, 1775)

(Pl. 1: 4,5,6)

Sphinx tabaniformis Rottemburg, 1775: 110. Type locality: Poland, Landsberg a.d. Warthe [Gorzów Wielkopolski]. Type material: lost.
Sphinx asiliformis [Denis & Schiffermüller], 1775: 305 [nec *Sphinx asiliformis* Rottemburg, 1775]. Type locality: Austria, Vienna. Type material: destroyed.
Sesia synagriformis Rambur, [1866]: 148. Type locality: Spain, Malaga. Type material: lectotype ♀, MNHP (des. by Špatenka, 1992b).

DIAGNOSIS.– Wingspan 20-38 mm; proboscis present; tegula black, or with yellow spot caudally; antenna black; forewing dark brown without transparent cells distally; abdominal segments 2, 4, 6 (7) with yellow rings (Fig. 12a), occasionally (in f. *rhingiaeformis* Hübner, 1790) yellow rings on all segments; in ssp. *synagriformis* antenna and forewing light brown; tegula with short yellow border caudally; metathorax with 2 small yellow spots (Fig. 12b) and yellow rings on all segments.

GENITALIA.– Fig. 12.

BIOLOGY AND HABITAT.– Open forests, windbreaks, poplar plantations; hostplants: *Populus* spp., *Salix* spp., (*Hippophae rhamnoides* L.); larva 2 years in trunk, in roots or in branches; pupation without cocoon in a gallery; adult V-VIII.

DISTRIBUTION.– Holarctic: nominotypical form in S, central, parts of N Europe; ssp. *synagriformis* in SE France, Sardinia (captured by Garrevoet and Pühringer) and in the Iberian Peninsula.

13. Paranthrene insolita Le Cerf, 1914

(Pl. 1: 7,8,9)

Paranthrene insolitus Le Cerf, 1914b: 422. Type locality: Syria, Akbés [Eibes]. Type material: holotype ♂, MNHP.

Paranthrene polonica Schnaider, [1939]: 143. Type locality: Ukraine, Lvov. Type material: lost.

Paranthrene novaki Toševski, 1987: 178. Type locality: Croatia, Muć Donji. Type material: holotype ♂, IPPB.

Paranthrene insolita hispanica Špatenka & Laštůvka, 1997: 20. Type locality: Spain, Andalusia, El Molinillo. Type material: holotype ♂, ZL.

DIAGNOSIS.– Wingspan 23-33 mm; proboscis short, yellowish; tegula and metathorax yellow bordered; antenna stout, longer than one half of forewing; forewing with 1-7 transparent cells distally; the transparency most developed in ssp. *polonica* (Fig. 13a) and reduced in nominotypical and particularly in ssp. *hispanica* (Fig. 13b); abdominal segments 2, 4-6 (7) in ssp. *polonica* and 2-6 (7) in ssp. *hispanica* with yellow rings.

GENITALIA.– Fig. 13; female genitalia not studied.

BIOLOGY AND HABITAT.– Open forests, forest-steppes, forest edges; hostplants: *Quercus* spp.; larva 2 years in a branch; pupation without cocoon (Schnaider, 1939; Auer, 1967; Bläsius, 1993); adult V-VII.

DISTRIBUTION.– W Palaearctic: central and S Europe (Toševski, 1987; Číla & Špatenka, 1987; Köhler, 1991; Petersen & Ernst, 1991; Scheuringer, 1991; Hamborg, 1991; Arheilger, 1992; Embacher, 1994; Cungs, 1998; Pühringer, 1998; Bettag & Bläsius, 1999a; Bąkowski, Ryrholm, pers. comm., etc.); ssp. *hispanica* in the Iberian Peninsula, ssp. *polonica* in the rest of Europe.

REMARKS.– In lepidopterological literature, article 31b of the ICZN remains generally unrespected. This fact is also reflected in this publication, except for *Paranthrene insolita*, in which case the grammatical error would be too flagrant and had to be corrected from *insolitus* to *insolita*.

14. Paranthrene diaphana Dalla Torre & Strand, 1925

(Pl. 1: 10)

Paranthrene tabaniformis var. *diaphana* Dalla Torre & Strand, 1925: 166. Type locality: Bosnia, Brčko. Type material: lectotype ♂, NHMV, type of ab. *diaphana* Schawerda, 1922 (des. by Toševski in Špatenka, 1992b, cf. Laštůvka, 1990f).

DIAGNOSIS.– Wingspan 24-27 mm; proboscis present; antenna yellowish brown with fine pectination in male; patagial collar yellow dorsally; tegula without yellow border; forewing with 1-3 transparent cells distally (between M1-Cu1); all abdominal segments with yellow rings (Fig. 14a).

GENITALIA.– Fig. 14.

BIOLOGY AND HABITAT.– Moist scrub, river banks, floodplain forests; hostplants: *Salix* spp. (*Salix alba* L.); *Populus alba* L.; larva 1-2 years in tumours on branches (Toševski, 1987 and pers. comm.); adult VI-VII.

DISTRIBUTION.– E Mediterranean, insufficiently known: Balkans, Anatolia.

Synanthedonini Niculescu, 1964

Synanthedoniinae [sic] Niculescu, 1964: 34. Type genus: *Synanthedon* Hübner, [1819].

TRIBE CHARACTERS.– Male antenna ciliate; R4-R5 of forewing and M3-Cu1 of hindwing stalked; valve and uncus with bifurcate hairs; united tegumen-uncus-gnathos complex, often with scopula; gnathos with crista gnathi and lateral wings. 6 genera with 93 species in the region.

Synanthedon **Hübner, [1819]**

Synanthedon Hübner, [1819]: 129.
Type species: *Sphinx oestriformis* Rottemburg, 1775 [= *Sphinx vespiformis* Linnaeus, 1761]; des. by Newman in Westwood, 1840.
Aegeria Fabricius, 1807, sensu Curtis, 1825.
Type species: *Sphinx culiciformis* Linnaeus, 1758; des. by Curtis, 1825 [incorrect type species designation].

GENUS CHARACTERS.– Proboscis present; longitudinal transparent area reaches as far as under the discal spot (up to Cu1); saccus usually shorter than 1:3 of valve; scopula long; aedeagus with internal and often also external sclerites apically, usually shorter than 3:4 of valve; lamella antevaginalis specialised, often sclerotised. Larva xylophagous, often with specific feeding pattern. 23 species in the region.

15. Synanthedon scoliaeformis (Borkhausen, 1789)
(Pl. 3: 1)

Sphinx scoliaeformis Borkhausen, 1789: 173. Type locality: Poland, Stettin. Type material: destroyed.

DIAGNOSIS.– Wingspan 25-35 mm; antenna yellowish white or with some white scales distally; eye with white border; labial palpus orange ventrally; discal spot of forewing conspicuously broad; abdomen with yellow rings on 2nd and 4th segments; anal tuft usually rusty.

GENITALIA.– Fig. 15.

BIOLOGY AND HABITAT.– Open forests, forest edges, groups of old birch trees; hostplants: *Betula pendula* Roth., *B. pubescens* Ehrh.; larva 2-3 years in trunk beneath bark; pupa in a tough cocoon in bark; adult V-VII.

DISTRIBUTION.– Eurasiatic: central and N Europe.

16. Synanthedon mesiaeformis (Herrich-Schäffer, 1846)
(Pl. 3: 2)

Sesia mesiaeformis Herrich-Schäffer, 1846: 74. Type locality: S Russia. Type material: holotype ♀, ZMHB.

DIAGNOSIS.– Wingspan 19-31 mm; antenna yellow distally; eye with white border; labial palpus orange ventrally; deep yellow rings on 2nd and 4th abdominal segments; ring on 4th segment distinctly broader than on 2nd; anal tuft black; hind tibia orange with black ring distally.

GENITALIA.– Fig. 16.

BIOLOGY AND HABITAT.– Floodplain forests, river banks, moist sandy places, with old alder trees; hostplant: *Alnus glutinosa* L.; larva 2 years beneath bark of old trees; pupation in bark in a cocoon; adult V-VII.

DISTRIBUTION.– W Palaearctic, disjunct: central and S France (Špatenka, pers. comm. and captured by Baumgarten), S Finland (Saramo, 1973), Estonia (Heidemaa & Kesküla, 1992), NE and E Poland (Buszko & Hołowiński, 1994; Bąkowski & Surmacki, 1995), S Hungary (Pazsiczky, 1941), Balkan Peninsula (Engelhard, 1975, etc.), Ukraine, S Russia.

17. Synanthedon spheciformis ([Denis & Schiffermüller], 1775)

(Pl. 3: 4)

Sphinx spheciformis [Denis & Schiffermüller], 1775: 306. Type locality: Germany, Frankfurt am Main. Type material: destroyed.

DIAGNOSIS.– Wingspan 23-31 mm; antenna yellowish white distally; eye without white border; labial palpus yellowish white ventrally; abdomen with narrow yellowish white ring on 2nd segment dorsally and on 4th segment ventrally.

GENITALIA.– Fig. 17.

BIOLOGY AND HABITAT.– Forest edges, clearings, river banks, meadows with groups of young alder trees, peat-bog edges, spring zones; hostplants: *Alnus* spp., *Betula* spp.; larva 2-3 years in the lower part of young trees, at first beneath bark, later in vertical gallery in the wood (about 15-30 cm long); pupation in the gallery without cocoon; adult V-VII.

DISTRIBUTION.– Eurasiatic: N and central, parts of S Europe.

18. Synanthedon stomoxiformis (Hübner, 1790)

(Pl. 3: 7,8; 9: 3)

Sphinx stomoxiformis Hübner, 1790: 93, pl. 3, fig. P. Type locality: Germany, Bavaria, Friedberg. Type material: lost.

Sesia stomoxyformis v. *amasina* Staudinger, 1856: 209. Type locality: Turkey, Amasya. Type material: lost.

Synanthedon stomoxiformis riefenstahli Špatenka, *in* de Freina, 1997: 70. Type locality: Spain, Prov. Alicante, Orcheta. Type material: holotype ♂, TW.

DIAGNOSIS.– Wingspan 21-31 mm; labial palpus and frons black; thorax black laterally below forewing; 4th abdominal segment red dorsally; segments 5, 6 partly red or with red scales laterally; segments 4-6 red ventrally; anal tuft and anal flaps black; broad wing margins, smaller transparent areas and reduction of red coloration ventrally in ssp. *riefenstahli* and in ssp. *amasina*.

GENITALIA.– Fig. 18.

BIOLOGY AND HABITAT.– Rocky and forest-steppe slopes, scrub, forest edges, sandy habitats; hostplants: *Frangula* spp., *Rhamnus* spp. (Schnaider et al., 1961; Obermajer, pers. comm.), *Sorbus, Corylus, Crataegus* (Špatenka et al., 1999); larva 2 years in the root, in a flat, broad chamber beneath bark; pupation in a tube from the root to the ground surface at trunk base; adult V-VII.

DISTRIBUTION.– W Palaearctic: S, parts of central and E Europe; ssp. *riefenstahli* in the Iberian Peninsula, ssp. *amasina* in southern Balkans, the nominotypical subspecies in other parts of Europe.

REMARKS.– The nominotypical subspecies is probably of East European (? Caspian) origin, populations with darker specimens have developed in isolated parts of the range independently in southwestern and southeastern Europe and Asia Minor.

19. Synanthedon culiciformis (Linnaeus, 1758)

(Pl. 3: 10,11,12)

Sphinx culiciformis Linnaeus, 1758: 493. Type locality: Europe. Type material: lost.

DIAGNOSIS.– Wingspan 19-29 mm; labial palpus orange ventrally; eye with narrow white border; thorax with orange red spot laterally below forewing (Fig. 19a); forewing with red or orange scales basally; 4th abdominal segment red; occasionally further segments (2nd, 5th, etc.) with narrow orange or red rings; anal tuft usually black, occasionally partly orange or yellowish.

GENITALIA.– Fig. 19; crista sacculi extremely large.

BIOLOGY AND HABITAT.– Open forests, forest edges, clearings, sandy habitats, peat-bog edges, embankments; hostplants: *Betula* spp., *Alnus* spp.; larva 1(2) years, usually in the stump or in sickly or wounded parts of trunk and branches; pupation in a cocoon of long chips; adult IV-VI(VII).

DISTRIBUTION.– Holarctic: N, central and E Europe, N and S Apennine and Balkan Peninsula.

REMARKS.– The specimens from southern Bulgaria and from Anatolia are partly different, being darker, with broader wing margins, almost without reddish scales on the forewing base and often with 2 red abdominal rings. This is possibly the result of glacial isolation and spatially independent evolution separate from most populations colonising Europe from the East as late as during the postglacial period. Obviously, even older is the isolation of the Azerbaidjanian *Synanthedon talischensis* (Bartel, 1906) considered to be a distinct biospecies. – *Populus* sp. is recorded from southern Italy as another hostplant, probably by misidentification with *Alnus cordata* Loisel.

20. Synanthedon formicaeformis (Esper, 1783)

(Pl. 3: 9)

Sphinx formicaeformis Esper, 1783: 216, pl. 32, figs 3, 4. Type locality: Germany, Frankfurt am Main. Type material: not traced.

Sesia serica Alphéraky, 1882: 21, pl. 1, fig. 29. Type locality: W China, Kuldsha, Kounguesse. Type material: holotype ♂, ZISP.

Synanthedon herzi Špatenka & Gorbunov, 1992: 377. Type locality: Russia, Sakhalin. Type material: holotype ♂, ZISP, **syn.n.**

DIAGNOSIS.– Wingspan 14-25 mm; antenna black, occasionally (in specimens from SW Europe and from E Palaearctic) white distally; labial palpus orange ventrally; eyes narrowly white bordered; thorax laterally black; apical area of forewing red or reddish; 4th and partly 5th (and/or 6th) abdominal segments red; anal tuft black, with whitish scales laterally; in ssp. *serica* also discal spot of forewing red, external transparent area broad (Fig. 20a) and abdomen with a large reddish suffusion.

GENITALIA.– Fig. 20.

BIOLOGY AND HABITAT.– Floodplain forests, forest edges, peat bogs, river banks, osier beds; hostplants: *Salix* spp.; larva 1(2) years in the trunk, in branches and twigs, often in tumours or in sickly parts; adult V-VIII.

DISTRIBUTION.– Eurasiatic (? Holarctic): nearly whole of Europe.

REMARKS.– No specific differences from *Synanthedon herzi* have been found.– The relationship to the nearly identical Nearctic taxon, *Synanthedon bolteri* (Edwards, 1883), is unknown.– Concerning *Synanthedon serica*, known in Europe from SE Russia and Kazakhstan, these are probably marginal, morphologically more distinct populations (subspecies) of *S. formicaeformis*.

21. Synanthedon polaris (Staudinger, 1877)

(Pl. 3: 3)

Sesia polaris Staudinger, 1877: 175. Type locality: Finland, Lappland, Kuusamo. Type material: holotype ♂, ZMHB.

Sesia rufibasalis Bartel, 1906a: 190. Type locality: Switzerland, Pontresina. Type material: not traced.

DIAGNOSIS.– Wingspan 19-23 mm; labial palpus orange ventrally; frons black; thorax usually with yellow spot laterally; forewing partly covered (at discal spot, basally and anal margin) with reddish scales; external transparent area narrow (Fig. 21a); abdomen with narrow yellow rings on 2nd, 4th and 6th segments (occasionally indistinct).

GENITALIA.– Fig. 21.

BIOLOGY AND HABITAT.– Light northern taiga, willow shrubs in arctic tundra and in alpine habitats; hostplants: *Salix* spp. (*Salix lapponum* L., *S. glauca* L., *S. phylicifolia* L. in N Europe, *Salix helvetica* Vill., *S. breviserrata* Flod. and *S. glaucosericea* Flod. in the Alps); larva 3 years in basal part of trunk and in roots; adult VII-VIII (Schantz, 1959; Priesner et al., 1989).

DISTRIBUTION.– Eurasiatic (or Holarctic) with arctic-alpine disjunction of its range; in Europe, subarctic and arctic regions of Fennoscandia (Fibiger & Kristensen, 1974; Ryrholm, pers. comm.) and several localities in the Alps: in Switzerland and South Tyrol (Bartel, 1906, as *Sesia rufibasalis*; Priesner et al., 1989; Pühringer et al., 1999).

REMARKS.– Probably conspecific with the Nearctic taxon, *Synanthedon arctica* (Beutenmüller, 1900).

22. Synanthedon flaviventris (Staudinger, 1883)

(Pl. 2: 16)

Sesia flaviventris Staudinger, 1883: 177. Type locality: Germany, Mecklenburg, Friedland. Type material: lectotype ♂, ZMHB (des. by Špatenka & Laštůvka, 1988).

DIAGNOSIS.– Wingspan 13-21 mm; labial palpus deep yellow; eye with narrow white border; thorax yellow laterally below forewing; external transparent area usually very narrow (Fig. 22a); abdomen with narrow yellow rings on 2nd, 4th and 6th segments dorsally and with whitish yellow segments 4-6(7) in male and 4-5 in female ventrally; anal tuft black in male and black with ochreous scales laterally in female; anal flaps in male black, partly yellowish distally.

GENITALIA.– Fig. 22.

BIOLOGY AND HABITAT.– Forests, forest edges, meadows with groups of willows, river bank growths, embankments, spring zones, old quarries; hostplants: *Salix* spp. (particularly *Salix caprea* L., *S. cinerea* L., *S. aurita* L., *S. repens* L., etc.); eggs are laid on slender (this year's) twig; larva 2 years in vertical tunnel of 3-6 cm length, producing a small characteristic gall; pupa in the tunnel without cocoon; adult VI-VII.

DISTRIBUTION.– Eurasiatic: little known, central and parts of N Europe.

23. *Synanthedon andrenaeformis* (Laspeyres, 1801)

(Pl. 3: 13,14,15)

Sesia andrenaeformis Laspeyres, 1801: 20. Type locality: Hungary, Ofen [Budapest]. Type material: destroyed.

Synanthedon perigordensis Garrevoet & Vanholder, 1996: 141. Type locality: France, dep. Dordogne, Plaisance/Issigeac. Type material: holotype ♂, TG.

DIAGNOSIS.– Wingspan 17-26 mm; labial palpus white or yellowish white ventrally; frons black; pericephalic hairs black; thorax with yellow spot laterally; segments 2 and 4 with narrow yellow rings dorsally; segments 4-6 (partly 7) whitish ventrally in male and 4 in female; anal tuft black, ochreous distally.

GENITALIA.– Fig. 23.

BIOLOGY AND HABITAT.– Scrub, open forests, rocky and forest-steppe slopes, occasionally parks, floodplain forests and river bank growths, roadsides; hostplants: *Viburnum lantana* L., *V. opulus* L. (occasionally); larva (1)2 years in the pith of twigs or branches; pupation without cocoon in a gallery; adult (V)VI-VII.

DISTRIBUTION.– W Palaearctic: scattered in central, S and E Europe.

24. *Synanthedon soffneri* Špatenka, 1983

(Pl. 2: 15)

Synanthedon soffneri Špatenka, 1983: 297. Type locality: Bohemia, Šumava Mts., Dobrá. Type material: holotype ♂, NMP.

DIAGNOSIS.– Wingspan 17-23 mm; labial palpus orange ventrally; frons black; thorax orange laterally; broad apical area of forewing; 1st-3rd abdominal segments orange yellow laterally; 4th segment orange yellow; anal tuft and anal flaps black.

GENITALIA.– Fig. 24.

BIOLOGY AND HABITAT.– Open forests, forest edges, clearings; hostplants: *Lonicera nigra* L., *L. xylosteum* L., *L. tatarica* L., *L. caerulea* L.; larva 2 (3) years in branch or twig; pupation without cocoon (Špatenka, 1983); adult V-VI.

DISTRIBUTION.– Insufficiently known, W Palaearctic or Eurasiatic: France (Dutreix, pers. comm.), Germany (Steffny, 1990), Switzerland (Bartsch & Pelz, 1997), Austria (Embacher, 1994), Bohemia, Slovakia, Russia (Kallies, Špatenka, pers. comm.).

25. *Synanthedon uralensis* (Bartel, 1906)

(Pl. 2: 14)

Sesia uralensis Bartel, 1906b: 169. Type locality: Kazakhstan, Uralsk. Type material: holotype ♂, BMNH.

DIAGNOSIS.– Wingspan 18-21 mm; labial palpus black, long tufted in male; frons black; thorax orange laterally; 4th abdominal segment red or orange-red dorsally; ventrally segments 4-7 and anal flaps orange in male and only 4-5 in female.

GENITALIA.– Fig. 25.

BIOLOGY AND HABITAT.– Steppes; hostplant: *Artemisia abrotanum* L.; larva 1 year in the woody stem similar as in *S. andrenaeformis* in *Viburnum;* adult V-VII (Gorbunov, pers. comm.).

DISTRIBUTION.– Siberian-E European: E Ukraine, SE European Russia, NW Kazakhstan, Mongolia.

26. *Synanthedon melliniformis* (Laspeyres, 1801)

(Pl. 2: 3,4)

Sesia melliniformis Laspeyres, 1801: 19. Type locality: S France. Type material: destroyed.
Aegeria danubica Králíček, 1975: 1. Type locality: Slovakia, Štúrovo. Type material: holotype ♂, MK
Synanthedon croaticus Kranjčev, 1979: 27. Type locality: Croatia, Podravina, Delekovec. Type material: holotype ♀, RK.

DIAGNOSIS.– Wingspan 17-22 mm; labial palpus yellow ventrally; eye with white border; thorax yellow laterally; external transparent area broad, particularly in male; abdominal segments 2, 4, 7 in male and 2, 4, 6 in female with narrow yellow rings; segments 4 and 5 in male and 4 in female yellow ventrally; anal tuft black in male, partly yellow medially and ventrally; anal flaps yellow; in female yellow, black laterally.

GENITALIA.– Fig. 26; unusually hooked crista sacculi; in female genitalia simple, unspecialised 8th segment, ostium bursae and lamella antevaginalis; antrum slender, tubular; unusual morphology in the genus.

BIOLOGY AND HABITAT.– Floodplain forests, marshy habitats, river bank growths; hostplants: *Populus* spp., *Salix* spp. (Tomala, 1913, as *S. flaviventris*; Kranjčev, 1979; Hamborg, 1993; Prola, pers. comm.); larva 1 year in sickly parts of trunk, branches or twigs; pupation in a cocoon beneath bark; adult V-VIII.

DISTRIBUTION.– Adriatico-Mediterranean: central and N Italy, Slovenia, E and SE Austria, Croatia, Hungary, Serbia, S Slovakia, S France (Laspeyres, 1801; Le Cerf, 1922; Kranjčev, 1979; Králíček, 1975a; Laštůvka & Laštůvka, 1988; Hamborg, 1993; Prola, Toševski, pers. comm.).

27. *Synanthedon martjanovi* Sheljuzhko, 1918

(Pl. 2: 13)

Synanthedon martjanovi Sheljuzhko, 1918: 104. Type locality: Russia, Minusinsk. Type material: holotype ♀, ZMUK.

DIAGNOSIS.– Wingspan 16-22 mm; labial palpus whitish yellow ventrally; eye with white border; thorax yellow laterally; external transparent area narrow in male, nearly square in female; abdominal segments 2 and 4 with narrow yellowish borders; segments 4-5 in male and 4-6 in female covered with whitish scales ventrally; anal tuft in male black dorsally and yellow ventrally, in female yellow medially, black laterally and ventrally.

GENITALIA.– Fig. 27 (after Gorbunov); no distinct differences from *S. melliniformis* have been found.

BIOLOGY AND HABITAT.– Open forests, forest edges, groups of aspen trees; hostplants: *Populus* spp. (*Populus tremula* L.); larva 1 year beneath bark in sickly parts of trunk or branches; pupation in a cocoon beneath bark (Ilinskij, 1962; Gorbunov & Tshistjakov, 1995; Gorbunov, pers. comm.).

DISTRIBUTION.– Siberian-E European; the occurrence in E Europe (S Russia) is insufficiently documented (Ilinskij, 1962; Gorbunov, pers. comm.).

28. *Synanthedon myopaeformis* (Borkhausen, 1789)

(Pl. 3: 16,17,18; 9: 6)

Sphinx myopaeformis Borkhausen, 1789: 169. Type locality: not mentioned. Type material: destroyed.

Sphinx typhiaeformis Borkhausen, 1789: 174. Type locality: Italy. Type material: destroyed.

Sesia cruentata Mann, 1859: 91. Type locality: Sicily, Palla-Gutta valley. Type material: holotype ♂, ZMHB.

Sesia myopiformis v. *graeca* Staudinger, 1870: 92. Type locality: Greece, Mt. Parnass. Type material: lectotype ♂, ZMHB (des. by Špatenka & Laštůvka, 1988).

DIAGNOSIS.– Wingspan 14-25 mm; eye with white border; thorax with orange-red spot laterally below forewing; in female 4th segment reddish ventrolaterally and black medially; anal tuft black; anal flaps in male whitish yellow; in nominotypical form antenna black, labial palpus white ventrally in male and brown to black in female, 4th segment red dorsally and in male 4th-6th segments whitish ventrally; in ssp. *graeca* 2nd-4th segments partly or entirely red dorsally, otherwise as in nominotypical subspecies; in ssp. *typhiaeformis* antenna white distally, labial palpus ochreous or orange in male and orange or brown in female, abdominal segments 2 (narrowly) and 4 with red rings dorsally (broader in female); in ssp. *cruentata* abdominal segments 2-4 entirely red dorsally, otherwise as in ssp. *typhiaeformis*.

GENITALIA.– Fig. 28, without differences between subspecies.

BIOLOGY AND HABITAT.– Open forests, orchards; hostplants: Maloidea (*Malus, Sorbus, Crataegus, Pyrus* spp., *Eriobotrya japonica* Thunb., etc.), ? *Hippophae rhamnoides* L.; larva 1 year beneath bark; pupation in a cocoon; adult V-IX.

DISTRIBUTION.– W Palaearctic: the nominotypical ssp. in central, W, E, parts of S Europe; ssp. *typhiaeformis* in SE France, Corsica, and Italy from Liguria to Campania; ssp. *cruentata* in Calabria, Sicily and Malta; ssp. *graeca* in Greece.

29. *Synanthedon vespiformis* (Linnaeus, 1761)

(Pl. 3: 5,6)

Sphinx vespiformis Linnaeus, 1761: 289. Type locality: [S Sweden]. Type material: lost.
Sphinx asiliformis Rottemburg, 1775: 108. Type locality: Poland, Landsberg an der Warthe [Gorzów Wielkopolski]. Type material: lost.

DIAGNOSIS.– Wingspan 16-27; eye with white border; thorax with yellow spot laterally; tegula with yellow border dorsally; discal spot of forewing red with narrow black border towards base; metathorax with yellow transverse spot dorsally (Fig. 29a); abdominal segments 2, 4, 6, 7 in male and 2, 4, (5), 6 in female with yellow rings dorsally; segments 4, (5, 6) in male and 4, (5), 6 in female yellow ventrally; anal tuft black in male and entirely or distally yellow in female; male anal flaps distally yellow (Fig. 29b).

GENITALIA.– Fig. 29; specialised gnathos and crista sacculi; in female specialised 8th segment and lamella antevaginalis.

BIOLOGY AND HABITAT.– Forests, clearings, parks, river banks; hostplants: *Quercus* spp. (primarily), occasionally or locally *Populus* spp., *Salix* spp., *Prunus* spp., *Juniperus* spp., *Fagus sylvatica* L., *Castanea sativa* Mill., *Ulmus* spp., *Loranthus europaeus* L.; larva 1-2 years beneath bark (often in stumps); pupation in a cocoon beneath bark; adult V-X.

DISTRIBUTION.– W Palaearctic: central, S and E Europe.

30. *Synanthedon codeti* (Oberthür, 1881)

(Pl. 2: 7,8)

Sesia codeti Oberthür, 1881: 67, pl. 11, fig. 5. Type locality: Algeria, Sebdou. Type material: lectotype ♀, MNHP (des. by Špatenka, 1992b).
Synanthedon ferdinandi Rungs, 1972: 671. Type locality: Morocco, Rabat. Type material: holotype ♂, MNHP.

DIAGNOSIS.– Wingspan 16-23 mm; eye with white border; tegula with yellow border; thorax yellow laterally; metathorax yellow dorsally; discal spot of forewing with red spot, black towards base; abdominal segments 2, 4, (6), 7 in male and 2, 4, 6 in female with yellow rings dorsally and 4-6(7) ventrally; anal tuft almost entirely black in male and yellow medially and black laterally in female; anal flaps in male black (Fig. 30a).

GENITALIA.– Fig. 30.

BIOLOGY AND HABITAT.– Open warm forests, forest edges, parks, forest-steppe slopes, hazel cultures; hostplants: *Quercus* spp. (Le Cerf, 1920; Bläsius, pers. comm.); *Corylus avellana* L. (Sarto, pers. comm.); *Platanus orientalis* L., *Carya pecan* Marsh., *Malus* sp., *Prunus* sp. (Rungs, 1972); larva 1 year in sickly parts of trunk; pupation in a cocoon beneath bark; adult V-VII.

DISTRIBUTION.– W Mediterranean: S and E Iberian Peninsula, S France (Bettag & Bläsius, 1999a), Sardinia (Angelo leg.).

31. Synanthedon theryi Le Cerf, 1916

(Pl. 2: 6)

Synanthedon theryi Le Cerf, 1916: 16; pl. 322, fig. 4678. Type locality: Algeria, Boufarik. Type material: holotype ♂, MNHP.

DIAGNOSIS.– Wingspan 16-21 mm; eye with white border; tegula in fore part yellow; thorax yellow laterally; metathorax yellow dorsally (Fig. 31a); discal spot of forewing almost entirely red; abdominal segments 2, 4, 6 (7) with broad yellow rings dorsally; nearly all segments yellow ventrally; anal tuft yellow medially and black laterally in male and almost entirely yellow distally in female; anal flaps in male yellow (Fig. 31b).

GENITALIA.– Fig. 31.

BIOLOGY AND HABITAT.– Sandy and coastal habitats, river banks; hostplants: *Tamarix gallica* L., *T. africana* Poiret (Le Cerf, 1920; Rungs, 1972); larva 1 year in trunk and in branches; pupation in a cocoon beneath bark (Bläsius, pers. comm.); adult V(VI).

DISTRIBUTION.– W Mediterranean: in Europe known only from S Andalusia (captured by Bläsius, Koschwitz, Lingenhöle, Petersen, etc.) and from the Balearic Islands (captured by T. & W. Garrevoet).

The following species group ("*tipuliformis*" group) constitutes a monophyletic unit and it has been described as subgenus *Tipulia* Králíček & Povolný, 1977.

32. Synanthedon conopiformis (Esper, 1782)

(Pl. 2: 10)

Sphinx conopiformis Esper, 1782: pl. 31, figs 1, 2; p. 213 (1783). Type locality: Austria, Vienna env. Type material: lectotype ♀, ZSBS (des. by Špatenka, 1992b).

Sesia nomadaeformis Laspeyres, 1801: 27. Type locality: S Germany. Type material: destroyed.

DIAGNOSIS.– Wingspan 16-24 mm; eye with white border; metathorax with yellow transverse spot (Fig. 32a); apical area of forewing with orange spots between veins; abdomen with narrow yellow rings on 2nd, 4th, 6th and 7th (in male) segments; only 4th segment narrowly bordered ventrally; anal tuft and anal flaps in male black.

GENITALIA.– Fig. 32; crista sacculi low, present.

BIOLOGY AND HABITAT.– Warm oak forests, forest-steppe habitats, solitary old oak trees; hostplants: *Quercus* spp.; larva 2 years in sickly parts of trunk and branches; pupation in a cocoon beneath bark; adult V-VII.

DISTRIBUTION.– W Palaearctic: central, S and E Europe.

33. *Synanthedon tipuliformis* (Clerck, 1759)

(Pl. 2: 11)

Sphinx tipuliformis Clerck, 1759: pl. 9, fig. 1. Type locality: not mentioned. Type material: lost.

Sphinx salmachus Linnaeus, 1758: 493. Type locality: Europe. Type material: lost (rejected by ICZN, Opinion 1288).

DIAGNOSIS.– Wingspan 11-21 mm; labial palpus yellow ventrally; eye with white border (Fig. 33b); metathorax dark; external transparent area higher than broad with straight external border; apical area with distinct ochreous spots between veins (Fig. 33a); abdominal segments 2, 4, 6 and 7(in male) with narrow yellow rings; anal tuft black.

GENITALIA.– Fig. 33; very homogeneous in the group and scarcely distinguishable from *S. spuleri*; crista sacculi absent; in female 8th segment, lamella antevaginalis and ostium bursae united in the group.

BIOLOGY AND HABITAT.– Gardens, currant cultures, scrub; hostplants: *Ribes* spp., *Grossularia* spp., *Euonymus europaea* L.; larva 1 year in the pith of twigs; pupation without cocoon in bordered gallery; adult V-VIII.

DISTRIBUTION.– W Palaearctic (autochthonous), introduced from Europe into many parts of the world.

REMARKS.– A common pest of currants.

34. *Synanthedon spuleri* (Fuchs, 1908)

(Pl. 2: 12)

Sesia spuleri Fuchs, 1908: 33. Type locality: Austria, Karwendel, Halltal. Type material: neotype ♂, TLM (des. by Králíček & Povolný, 1977).

Aegeria schwarzi Králíček & Povolný, 1977: 91. Type locality: Czech Republic, Chřiby Mts., Buchlov. Type material: holotype ♀, MK.

DIAGNOSIS.– Wingspan 14-24 mm; labial palpus yellow ventrally; eye with white border; metathorax dark; external transparent area nearly square, slightly convex towards apical area; apical area dark without or with indistinct spots (Fig. 34a); abdominal segments 2, 4, 6 and 7(in male) with yellow rings; anal tuft entirely black or with several yellow hairs mediodistally in male or laterally in female; some specimens hardly distinguishable from *S. tipuliformis*.

GENITALIA.– Fig. 34 (any differences in figs. 33 and 34 are due to variability rather than specific differences).

BIOLOGY AND HABITAT.– Open forests, river banks, sandy places and pastures with *Juniperus*; hostplants: *Acer* spp., *Betula pendula* Roth., *Carpinus betulus* L., *Corylus avellana* L., *Fagus sylvatica* L., *Juniperus communis* L. (in tumours caused by *Gymnosporangium clavariiforme* Jacq.), *Populus* spp., *Quercus* spp., *Salix* spp., *Ulmus* spp. (Fuchs, 1908; Hamborg, 1993; Bläsius, Kallies, Králíček, Špatenka, pers. comm.); ? *Diospyros kaki* L. (Zuccherelli, 1969), individual populations probably specialised; larva 1(2) years in sickly parts of trunk and branches; pupation in a cocoon beneath bark; adult V-VIII.

DISTRIBUTION.– W Palaearctic: central, W, S and E Europe.

35. Synanthedon geranii Kallies, 1997

(Pl. 9: 4,5)

Synanthedon geranii Kallies, 1997: 60. Type locality: Greece, Mt. Parnassos, N of Arahova. Type material: holotype ♂, TW.

DIAGNOSIS.– Wingspan 17-22 mm; fore coxa entirely yellow laterally; discal spot of forewing broad, triangular; apical area broad with distinct orange spots between veins; external transparent area higher than broad (narrow); discal spot with yellow scales ventrally; discal spot of hindwing large; yellow rings on 2nd, 4th, 6th and 7th (in male) segments; anal tuft in male black with yellow hairs medially and ventrally, in female orange-yellow distally.

GENITALIA.– Fig. 35.

BIOLOGY AND HABITAT.– Shady limestone screes and rocks in mountains near the upper forest border (1100-1800 m); hostplant: *Geranium macrorrhizum* L.; larva 1 year in the rhizomes above the ground surface; pupa in a cocoon in the rhizome, usually in its tip; adult V-VII (Kallies, 1997).

DISTRIBUTION.– Greece: Mt. Parnass and Mt. Timfi.

36. Synanthedon loranthi (Králíček, 1966)

(Pl. 2: 17,18)

Aegeria loranthi Králíček, 1966: 231. Type locality: Moravia, Hodonín. Type material: holotype ♂, MMB.

Aegeria cryptica Králíček & Povolný, 1977: 85. Type locality: Slovakia, Štúrovo. Type material: holotype ♂, MK.

DIAGNOSIS.– Wingspan 14-22 mm; fore coxa laterally up to 2:3 of its length yellow (Fig. 36b); external transparent area broader than high (Fig. 36a); discal spot of hindwing small (Fig. 36a); yellow rings on 2nd, 4th and 7th segments in male and 2nd, 4th and (6th) in female.

GENITALIA.– Fig. 36; saccus usually even or only slightly concave at the end; bare area on ventral margin of valve indistinct, small; crista sacculi not indicated (Králíček, 1966; Laštůvka, 1983c); in female genitalia no differences from *S. cephiformis* have been found.

BIOLOGY AND HABITAT.– Forests with *Viscum* spp. or *Loranthus*, parks, forest-steppe habitats; hostplants: *Viscum* spp., *Loranthus europaeus* L.; larva 1(2) years beneath bark and/or in wood; pupation in a cocoon beneath bark; adult V-VIII (Králíček, 1966; Laštůvka, 1983c).

DISTRIBUTION.– European: central, W and S Europe (Laštůvka, 1990f; Blum, 1990; Kristal, 1990; Cungs, 1998; Hołowiński & Miłkowski, 1999).

37. Synanthedon cephiformis (Ochsenheimer, 1808)

(Pl. 2: 19, 21)

Sesia cephiformis Ochsenheimer, 1808: 169. Type locality: Austria, Vienna env. Type material: destroyed.

Aegeria gaderensis Králíček & Povolný, 1977: 83. Type locality: Slovakia, Velká Fatra Mts., Gaderská valley. Type material: holotype ♂, MK.

DIAGNOSIS.– Wingspan 16-24 mm; fore coxa entirely yellow laterally (Fig. 37b); external transparent area nearly square or narrower (Fig. 37a); discal spot of hindwing large (Fig. 37a); abdomen with yellow rings on 2nd, 4th, 6th and 7th (in male) segments; ring on 6th segment occasionally absent in male; anal tuft in male black with yellow hairs mediodistally and ventrally; in female largely yellow distally.

GENITALIA.– Fig. 37; saccus usually slightly broadened and concave at the end; distinct bare area on the ventral margin of valve; crista sacculi often indicated as a line of hairs; no specific differences from *S. loranthi* have been found in female genitalia.

BIOLOGY AND HABITAT.– Fir and mixed forests; hostplants: *Abies alba* Miller, *A. borisii-regis* Mattf., *A. cephalonica* Loudon; tumours produced by *Melampsorella caryophyllacearum* DC., occasionally other diseased or damaged places; larva 1(2) years in tumour tissue; pupation in a cocoon in bark; adult VI-VIII.

DISTRIBUTION.– European: central, parts of W Europe, Balkan Peninsula.

REMARKS.– Mountain specimens, described as *Aegeria gaderensis* and occurring throughout the range, show no specific differences (cf. Špatenka et al., 1999).– The Greek specimens from *Abies cephalonica* and *A. borisii-regis* with large external transparent area could be confused with *S. loranthi*.

Bembecia Hübner, [1819]

Bembecia Hübner, [1819]: 128.
Type species: *Sphinx ichneumoniformis* [Denis & Schiffermüller], 1775; des. by Newman in Westwood, 1840.
Dipsosphecia Spuler, 1910: 316.
Type species: *Sphinx ichneumoniformis* [Denis & Schiffermüller], 1775; by original designation.

GENUS CHARACTERS.– Proboscis reduced or absent; discal spot of forewing orange or yellow on the outside; often yellow scapular spot at forewing base; aedeagus with a number of internal sclerites apically; antrum usually long. Larva in the roots of Fabaceae. 21 species in the region.

REMARKS.– The absence of transparency in different species groups is not relevant in their classification (Laštůvka, 1992a). – The species of this genus can be classified into several groups on the basis of their genitalia morphology.

38. Bembecia hymenopteriformis (Bellier, 1860)

(Pl. 2: 28,29)

Sesia hymenopteriformis Bellier, 1860: 681. Type locality: Sicily, Palermo env. Type material: lectotype ♂, MNHP (des. by Špatenka, 1992b).

DIAGNOSIS.– Wingspan 15-22 mm; ground coloration in male greyish brown; antenna black; external transparent area small, of 3 slightly elongate cells; apical area with ochreous spots; abdomen with narrow white rings on 2nd, 4th and 6th segments and with ochreous scales; female black; usually only external transparent area present, consisting of 1-3 small cells; abdomen with white rings on 2nd, 4th and 6th segments.

GENITALIA.– Fig. 38; specialised in both sexes.

BIOLOGY AND HABITAT.– Xerothermic, rocky, or sandy places, roadsides, pastures; hostplants: *Lotus* spp., *Anthyllis vulneraria* L. (Le Cerf, 1920); larva 1 year in root; pupa deep (3-8 cm) in root; adult V-IX(X), in extreme xerothermic habitats the adults occur in two waves V-VI and VIII-X.

DISTRIBUTION.– W Mediterranean: S Spain, Sicily, S and central Italy.

REMARKS.– The position of this species is quite isolated in the genus.

"ichneumoniformis"-group

39. Bembecia lomatiaeformis (Lederer, 1853)

(Pl. 4: 10; 9: 15,16)

Sesia lomatiaeformis Lederer, 1853: 89. Type locality: Turkey, Diarbekir. Type material: lectotype ♂, ZMHB (des. by Špatenka & Laštůvka, 1988).

DIAGNOSIS.– Wingspan 28-35 mm; antenna black; metathorax with long whitish hairs; transparent areas large in male and small or occasionally reduced in female (Fig. 39a); male abdominal segments 2, 4, 6 and 7 yellow dorsally, segments 2, 3, 5 with narrow yellow rings; female with yellow rings on 2nd, 4th (and 6th) segments.

GENITALIA.– Fig. 39.

BIOLOGY AND HABITAT.– Treeless habitats in mountains (pastures, roadsides); hostplant: *Astragalus angustifolius* Lam.; larva 2 years in root; pupation in a cocoon in upper part of root or in a tube outside the root; adult VI-VIII (Petersen & Bartsch, 1998; reared by present authors).

DISTRIBUTION.– E Mediterranean: Greece, Timfi and Taygetos Mts. (Hofmann, Lingenhöle, pers. comm.).

40. Bembecia sareptana (Bartel, 1912)

(Pl. 4: 13; 9: 10)

Dipsosphecia sareptana Bartel, 1912: 395. Type locality: Russia, Sarepta [Krasnoarmejskoe, Volgograd env.]. Type material: holotype ♀, ZMHB.

DIAGNOSIS.– Wingspan 18-27 mm; antenna black; external transparent area round, of 4-5 cells; discal spot of forewing partly red; apical area and anal margin orange-red; hindwing basally, cilia near base and hind tibia orange-red; segments 2, 4, 6, 7 in male and 2, 4, 6 in female with yellow rings; anal tuft black, in male partly with yellow hairs medially.

GENITALIA.– Fig. 40.

BIOLOGY AND HABITAT.– East European steppes; hostplant: unknown.

DISTRIBUTION.– SE European Russia.

REMARKS.– An insufficiently known taxon with unresolved relationship to *Bembecia strandi* (Kozhantshikov, 1936) and several other Central Asiatic species.

41. Bembecia volgensis Gorbunov, 1994

(Pl. 2: 27,31)

Bembecia volgensis Gorbunov, 1994: 563. Type locality: Russia, 160 km south of Ulyanovsk, Ryabina. Type material: holotype ♂, ZISP.

DIAGNOSIS.– Wingspan 15-18 mm; antenna black in male, brownish distally in female; apical area of forewing narrow, dark distally and pale yellow proximally, not distinctly delimited from the broad external transparent area; discal spot broad, dark, pale yellow only distally; transparent areas covered more or less with yellowish-white scales; longitudinal transparent area absent in female; abdominal segments 2, 4, 6 (7 in male) with broad yellow rings; segments 3 and 5 occasionally with some yellow scales; male anal tuft black with yellow hairs medially, in female black with individual yellow scales.

GENITALIA.– Fig. 41.

BIOLOGY AND HABITAT.– Limestone steppes with denuded saline soils; hostplant: unknown; adult: VI-VIII.

DISTRIBUTION.– S Russia, Ulyanovsk and Rostov districts.

REMARKS.– The paratype female (see the original description) resembles *B. megillaeformis* and according to its genitalia it really belongs to the *megillaeformis*-group, but not to this species.

42. Bembecia abromeiti Kallies & Riefenstahl, 2000

(Pl. 4: 11)

Kallies & Riefenstahl, 2000: 359. Type locality: NE Mallorca, coast near Cala Canyamel, S of Capdepera. Type material: holotype ♂, ZIMH.

DIAGNOSIS.– Wingspan 15-28 mm; antenna black; ground coloration black; tegula with orange reddish margin; external transparent area large, of 7 cells; apical area nearly absent; longitudinal transparent area up to Cu2; discal spot of forewing black, only with a few orange scales; abdominal segments 2-6 (7) with orange reddish rings.

GENITALIA.– Fig. 42.

BIOLOGY AND HABITAT.– Coastal rocky cliffs and sandy beaches, mountain rocky habitats; hostplants: *Astragalus balearicus* Chater, *Lotus cytisoides* L.; larva 1 year in root; pupa in a cocoon, usually in the upper part of an emergence tube; adult V-VII (Kallies & Riefenstahl, 2000).

DISTRIBUTION.– Mallorca.

43. Bembecia ichneumoniformis ([Denis & Schiffermüller], 1775)

(Pl. 4: 1,2)

Sphinx ichneumoniformis [Denis & Schiffermüller], 1775: 44. Type locality: Slovakia, Bratislava. Type material: neotype ♂, ZMHB (des. by Špatenka & Laštůvka, 1990).

DIAGNOSIS.– Wingspan 12-29 mm; antenna entirely black or with indistinct ochreous part in male and basally ochreous with black apex in female; external transparent area large, broader than apical area (Fig. 43a); apical area between veins orange; hind tibia and tarsus deep yellow; all segments with yellow rings of the same width, the ring on 5th segment occasionally narrower or absent; anal tuft in female usually yellow with short black hairs medially and laterally.

GENITALIA.– Fig. 43; gnathos simple; crista sacculi straight; valve usually pointed apically; in female antrum long, sclerotised; ostium bursae distinct, slightly concave.

BIOLOGY AND HABITAT.– Various xerophilous and/or mesophilous treeless habitats; hostplants: *Lotus* spp., *Dorycnium* spp., *Hippocrepis* spp., *Anthyllis vulneraria* L., *Tetragonolobus* spp., *Psoralea bituminosa* L.; larva 1-2 years in root; pupation in a cocoon in upper part of root; adult VI-IX(X).

DISTRIBUTION.– W Palaearctic: central, S, E and partly N Europe.

44. *Bembecia albanensis* (Rebel, 1918)

(Pl. 4: 4,5,6)

Sesia ichneumoniformis ssp. *albanensis* Rebel, 1918: 86. Type locality: Albania, Bilalas. Type material: holotype ♀, NHMV.

Dipsosphecia kalavrytana Sheljuzhko, 1924: 183. Type locality: Greece, Peloponnes, Kalavryta. Type material: holotype ♂, ZMUK.

Dipsosphecia megillaeformis var. *tunetana* Le Cerf, 1920: 305. Type locality: Tunisia, Tunis. Type material: lectotype ♂, MNHP (des. by Špatenka, 1992b).

Bembecia psoraleae Bartsch & Bettag, 1997: 29. Type locality: Spain, Serrania de Ronda, Cortes de la Frontera. Type material: holotype ♀, SMNS.

DIAGNOSIS.– Wingspan 14-23 mm; antenna black, in female sometimes ochreous basally; external transparent area small, narrower than apical area, of 3(4) cells (Fig. 44a), usually absent in ssp. *kalavrytana*; apical area (between veins) orange-yellow or deep yellow; all segments with yellow rings of the same width, exceptionally ring on 5th segment narrower; anal tuft in female yellow with long black hairs laterally.

GENITALIA.– Fig. 44; gnathos simple; crista sacculi slightly hooked at end; valve usually rounded apically; in female antrum shorter; ostium with distinct incision.

BIOLOGY AND HABITAT.– Grassy, sandy and rocky treeless places, pastures, embankments; hostplants: *Ononis spinosa* L., *O. repens* L., *O. arvensis* L., *Psoralea bituminosa* L., *Hedysarum coronarium* L. (Le Cerf, 1920; Špatenka & Laštůvka, 1990; Bläsius, Petersen, Toševski, pers. comm.); larva 1(2) years in root; pupa in a cocoon, occasionally with short tube; adult VI-IX.

DISTRIBUTION.– W Palaearctic: W, central, S and E Europe; ssp. *kalavrytana* in Greece and Crete, ssp. *tunetana* in S Italy, Sicily and Malta (reared by authors and Petersen, cf. Bettag & Bläsius, 1999b), ssp. *psoraleae* in Iberian Peninsula and S France (Bettag & Bläsius, 1999a), ssp. *albanensis* in other parts of the European range.

REMARKS.– Several taxa close to *B. albanensis* have been described, showing poorly differentiated genitalia morphology. In view of the unclear situation, all such taxa are treated as subspecies in this work.

45. *Bembecia pavicevici* Toševski, 1989

(Pl. 4: 3; 9: 8,9)

Bembecia pavicevici Toševski, 1989: 85. Type locality: Macedonia, Konsko, Gevgelia. Type material: holotype ♂, IT.

Bembecia pavicevici dobrovskyi Špatenka, *in* de Freina, 1997: 126. Type locality: Greece, Taygetos gorge, Exohori. Type material: holotype ♂, TW.

DIAGNOSIS.– Wingspan 19-27 mm; antenna with ochreous-white spot before apex in male and orange-yellow with black apex in female; external transparent area small, nearly round, usually narrower than apical area (Fig. 45a); black coloration of veins in apical area broader towards margin; abdominal segments 2, 4, 6 and 7 (in male) broadly yellow dorsally, 3 and 5 with narrow yellow rings (Fig. 45b); 4th-6th (7th) segments with narrow yellow rings ventrally; hind tibia with broad dark part distally (more than 1/3 its length); ssp. *dobrovskyi* more robust, with broader and darker margins of wings.

GENITALIA.– Fig. 45.

BIOLOGY AND HABITAT.– Xerothermic shrubby places, rocky habitats, roadsides; hostplant: *Coronilla emerus* L.; larva 2 years in root; pupa in a cocoon in root or in a short tube from the root (Toševski, 1992; Petersen & Bartsch, 1998; reared by present authors); adult VI-VIII.

DISTRIBUTION.– Balkan Peninsula, northwards to Istria (Toševski, 1992; Gelbrecht, Kallies, pers. comm.); ssp. *dobrovskyi* in southern Peloponnes.

46. *Bembecia fibigeri* Laštůvka & Laštůvka, 1994

(Pl. 4: 14,15)

Bembecia fibigeri Laštůvka & Laštůvka, 1994: 233. Type locality: Spain, Aragon, Vivel del Rio Martín. Type material: holotype ♂, ZL.

DIAGNOSIS.– Wingspan 19-26 mm; antenna black in male and black with whitish scales before apex in female; external transparent area small, round or narrower than apical area; discal spot with large orange or deep yellow coloration; apical area and anal margin orange or yellow (Fig. 46a); hind tibia yellow with indistinct dark ring distally; broad yellow rings on abdominal segments 2, 4, 6, 7 in male and 2, 4, 5, 6 in female dorsally, narrower rings on 3rd and 5th segments in male and 3rd in female (Fig. 46b); usually all segments bordered ventrally in male and 4-6 in female.

GENITALIA.– Fig. 46.

BIOLOGY AND HABITAT.– Rocky, stony or grassy treeless or shrubby habitats; hostplants: *Ononis rotundifolia* L. (Laštůvka & Laštůvka, 1994); *O. fruticosa* L. (Bettag, Bläsius, Blum, pers. comm.; Bettag & Bläsius, 1999a); larva 1(2) year in root; pupa in a cocoon in upper part of root; adult VII-VIII.

DISTRIBUTION.– Insufficiently known: S France, NE Spain (Laštůvka & Laštůvka, 1994; Bettag & Bläsius, 1999a; Drouet, pers. comm.).

47. Bembecia scopigera (Scopoli, 1763)

(Pl. 4: 7,8)

Sphinx scopigera Scopoli, 1763: 188. Type locality: Slovenia, Ljubljana env. Type material: neotype ♂, ZMHB (des. by Špatenka & Laštůvka, 1990).

DIAGNOSIS.– Wingspan 17-27 mm; antenna entirely black or with ochreous scales before apex in male and ochreous-brown with black apex in female; external transparent area large, broader than apical area, of 5 cells (Fig. 47a); apical area, anal margin and hind tibia orange; all abdominal segments with yellow rings; ring on 3rd segment narrower than on 2nd, often interrupted laterally (Fig. 47b); narrow or indistinct rings ventrally.

GENITALIA.– Fig. 47; crista gnathi doubled proximally; crista sacculi split distally; in female, antrum short, distinctly broadened medially.

BIOLOGY AND HABITAT.– Grassy and rocky, occasionally sandy treeless habitats, embankments, crops of *Onobrychis*; hostplants: *Onobrychis* spp. (*O. viciifolia* Scop., *O. hypargyrea* Boiss., *O. ebenoides* Boiss. & Spruner, *O. arenaria* Kit., *O. pallasii* Willd.), *Hedysarum candidum* Bieb. (Gorbunov, pers. comm.); larva 1(2) years in root; pupa in cocoon in root; adult VI-IX (Bournier & Khial, 1969; Špatenka & Laštůvka, 1990; reared by present authors).

DISTRIBUTION.– W Palaearctic: S, parts of central, E and SE Europe.

48. Bembecia priesneri Kallies, Petersen & Riefenstahl, 1998

(Pl. 5: 1,2)

Bembecia priesneri Kallies, Petersen & Riefenstahl, 1998: 58. Type locality: Turkey, Cappadocia, Prov. Nevşehir, Ürgüp-Avanos. Type material: holotype ♂, AK.

DIAGNOSIS.– Wingspan 18-23 mm; antenna with yellow scales and black apex in male, orange yellow with black apex in female; external transparent area small, of 4 cells, broader than high or nearly round, broad as apical area; apical area yellow, with conspicuously dark veins only distally; longitudinal transparent area present; discal spot of forewing black with reddish orange spot in male and nearly entirely reddish orange in female; anal margin yellow in male and orange in female; broad yellow rings on abdominal segments 2, 4-6 and 7 (in male), 3rd segment with narrow ring in male and with yellow spot in female; segments 4-6 (7) with yellow rings ventrally, yellow margin on 2nd segment interrupted medially (Fig. 48a, b).

GENITALIA.– Fig. 48; gnathos with small additional lateral wings.

BIOLOGY AND HABITAT.– Stony places, erosion gullies, banks of streams; hostplant: *Ononis* sp., larva 1 year in root; pupa in cocoon with a short tube; adult VIII-IX (Kallies et al., 1998).

DISTRIBUTION.– Anatolia, in Europe known only from Rhodes (Kallies, pers. comm.).

49. Bembecia iberica Špatenka, 1992

(Pl. 2: 24, 25)

Bembecia iberica Špatenka, 1992a: 429. Type locality: Spain, Aragon, Sra de Gúdar, Alcalá env. Type material: holotype ♀, TW.

DIAGNOSIS.– Wingspan 12-24 mm; antenna black (more often in male) or orange-brown with black apex (more often in female) dorsally and orange-brown with black apex ventrally; external transparent area small, narrower than apical area, of 3 slightly elongate cells (Fig. 49a); discal spot with reddish orange coloration; apical area and anal margin partly orange; segments 2, 4, 6 and 7 (in male) with broad (4th nearly entire) and 3, 5 with narrower yellow rings dorsally; 4-6 (7) segment with yellow ring ventrally, often entire 2nd segment yellow and 3rd segment usually without ring (Fig. 49b).

GENITALIA.– Fig. 49; gnathos specialised; crista sacculi split distally.

BIOLOGY AND HABITAT.– Treeless grassy and stony habitats: embankments, roadsides, pastures; hostplants: *Hippocrepis* spp., *Tetragonolobus* spp., *Lotus* spp., *Anthyllis vulneraria* L. (reared by present authors), *Melilotus* sp. (one record by Bläsius); larva 1 year in root; pupa in a cocoon in root or between basal parts of stems; adult VI-VIII.

DISTRIBUTION.– W Mediterranean: Iberian Peninsula, W and S France, NW Italy, Corsica, Sardinia (Špatenka, 1992a, and pers. comm.).

"megillaeformis"-group

50. Bembecia blanka Špatenka, 2001

(Pl. 5: 3)

Bembecia blanka Špatenka, 2001. Type locality: Crete, Psiloritis Mts., Fourfouras. Type material: holotype ♂, TW.

DIAGNOSIS.– Wingspan 12-19 mm; external transparent area round, of 4-5 cells, as broad as apical area; longitudinal transparent area long, up to discal spot in male, absent in female; apical area between veins, discal spot distally and hind margin of forewing yellow; abdominal segments 2-7 with yellow rings; the ring on 3rd and 5th segments narrower and interrupted laterally, sometimes quite absent, especially in female; anal tuft brownish black, ochreous medially and laterally; yellow rings not continuous ventrally in male and broadly yellow on segments 4-6 in female.

GENITALIA.– Fig. 50.

BIOLOGY AND HABITAT.– Stony shrubby xerothermic habitats on limestone up to 2000 m; hostplant: *Trifolium uniflorum* L.; larva 1 year in root; pupa in a cocoon in upper part of root; adult V-VII (Špatenka, Lingenhöle, pers. comm.).

DISTRIBUTION.– Crete.

51. Bembecia fokidensis Toševski, 1991

(Pl. 2: 20)

Bembecia fokidensis Toševski, 1991: 169. Type locality: Greece, Domokos. Type material: holotype ♂, IT.

DIAGNOSIS.– Wingspan 15-20 mm; antenna black in male and ochreous-orange with black apex in female; external transparent area broad, of 4-5 cells in male and very small, of 3 cells in female; apical area orange between veins (Fig. 51a); discal spot of forewing dark brown with reddish orange spot; hind tibia in lateral view with broad black bands proximally and distally; segments 2, 4, 6 and 7 with yellow rings dorsally; abdomen dark ventrally or with narrow distal border on 4th segment.

GENITALIA.– Fig. 51.

BIOLOGY AND HABITAT.– Scrub and treeless habitats from 100 m up to mountain meadows at 1600 m; hostplant: *Trifolium fragiferum* L. (Bettag & Blum, pers. comm.); larva 1 year in root; adult V-VII.

DISTRIBUTION.– Greece (Toševski, 1991; Lingenhöle, Špatenka, pers. comm.).

52. Bembecia megillaeformis (Hübner, [1813])

(Pl. 5: 4,5)

Sphinx megillaeformis Hübner, [1813]: pl. 24, fig. 114. Type locality: not mentioned. Type material: lost.

DIAGNOSIS.– Wingspan 16-27 mm; antenna black, occasionally brown with black apex in female; external transparent area round, of 5 cells (Fig. 52a); discal spot of forewing dark brown with reddish orange spot; apical area dark, orange between veins; hind tibia orange or deep yellow (in male), distal dark ring narrow or indistinct; male segments 4, 6 and 7 with broader and 2 and 5 (occasionally 3) with narrow yellow rings; in female rings on 2nd, 4th and 6th segments; segments 4-6 (7) bordered ventrally (Fig. 52b); anal tuft in female black with rusty scales.

GENITALIA.– Fig. 52; male genitalia very homogeneous and characteristic in *Bembecia megillaeformis*-group; in the European species of this group usually indistinguishable; apparent differences in figs 50 - 55 reflect intrinsic variability; female genitalia scarcely distinguishable within the species group: lamella antevaginalis broad; antrum short, membranous.

BIOLOGY AND HABITAT.– Stony or rocky, treeless or forest-steppe habitats; hostplants: *Genista tinctoria* L., *Colutea arborescens* L., occasionally *Chamaecytisus* spp., *Corothamnus procumbens* Waldst. & Kit., *Astragalus glycyphyllos* L. (Laštůvka, 1990b; reared by Bartsch, Stadie, present authors, etc.); larva 1(2) years in root; pupa in a cocoon with a short tube from the root; adult VI-VIII.

DISTRIBUTION.– W Palaearctic: scattered in W, central, SE and E Europe; from the end of the 19th century this species has been retreating from central Europe and is now lost from Germany, Poland and the Czech Republic.

REMARKS.– Specimens from around Angers in W France have been described as ssp. *luqueti* Špatenka, 1992.

53. Bembecia puella Laštůvka, 1989

(Pl. 5: 6,7)

Bembecia puella Laštůvka, 1989b: 85. Type locality: Slovakia, Plešivec. Type material: holotype ♀, ZL.

Bembecia daghestanica Gorbunov, 1991: 129. Type locality: Russia, Daghestan, NE Caucasus, Upper Gunib. Type material: holotype ♂, OG; **syn.n.**

DIAGNOSIS.– Wingspan 15-24 mm; antenna black; discal spot of forewing with small or indistinct orange spot; external transparent area extremely large (Fig. 53a); apical area almost absent in male, very narrow in female; segments 2, 4, 6, 7 with broad and 3, 5 with narrow yellow rings in male and 2, 4, 6 with broad yellow rings in female; in female 4th segment bordered ventrally and anal tuft entirely black (Fig. 53b).

GENITALIA.– Fig. 53; see *B. megillaeformis*; small distal wings of gnathos usually present.

BIOLOGY AND HABITAT.– Scrub, forest edges, often partly in shade; host-plants: *Astragalus glycyphyllos* L., *A. sigmoideus* Bunge, ? *A. utriger* Pallas; larva 1(2) years in root; pupa in a cocoon between bases of stems; adult VI-VIII (Laštůvka, 1989b, Gorbunov, Špatenka, pers. comm.).

DISTRIBUTION.– S Slovakia, Hungary, W Romania, NE Bulgaria, Greece (captured by Lingenhöle), Ukraine (Crimea), S Russia.

REMARKS.– No differences between *B. puella* and *B. daghestanica* have been found which would justify the treatment of *B. daghestanica* as a species.

54. Bembecia sirphiformis (Lucas, 1849)

(Pl. 4: 17,18)

Sesia sirphiformis Lucas, 1849: 367. Type locality: Algeria, El Kala [Lacalle], Tonga-See. Type material: holotype ♀, MNHP.

Sesia astragali Joannis, 1909: 183. Type locality: France, Vaucluse, La Bonde. Type material: lectotype ♀, MNHP (des. by Špatenka, 1992b).

DIAGNOSIS.– Wingspan 10-28 mm; antenna black in male and orange with black apex in female (exceptionally in male); discal spot of forewing brown, orange or yellow laterally; external transparent area of 4-5 cells (Fig. 54a, variation); apical area yellow; hind tibia and 1st tarsal segment conspicuously hairy (Fig. 54b); abdominal segments 2, 4, 6 (7) with broad and 3, 5 with narrow (or without) yellow rings dorsally; segments 4-6 (7) bordered ventrally.

GENITALIA.– Fig. 54.

BIOLOGY AND HABITAT.– Xerothermic, sandy, stony or rocky habitats; host-plants: *Astragalus monspessulanus* L. (Joannis, 1909), *Astragalus granatensis* Lam., *Colutea arborescens* L. (Bläsius, Hamborg, Špatenka, pers. comm.); *Acanthyllis armata* Battand (Le Cerf, 1920); larva 1(2) years in root; pupa in a cocoon 3-10 cm deep in the root with a long emergence tube; adult VI-VIII.

DISTRIBUTION.– W Mediterranean: Spain, S France, NW Italy.

55. Bembecia sanguinolenta (Lederer, 1853)

(Pl. 4: 16)

Sesia sanguinolenta Lederer, 1853: 81; replacement name for *Sesia tengyraeformis* Herrich-Schäffer, 1851; see remarks.

Sesia tengyraeformis Herrich-Schäffer, 1851: 47, pl. 10, fig. 59 [nec *Sesia tenthrediniformis* v. *tengyraeformis* Boisduval, 1840]. Type locality: not mentioned. Type material: lectotype ♂, ZMHB (des. by Špatenka & Laštůvka, 1988).

DIAGNOSIS.– Wingspan 21-33 mm; antenna, thorax and abdomen black or brownish black, without yellow rings; entire forewing opaque, rusty red, brownish black apically.

GENITALIA.– Fig. 55.

BIOLOGY AND HABITAT.– Xerothermic grassy, stony or rocky treeless or shrubby habitats, waste places; hostplants: *Astragalus dipsaceus* Bunge, ? *A. monspessulanus* L., *Astragalus* sp.; larva 2 years in root; pupa in a cocoon in the upper part of root; adult VI-VIII (Špatenka, pers. comm.).

DISTRIBUTION.– E Mediterranean: in Europe known from NE Greece and SW Bulgaria (captured by Bruer, Kolev, present authors).

REMARKS.– Lederer (1853) established "*sanguinolenta*" as a new name for *Sesia tengyraeformis* Herrich-Schäffer, 1851. Types of this species can therefore originate only from Herrich-Schäffer´s material. The collection of Staudinger comprises numerous type specimens of Herrich-Schäffer. The designated lectotype has a Staudinger label "origin" and was probably used by Herrich-Schäffer for the original drawing. See also remark on the genus *Bembecia*.

"*uroceriformis*"-group

56. Bembecia flavida (Oberthür, 1890)

(Pl. 9: 7)

Sesia flavida Oberthür, 1890: 24. Type locality: Algeria, Constantine. Type material: lectotype ♀, MNHP (des. by Špatenka, 1992b).

DIAGNOSIS.– Wingspan 11-20 mm; antenna black in male and ochreous with black tip in female; apical area of forewing narrow, yellow, without dark veins; external transparent area in male round or broader than high, of 4-5 cells; in female smaller, of 3 cells; longitudinal transparent area in female absent; male abdominal segment 2 with narrow and 4 with broad rings, 6, 7 entirely yellow; segment 4 with broad and 5-7 with narrow yellow margins ventrally; female segments 2, 4, 6 with broad yellow rings, occasionally 3 and 5 with indistinct margins.

GENITALIA.– Fig. 56.

BIOLOGY AND HABITAT.– Rocky limestone places with a rich mediterranean treeless vegetation; hostplant: unknown, at the localities in Sicily *Ononis* sp. (yellow), *Anthyllis vulneraria* L. and *Lotus* sp. occur; adult: VI, VII.

DISTRIBUTION.– W Mediterranean-North African: SE Sicily (captured near Sortino in 1995 and 2000 by present authors).

57. Bembecia himmighoffeni (Staudinger, 1866)
(Pl. 2: 22,26)

Sesia himmighoffeni Staudinger, 1866: 51. Type locality: Spain, Catalonia, Barcelona. Type material: lectotype ♀, ZMHB (des. by Špatenka & Laštůvka, 1988).
Bembecia baumgartneri Špatenka, 1992a: 427. Type locality: Croatia, Istria, Poreč. Type material: holotype ♂, TW.

DIAGNOSIS.– Wingspan 10-23 mm; antenna black, usually with whitish scales before apex in female; external transparent area nearly square or higher than broad, narrower than apical area, indistinctly limited towards yellow apical area or apical area dark (Fig. 57a); discal spot brown with deep yellow spot; all segments usually with yellow rings, 2, 4, 6, 7 broader, 4 often nearly entirely yellow, in large specimens (females) often all segments with broad rings.

GENITALIA.– Fig. 57; ground morphology very similar to that of *B. uroceriformis*; aedeagus and antrum slightly more slender.

BIOLOGY AND HABITAT.– Shrubby, rocky or stony habitats, roadsides, embankments; hostplants: *Coronilla minima* L., *Lotus corniculatus* L.; larva 1 year in root; pupa deep (4-8 cm) in root with a long emergence tube to ground surface (reared by present authors); adult VI-VIII.

DISTRIBUTION.– W Mediterranean: Iberian Peninsula, S France, NW and central Italy, Istria.

REMARKS.– The dark specimens (Fig. 57b), described as *Bembecia baumgartneri*, were later given by Špatenka et al. (1999) as subspecies of *B. himmighoffeni*. However, such individuals occur throughout the range of *B. himmighoffeni* and they are, thus, geographically not definable.

58. Bembecia uroceriformis (Treitschke, 1834)
(Pl. 4: 9,12)

Sesia uroceriformis Treitschke, 1834: 121. Type locality: Hungary. Type material: lectotype ♂, TMB (des. by Toševski, 1992).

DIAGNOSIS.– Wingspan 16-26 mm; antenna black with white spot before apex in female; external transparent area narrower than apical area, round or oval-shaped, in male of 3 slightly elongated and 1-2 very small cells (Fig. 58a), in female of 3 cells; apical area in male with yellow spots between veins; in female almost entirely yellow; in male yellow abdominal rings 2, 4, 6, 7 broad and 3, 5 narrower, in female 2, 4, (5), 6 broad, 3, (5) narrow and/or absent.

GENITALIA.– Fig. 58; see *B. himmighoffeni*.

BIOLOGY AND HABITAT.– Xerothermic, often stony or rocky habitats; hostplants: *Chamaecytisus* spp. (Laštůvka, 1990d), *Ulex* spp. (Le Cerf, 1920), *Corothamnus procumbens* Waldst. & Kit. (Špatenka, pers. comm.), *Coronilla emerus* L. (Toševski, 1992), *Spartium junceum* L., *Calicotome* spp. (reared by present authors); larva 1-2 years in root; pupa in a cocoon in root, occasionally with a short emergence tube; adult VI-VII.

DISTRIBUTION.– Holomediterranean: S, parts of W and central Europe.

Pyropteron Newman, 1832

Pyropteron Newman, 1832: (73)75.
Type species: *Sphinx chrysidiformis* Esper, 1782; by monotypy.
Synansphecia Căpuşe, 1973: 166 (as subgenus of *Chamaesphecia* Spuler, 1910), **syn n.**
Type species: *Sesia triannuliformis* Freyer, 1842; by original designation.

GENUS CHARACTERS.– Antenna usually dark ventrally; proboscis present; crista sacculi high, in distal part bent ventrally; saccus and aedeagus long; aedeagus with a number of small cornuti apically; antrum thin, membranous or sclerotised. Larva in the roots of herbaceous plants. 15 species in the region.

REMARKS.– The genera *Pyropteron* and *Synansphecia* are synonymized here, because there does not exist any clear morphological limit between these two taxa (cf. Laštůvka, 1990a). The West Mediterranean species of *Synansphecia* have been revised by Kallies (1999).

59. Pyropteron chrysidiformis (Esper, 1782)

(Pl. 5: 10,11,12,13)

Sphinx chrysidiformis Esper, 1782: 210, pl. 30, fig. 2. Type locality: S France, Languedoc. Type material: holotype ♂, SMW.
Pyropteron chrysidiformis var. *sicula* Le Cerf, 1922: 27, pl. 539, figs 4526, 4527. Type locality: Sicily. Type material: lectotype ♂, MNHP (des. by Špatenka, 1992b).

DIAGNOSIS.– Wingspan 17-26 mm; antenna black, often with ochreous subapical spot in female and occasionally in male; labial palpus with long black scales; 1st hind tarsal segment with bristled scales; in nominotypical subspecies scapular spot of forewing usually present (absent occasionally in dark specimens) and 2nd segment without ring; in ssp. *sicula* scapular spot usually reduced or absent; abdominal segments 2, 4, 6 and 7 (in male) with white or yellowish rings; very variable species.

GENITALIA.– Fig. 59.

BIOLOGY AND HABITAT.– Sandy places, meadows, roadsides, embankments, field edges, waste places; hostplants: *Rumex* spp.; larva 1 year in root; pupa in upper part of root or in basal part of old stem; adult V-VIII.

DISTRIBUTION.– Atlantico-Mediterranean: nominotypical ssp. in SW, W and parts of central Europe, eastwards to Istria, doubtful record from SW Romania (Popescu-Gorj, 1962); ssp. *sicula* in Sicily, central and S Italy.

REMARKS.– The taxonomic relationship of the taxa "*chrysidiformis*" and "*sicula*" (sympatric ?) is insufficiently known (cf. Laštůvka, 1990b).

60. Pyropteron minianiformis (Freyer, 1843)

(Pl. 5: 14,15)

Sesia minianiformis Freyer, 1843: 35, pl. 404, fig. 3. Type locality: [Turkey, Istanbul]. Type material: lost.

DIAGNOSIS.– Wingspan 12-24 mm; antenna black; tegula yellow bordered (distinct in female); vertex with ochreous hairs; forewing scapular spot present; segments 2, 4, 6 and 7(in male) with yellow rings (Fig. 60a).

GENITALIA.– Fig. 60; crista gnathi medialis double.

BIOLOGY AND HABITAT.– Embankments, village edges, pastures; hostplants: *Rumex* spp. (*Rumex conglomeratus* Murr., *R. crispus* L., *R. maritimus* L., etc.); larva 1 year in root; pupa in root or stem base (Engelhard, 1978; Laštůvka & Laštůvka, 1980); adult V-VIII.

DISTRIBUTION.– E Mediterranean: Balkan Peninsula, Crete, Crimea.

REMARK.– The date of Freyer's original description is indicated as 1842, 1843 or 1845 by different authors. We have followed Olivier (2000).

61. *Pyropteron doryliformis* (Ochsenheimer, 1808)

(Pl. 5: 16,17,18)

Sesia doryliformis Ochsenheimer, 1808: 141. Type locality: Portugal. Type material: destroyed.

Sesia icteropus Zeller, 1847b: 403. Type locality: Sicily, Syracuse. Type material: lost.

DIAGNOSIS.– Wingspan 15-26 mm; antenna brown with black apex, ochreous ventrally; frons brown, occasionally with indistinct yellow borders before eyes; labial palpus with long black scales in male and orange in female; external transparent area narrow, of (4)-5 cells; apical area without light spots; abdomen covered with reddish rusty scales in female, with yellow or orange ones in male, segments 2, 4, 6 with white rings; ssp. *icteropus* more slender and yellowish.

GENITALIA.– Fig. 61.

BIOLOGY AND HABITAT.– Roadsides, waste places, pastures; hostplants: *Rumex* spp. (*Rumex conglomeratus* Murr., *R. acetosa* L., etc.); larva 1 year in root; adult V-VII.

DISTRIBUTION.– W Mediterranean: nominotypical form in S Iberian Peninsula; ssp. *icteropus* in Sardinia, Sicily and S Italy.

REMARKS.– The type locality of *Sesia doryliformis* var. *teriolensis* Staudinger, 1894 (Bolzano env.) lies distinctly outside the hitherto known range of *Pyropteron doryliformis*, and a mistake of the type specimen in ZMHB [= *P. doryliformis icteropus*] cannot be excluded.

62. *Pyropteron triannuliformis* (Freyer, 1843) **comb. n.**

(Pl. 6: 4,5)

Sesia triannuliformis Freyer, 1843: 35, pl. 404, fig. 2. Type locality: Turkey, Istanbul. Type material: lost.

Chamaesphecia balcanica Zukowsky, 1929: 21. Type locality: Albania, Hudowa. Type material: lectotype ♂, LNM (des. by Špatenka, 1992b).

Bembecia (*Synansphecia*) *ljiljanae* Toševski, 1986b: 191. Type locality: Macedonia, Crnićani, Bogdanci. Type material: holotype ♀, IT.

DIAGNOSIS.– Wingspan 11-26 mm; antenna black, usually with white subapical spot in female; labial palpus in male smooth; eye with white border; external transparent area usually of 5 cells, round or oval; longitudinal transparent area at least indicated in female; discal spot of hindwing broad to M2 (Fig. 62a); abdominal segments 2, 4, 6 with white rings; often dorsal spot line present; male anal tuft divided into three parts.

GENITALIA.– Fig. 62; gnathos broad distally, specialised, not rounded.

BIOLOGY AND HABITAT.– Sandy and rocky places, waste places, pastures, embankments; hostplants: *Rumex* spp. (*Rumex acetosella* L., *R. acetosa* L., *R. crispus* L., etc.); larva 1 year in root; pupa in root or in basal part of old (dry) stem; adult V-VIII.

DISTRIBUTION.– E Mediterranean: central, SE and E Europe.

REMARKS.– The taxonomy of *P. triannuliformis* was discussed by Laštůvka (1990e).– The date of Freyer´s original description is indicated as 1842, 1843 or 1845 by different authors. We have followed Olivier (2000).

63. Pyropteron meriaeformis (Boisduval, 1840) **comb. n.**

(Pl. 6: 6)

Sesia meriaeformis Boisduval, 1840: 42. Type locality: Andalusia [Granada]. Type material: lost.
Sesia corsica Staudinger, 1856: 274. Type locality: Corsica. Type material: lectotype ♂, ZMHB (des. by Špatenka & Laštůvka, 1988).

DIAGNOSIS.– Wingspan 11-18 mm; ground coloration usually dark brown, body and forewing borders more or less covered with ochreous scales; antenna black, usually subapically with white scales in male and with white spot in female; eye with white border; external transparent area very small, usually of 3 cells, narrower than apical area; longitudinal transparent area absent; abdominal segments 2 (often indistinct), 4, 6 (occasionally 7 in male) with narrow white rings.

GENITALIA.– Fig. 63; see *P. hispanica.*

BIOLOGY AND HABITAT.– Sandy, rocky habitats, pastures, rocks, roadsides, abandoned fields; hostplants: *Rumex* spp. (small species, of *Rumex acetosella*-group); larva 1 year in root; adult V-VII.

DISTRIBUTION.– Atlantico-Mediterranean: Iberian Peninsula, France, Italy, Sardinia, Corsica, Sicily.

64. Pyropteron hispanica (Kallies, 1999) **comb. n.**

(Pl. 6: 2,3)

Synansphecia hispanica Kallies, 1999: 92. Type locality: Spain, prov. Almería, Sierra Baza. Type material: holotype ♂, ZMHB.
Synansphecia atlantis auct., nec Schwingenschuss, 1935

DIAGNOSIS.– Wingspan 16-21 mm; ground coloration white/greyish black; antenna with white subapical spot usually in both sexes; labial palpus nearly smooth; eye with white border (or frons entirely white); tegula with yellow border; external transparent area of 5 cells, round, slightly higher or broader; longitudinal transparent area present (at least indicated in female); discal spot of hindwing of the same width up to M2 (Fig. 64a); segments 2, 4, 6 with white rings.

GENITALIA.– Fig. 64; male genitalia very homogeneous and usually not distinguishable in the species group (*P. hispanica, P. meriaeformis, P. koschwitzi, P. muscaeformis*); gnathos simple and rounded distally in the group; female genitalia not significantly different from those of *P. meriaeformis.*

BIOLOGY AND HABITAT.– Embankments, pastures, waste places; hostplants: *Rumex* spp. (*Rumex scutatus* L.); larva 1 year in root; adult V-VII.

DISTRIBUTION.– W Mediterranean: S and SE Spain, S France (recorded near Aigues Mortes by the present authors in 1994).

REMARKS.– Previously, *Pyropteron hispanica* was reported as *Synansphecia atlantis* (Schwingenschuss, 1935), but this species is known so far only from Morroco (Kallies, 1999).

65. *Pyropteron koschwitzi* (Špatenka, 1992) **comb. n.**

(Pl. 6: 7)

Synansphecia koschwitzi Špatenka, 1992a: 437. Type locality: Spain, prov. Toledo, Aranjuez. Type material: holotype ♂, TW.

DIAGNOSIS.– Wingspan 15-21 mm; antenna black, with white spot in female, exceptionally in male; eye with white border; labial palpus slightly tufted in male; ground coloration greyish brown; external transparent area round, of 4-5 cells; tegula with light ochreous border; longitudinal transparent area absent in female; apical area with greyish spots; discal spot of hindwing usually narrows to M2; abdominal segments 2, 4, 6 with white rings.

GENITALIA.– Fig. 65; see *P. hispanica,* female genitalia similar to *P. muscaeformis.*

BIOLOGY AND HABITAT.– Xerothermic salty, sandy habitats; hostplant: *Limonium toletanum* Erben (Špatenka, pers. comm.); larva 1 year in root; adult IV-VI.

DISTRIBUTION.– Known from the type locality and from NE Spain, Fraga (recorded by present authors).

REMARKS.– The species is close to the N African *Pyropteron borreyi* (Le Cerf, 1922), which was incorrectly reported in the first edition of this book from southern Spain (misidentification of *P. hispanica*).

66. *Pyropteron muscaeformis* (Esper, 1783) **comb. n.**

(Pl. 6: 10,11)

Sphinx muscaeformis Esper, 1783: 217, pl. 32, fig. 5. Type locality: Germany, Frankfurt am Main. Type material: not traced.
Sesia philanthiformis Laspeyres, 1801: 31. Type locality: Germany. Type material: destroyed.
Chamaesphecia aestivata Králíček, 1969: 115. Type locality: Czech Republic, Hodonín. Type material: holotype ♂, MMB.

DIAGNOSIS.– Wingspan 13-19; antenna black, with white subapical spot in female and occasionally in male; eye with white border, frons brown or ochreous-yellow; labial palpus slightly tufted in male; ground coloration usually brownish or yellowish brown; tegula with yellow border; external transparent area round or oval, broader than apical area, consisting of 5 cells; longitudinal transparent area absent in female; ochreous or yellowish spots in apical area; discal spot of hindwing usually tapers to M2 (Fig. 66a); abdominal segments 2, 4, 6 with white rings.

GENITALIA.– Fig. 66; see *P. hispanica*; in female genitalia ductus bursae more or less straight in fertilised and strongly bent in unfertilised female (Fig. 66b, c).

BIOLOGY AND HABITAT.– Sandy or rocky places, roadsides, coastal habitats; hostplants: *Armeria* spp. (*Armeria maritima* Miller); larva 1 year in root; pupa in upper part of root; adult V-VIII(IX).

DISTRIBUTION.– Scattered, European.

REMARKS.– More yellow-coloured specimens from western France have been described as *Sesia muscaeformis occidentalis* Joannis, 1908.

67. Pyropteron kautzi (Reisser, 1930) **comb. n.**

(Pl. 6: 8)

Chamaesphecia kautzi Reisser, 1930: 101. Type locality: Andalusia, Sierra Nevada, Monte del Lobo. Type material: lectotype ♀, LNK (des. by Špatenka, 1992b).

DIAGNOSIS.– Wingspan 16-22 mm; antenna black without white subapical spot in both sexes; frons with bronze sheen and yellow or white before eyes; labial palpus deep yellow and black basally and nearly entire black distally in male and orange basally and brownish black distally in female; transparent areas small, covered with whitish scales; external transparent area of 5 cells in male and smaller, of 3-4 cells in female; longitudinal transparent area absent in female; apical area with yellow or orange spots; abdomen with oily sheen, dark brown, with yellow ring on 2nd, 4th and 6th segments (Fig. 67a).

GENITALIA.– Fig. 67.

BIOLOGY AND HABITAT.– Rocky habitats on the top of Sierra Nevada; hostplant: *Erodium cheilanthifolium* Boiss.; larva 2 (3) years in root (Pühringer & Pöll, 2001a; Pühringer, pers. comm.); adult VII.

DISTRIBUTION.– Known only from the type material and several other specimens (Pühringer & Pöll, 1999 and 2001a) from the Sierra Nevada.

68. Pyropteron aistleitneri (Špatenka, 1992) **comb. n.**

(Pl. 6: 1)

Synansphecia aistleitneri Špatenka, 1992a: 436. Type locality: Spain, Andalusia, Sra de Guillimona. Type material: holotype ♀, TW.

DIAGNOSIS.– Wingspan 16-23 mm; male antenna entirely dark, in female with white subapical spot; ground coloration extremely dark, nearly black; external transparent area large, broad in male and slightly higher than broad in female, of 5 cells; apical area with distinct greyish spots; discal spot of hindwing large, broad to M2 and narrowing between M2 and M3; white abdominal rings not broadened laterally; abdomen black ventrally; hind tibia only white laterally in the fore part, not distally.

GENITALIA.– Fig. 68 (paratype, after Špatenka), female genitalia not studied.

BIOLOGY AND HABITAT.– Rocky or stony habitats in the mountains of southern Spain; bionomics unknown; adult VII (Pühringer & Pöll, 2001b).

DISTRIBUTION.– Andalusia.

69. Pyropteron leucomelaena (Zeller, 1847) **comb. n.**

(Pl. 6: 14,15)

Sesia leucomelaena Zeller, 1847a: 12, part. Type locality: Turkey, Macri. Type material: lectotype ♂, BMNH (des. by Špatenka, 1992b); see note.
Sesia cretica Rebel, 1916: 143. Type locality: Crete, Kristallenia. Type material: lectotype ♀, NHMV (des. by Špatenka, 1992b).

DIAGNOSIS.– Wingspan 9-18 mm; antenna black, in female (exceptionally in male) with white subapical spot; frons white or yellowish white; tegula with yellow border; thorax laterally below forewing with yellowish white to deep yellow spot; external transparent area large in male, of 5 (6) cells; apical area narrow (Fig. 69a); in female small, round or often higher than broad (2-5 cells); longitudinal transparent area present in male, usually absent in female; abdominal segments 4 and 6 with white rings.

GENITALIA.– Fig. 69; crista sacculi extremely high distally and often without hairs in middle.

BIOLOGY AND HABITAT.– Xerothermic treeless, grassy, sandy or rocky habitats, also waste places; hostplant: *Poterium minor* Scop.; larva 1 year in root (Toševski, 1986a); oviposition on *Geranium sanguineum* L. observed by present authors in Sicily; adult V-VIII.

DISTRIBUTION.– Holomediterranean: S Europe.

REMARKS.– Zeller (1847a) based the description of *Sesia leucomelaena* on two specimens, the female of which was recognised to be a different species, viz. *Sesia annellata* Zeller, 1847b. Consequently the designation of the paralectotype was incorrect (cf. Špatenka, 1992b).

70. Pyropteron affinis (Staudinger, 1856) **comb. n.**

(Pl. 6: 13)

Sesia affinis Staudinger, 1856: 278. Type locality: Italy, Bolzano. Type material: lectotype ♀, ZMHB (des. by Špatenka & Laštůvka, 1988).

DIAGNOSIS.– Wingspan 11-20 mm; antenna without white subapical spot; ground coloration silvery grey; eye with white border; tegula with yellowish white border; external transparent area small, broader than high, exceptionally round, of 3 cells (Fig. 70a)(exceptionally of 2 further very small cells over R4 and under M3); longitudinal transparent area at least basally present in male, usually absent in female; abdominal segments (2) 4 (5) 6 (7) with white rings.

GENITALIA.– Fig. 70; gnathos specialised (doubled) distally; in female genitalia antrum long, bent, with less sclerotised part in middle.

BIOLOGY AND HABITAT.– Xerothermic stony or rocky places, often on limestone; hostplants: *Helianthemum* spp., *Fumana* spp. (Laštůvka, 1990d); larva 1 year in root; infested plant with a pile of reddish brown "sawdust" on the ground; adult V-VII.

DISTRIBUTION.– W Palaearctic: S, parts of central and W Europe.

71. Pyropteron umbrifera (Staudinger, 1870) **comb. n.**

(Pl. 9: 14)

Sesia umbrifera Staudinger, 1870: 96. Type locality: Greece, Corfu. Type material: lectotype ♂, ZMHB (des. by Špatenka & Laštůvka, 1988).

DIAGNOSIS.– Wingspan 20-27 mm; antenna brown with darker apex; frons brown; labial palpus brown; external transparent area small, narrow, of 3-4 cells; longitudinal transparent area absent; discal spot of hindwing broad, connected by a scale band with broad brown wing margin; brown abdomen with white ring on 4th segment.

GENITALIA.– Fig. 71; valve broad; crista sacculi short; gnathos more or less straight distally; in female genitalia antrum relatively broad, tubular, distinctly shorter than apophyses anteriores.

BIOLOGY AND HABITAT.– Salt marshes; hostplant: *Limonium vulgare* Mill. (*Limonium narbonense* Mill. according to Petersen & Bartsch, 1998); larva 1 year in root; adult V-VII (Petersen & Bartsch, 1998; reared by Špatenka, and present authors).

DISTRIBUTION.– Greece: Corfu, western coast and eastern Peloponnes (recorded by Petersen, Špatenka, and present authors).

72. Pyropteron cirgisa (Bartel, 1912) **comb. n.**

(Pl. 9: 13)

Chamaesphecia cirgisa Bartel, 1912: 408, pl. 50k. Type locality: Kazakhstan, Uralsk. Type material: holotype ♀, BMNH.

DIAGNOSIS.– Wingspan 19-28 mm; antenna brown, with black apex and with ochreous scales subapically, occasionally ochreous outside; eye with white border or frons entirely white; labial palpus only slightly tufted; external transparent area large, narrow (higher than broad), or quadrate, of 5-6 cells; discal spot of hindwing broad to M3; abdominal segments 2, 4 with white rings; dorsal spot line on abdomen present.

GENITALIA.– Fig. 72; gnathos straight, not rounded distally; crista sacculi extremely high distally; in female genitalia antrum short, straight.

BIOLOGY AND HABITAT.– E European and W Asiatic salt steppes; hostplant: *Limonium gmelini* Willd. (Zukowsky, 1915); larva probably 1 year in root; adult VI-VII.

DISTRIBUTION.– Caspian: SE European Russia, W Kazakhstan, S Ukraine, E Romania, Turkey (Lingenhöle, pers. comm.).

73. Pyropteron mannii (Lederer, 1853) **comb. n.**

(Pl. 6: 9,12)

Sesia mannii Lederer, 1853: 88. Type locality: Turkey, Brussa [Bursa]. Type material: lectotype ♂, ZMHB (des. by Špatenka & Laštůvka, 1988).

DIAGNOSIS.– Wingspan 11-18 mm; antenna black; frons yellow; labial palpus yellow with long black scales in male and orange in female; external transparent area extremely narrow and high, of 5 cells in male and 3-5 cells in female (Fig. 73a); hind tibia in middle ochreous-yellow in male, orange in female; abdominal segments 2, 4, 6 with narrow white rings.

GENITALIA.– Fig. 73; gnathos simple, rounded distally; crista sacculi high distally; similar to *P. leucomelaena*; in female genitalia antrum short.

BIOLOGY AND HABITAT.– Xerothermic scrub, bushes and forest edges, roadsides; hostplant: *Geranium rotundifolium* L.; larva 1 year in root; at the time of flight the above-ground part of the hostplant is usually already dry or absent (Laštůvka & Laštůvka, 1980); adult VI-VII.

DISTRIBUTION.– E Mediterranean (known range very small): in Europe, SE Bulgaria, European Turkey.

Dipchasphecia Căpuşe, 1973

Chamaesphecia subgen. *Dipchasphecia* Căpuşe, 1973: 161.

Type species: *Dipsosphecia roseiventris* Bartel, 1912; by original designation.

GENUS CHARACTERS.– Antenna whitish or white-yellow on the outside; proboscis present; transparent areas covered with white scales; crista sacculi reduced to a short projection in middle of valve; aedeagus with one small cornutus apically; gnathos elongate, flat, without crista medialis; antrum sclerotised, slightly broadened distally (Fig. 74, cf. Laštůvka, 1990a). Larva in roots of Plumbaginaceae and Caryophyllaceae. 1 species in the region.

74. Dipchasphecia lanipes (Lederer, 1863)

(Pl. 6: 16)

Sesia lanipes Lederer, 1863: 20. Type locality: Bulgaria, Sliwno [Sliven]. Type material: lectotype ♂, ZMHB (des. by Špatenka & Laštůvka, 1988).

DIAGNOSIS.– Wingspan 12-23 mm; antenna black, often whitish outside; frons white at least before eyes; labial palpus slightly tufted in male; transparent areas small, covered with whitish scales; external transparent area of 3-5 cells; the longest cell between R5 and M1, further cells get shorter from distal side towards Cu2 (Fig. 74a); longitudinal transparent area present in male, absent in female; abdominal segments 2, 4, 6 with white rings.

GENITALIA.– Fig. 74.

BIOLOGY AND HABITAT.– Shrubby slopes; hostplant: unknown; adult VI-VII.

DISTRIBUTION.– E Mediterranean: SE Bulgaria.

REMARKS.– The relationship to *Dipchasphecia intermedia* Špatenka, 1997 given by Špatenka (1997) also from SE Bulgaria needs clarification.

Chamaesphecia Spuler, 1910

Chamaesphecia Spuler, 1910: 311.
Type species: *Sphinx empiformis* Esper, 1783; by original designation.

GENUS CHARACTERS.– Antenna often ochreous or yellow on the outside; proboscis present; crista sacculi short, straight; the part of valve with bifurcate hairs is bordered by a crest (Fig. 76); scopula reduced or absent; aedeagus with 2 small cornuti; gnathos elongate, flat; antrum funnel-shaped; periostial region with microscopic spines. Larva in root and/or stem of herbaceous plants. 32 species in the region.

REMARKS.– There are two phylogenetic lines in the genus, classified as subgenera *Scopulosphecia* Laštůvka, 1990 (from *C. mysiniformis* to *C. masariformis*) and *Chamaesphecia* s. str. (Laštůvka, 1990a).

75. Chamaesphecia mysiniformis (Boisduval, 1840)

(Pl. 7: 1,2)

Sesia mysiniformis Boisduval, 1840: 42. Type locality: Andalusia [Granada]. Type material: lectotype ♂, MNHP (des. by Špatenka, 1992b).
Chamaesphecia rondouana Le Cerf, 1922: 32, pl. 540, figs 4534, 4536. Type locality: France, Pyrenees, Gédre. Type material: holotype ♂, MNHP.

DIAGNOSIS.– Wingspan 11-24 mm; coloration from silvery whitish to dark brown; antenna occasionally yellow outside; eye white bordered or frons white; external transparent area broader than high, of 3 (-5) long cells (Fig. 75a); segments 2, 4, 6 with white rings; dorsal line on abdomen.

GENITALIA.– Fig. 75.

BIOLOGY AND HABITAT.– Rocky slopes, pastures, waste places, from sea-level to about 2500 m; hostplants: *Marrubium* spp. (*M. vulgare* L., *M. supinum* L.); *Stachys* spp., *Ballota hirsuta* Bentham; *Sideritis* spp. (reared by Bläsius, Blum, Špatenka, present authors, cf. also Sagliocco & Coupland, 1995); larva 2 years in root; adult V-VIII.

DISTRIBUTION.– W Mediterranean: Iberian Peninsula, S France, NW Italy.

76. Chamaesphecia anatolica Schwingenschuss, 1938

(Pl. 7: 3; 9: 11)

Chamaesphecia anatolica Schwingenschuss, 1938: 175. Type locality: Turkey, Akşehir. Type material: lectotype ♂, NHMV (des. by Špatenka, 1992b).

DIAGNOSIS.– Wingspan 17-21 mm; antenna yellow outside; labial palpus white basally, yellowish distally; frons yellow; external transparent area round or oval, of 3-4 cells (Fig. 76a); narrow white rings on 2nd, 4th and 6th segments; body densely covered with ochreous yellow scales; discal spot of hindwing reaches M2 (Fig. 76b).

GENITALIA.– Fig. 76.

BIOLOGY AND HABITAT.– Xerothermic stony and rocky places, roadsides, dry riverbeds; hostplants: *Nepeta nuda* L., *Nepeta spruneri* Boiss., *N. parnassica* Heldr. & Sart. (reared by present authors and Petersen & Bartsch, 1998); larva 2 years in root; adult VI-VII.

DISTRIBUTION.– E Mediterranean: N Hungary, SW Romania, NE Serbia, Greece, Anatolia (Lingenhöle, Toševski, pers. comm.).

77. Chamaesphecia chalciformis (Esper, [1804])

(Pl. 6: 17)

Sphinx chalciformis Esper, [1804]: 44, pl. 47, figs 1, 2. Type locality: Hungary, Ofen [Budapest]. Type material: not traced.

Sphinx chalcidiformis Hübner, [1806]: 90, pl. 19, fig. 93. Type locality: Hungary. Type material: lost.

Sesia prosopiformis Ochsenheimer, 1808: 146. Type locality: Hungary. Type material: destroyed.

DIAGNOSIS.– Wingspan 12-22 mm; antenna black; frons black; labial palpus with long black scales in male and nearly smooth, white ventrally in female; forewing, hind and middle tibia and anal tuft partly red; abdomen without rings, black; 1st hind tarsal segment not distinctly haired.

GENITALIA.– Fig. 77; variability of crista sacculi demonstrated.

BIOLOGY AND HABITAT.– Dry stony places, shrubby slopes, roadsides, forest edges, pastures; hostplants: *Origanum vulgare* L. (Popescu-Gorj et al., 1958), *Origanum* sp. (Špatenka, pers. comm.); larva 1 year in root; adult V-VII.

DISTRIBUTION.– E Mediterranean-Asiatic: SE central Europe, Balkan Peninsula, Crete, S Ukraine, S Russia, Kazakhstan.

78. Chamaesphecia schmidtiiformis (Freyer, 1836)

(Pl. 6: 18)

Sesia schmidtiiformis Freyer, 1836: 140, pl. 182, fig. 1. Type locality: Croatia, Fiume [Rijeka]. Type material: lost.

DIAGNOSIS.– Wingspan 17-26 mm; antenna and frons black; labial palpus with long black scales in male and nearly smooth, white ventrally in female; forewing, hind and middle tibia and anal tuft partly red; 1st hind tarsal segment not distinctly hairy; abdomen with white ring on 4th segment.

GENITALIA.– Fig. 78. The genitalia of *C. chalciformis* and *C. schmidtiiformis* are rather variable, but no significant differences have been found between them.

BIOLOGY AND HABITAT.– Grassy, stony or rocky slopes and places, roadsides, pastures; hostplants: *Salvia* spp. (*Salvia verticillata* L., *S. sclarea* L., *S. syriaca* L.); larva 1 year in root; pupa in upper part of root, in a gallery with short emergence tube (Toševski, 1986a, Špatenka, pers. comm.); adult V-VI.

DISTRIBUTION.– E Mediterranean: N Italy, Balkans, S Ukraine, S Russia.

REMARKS.– The taxonomic status of some forms, particularly of ab. *albotarsata* Staudinger & Rebel, 1901 is unknown.

79. Chamaesphecia anthrax Le Cerf, 1916

(Pl. 8: 4)

Chamaesphecia anthrax Le Cerf, 1916: 15, pl. 321, fig. 4667. Type locality: Algeria, Sebdou. Type material: lectotype ♂, MNHP (des. by Špatenka, 1992b).

DIAGNOSIS.– Wingspan 15-19 mm; antenna occasionally with yellow scales outside; frons black (Fig. 79a); labial palpus with long black scales in male and nearly smooth, white basally in female; thorax black laterally; external transparent area round, of 4-5 cells; apical area broad, without distinct spots (Fig. 79b); longitudinal transparent area present in male and absent in female; abdominal segments 4 and 6 with white rings.

GENITALIA.– Fig. 79; very similar or indistinguishable in the species group (*C. anthrax* to *C. alysoniformis* and/or *C. albiventris*).

BIOLOGY AND HABITAT.– Dry, grassy or rocky semiruderal places, roadsides, pastures; hostplant: *Nepeta apuleii* Ucria; larva 2 years in root; pupa in root or in basal part of old stem (reared by present authors); adult VI-VII.

DISTRIBUTION.– W Mediterranean: known so far only from several localities in central and S Spain (captured by present authors, Bläsius, Lingenhöle, Petersen, etc.), from Algeria and Morocco.

80. Chamaesphecia maurusia Püngeler, 1912

(Pl. 9: 12)

Chamaesphecia maurusia Püngeler, *in* Seitz, 1912: 412, pl. 50m. Type locality: Algeria, Teniet-el-Had. Type material: holotype ♀, ZMHB.

DIAGNOSIS.– Wingspan 14-20 mm; antenna and frons black; labial palpus with long black scales in male and nearly smooth, white basally in female; thorax with yellowish spot laterally; forewing black, external transparent area distinctly higher than broad (not round), of 5 cells (Fig. 80a); apical area broad; longitudinal transparent area absent in female; abdominal segments 4, 6 in male and 2, 4, 6 in female with white rings.

GENITALIA.– Fig. 80.

BIOLOGY AND HABITAT.– Pastures, roadsides; hostplant: *Nepeta tuberosa* L.; larva 2 years in root (reared by present authors); adult: VI.

DISTRIBUTION.– W Mediterranean: in Europe known so far from one locality in Sicily, near Palermo (found by present authors).

REMARKS.– The scanty material and the limited knowledge of individual and geographical variation make it impossible to elucidate the taxonomic relationship of the taxa *C. anthrax, C. maurusia* and *C. powelli* Le Cerf, 1916. Therefore *C. anthrax* and *C. maurusia* are treated here as two separate species.

81. Chamaesphecia alysoniformis (Herrich-Schäffer, 1846)

(Pl. 8: 6)

Sesia alysoniformis Herrich-Schäffer, 1846: 73, pl. 8, fig. 46. Type locality: not mentioned. Type material: lost.

DIAGNOSIS.– Wingspan 10-16 mm; antenna black, yellowish ventrally; frons dark; labial palpus nearly smooth, whitish basally, yellowish distally; thorax laterally below forewing yellow; metathorax with transverse yellow spot dorsally (Fig. 81a); transparent areas very small, covered with yellowish scales; external transparent area of 1-3 cells; longitudinal transparent area absent; abdominal segments 2, 4, 6 with yellow rings.

GENITALIA.– Fig. 81; see *C. anthrax*.

BIOLOGY AND HABITAT.– Dry or moist grassy or shrubby places, roadsides, pastures; hostplants: *Mentha* spp. (*Mentha longifolia* L.: Laštůvka & Laštůvka, 1980), *Salvia* sp. (Petersen & Bartsch, 1998); larva 1 year in root; adult V-VII.

DISTRIBUTION.– E Mediterranean: Balkan Peninsula, Crete.

82. Chamaesphecia aerifrons (Zeller, 1847)

(Pl. 8: 5)

Sesia aerifrons Zeller, 1847b: 415. Type locality: Sicily, Siracusa. Type material: lectotype ♂, ZMHB (des. by Špatenka & Laštůvka, 1988).

Sesia aerifrons var. *sardoa* Staudinger, 1856: 281. Type locality: Sardinia, Barbagia Ollolai. Type material: lectotype ♂, ZMHB (des. by Špatenka & Laštůvka, 1988).

DIAGNOSIS.– Wingspan 12-19 mm; antenna black, yellow outside in ssp. *sardoa*; frons brownish, with metallic sheen, without white borders before eyes; labial palpus white basally, dark apically, nearly smooth (Fig. 82a); thorax laterally below forewing yellow; external transparent area small, of 3, usually slightly elongate cells (Fig. 82b); longitudinal transparent area absent or occasionally short in male; abdominal segments 4 and 6 with white rings (also 2 in ssp. *sardoa*).

GENITALIA.– Fig. 82; see *C. anthrax*.

BIOLOGY AND HABITAT.– Various habitats: dry to moist grassy, rocky or shrubby places, forest edges, river banks, pastures, abandoned fields; hostplants: *Calamintha nepeta* L., *Satureja* spp., *Thymus* spp., *Lavandula* spp., *Mentha* spp., *Origanum vulgare* L. (Le Cerf, 1920; reared by Bläsius, Blum, Bettag and present authors); larva 2 years in root; adult V-VII.

DISTRIBUTION.– Atlantico-Mediterranean: SW Europe, in central Europe known only from SW Germany (Kaiserstuhl: Bläsius, 1992) and from SW Switzerland (Whitebread, pers. comm.); in Sardinia and Corsica as ssp. *sardoa*; in other parts of the range as nominotypical subspecies.

83. Chamaesphecia albiventris (Lederer, 1853)

(Pl. 2: 30)

Sesia albiventris Lederer, 1853: 82. Type locality: Turkey, Brussa [Bursa]. Type material: lectotype ♂, ZMHB (des. by Špatenka & Laštůvka, 1988).

DIAGNOSIS.– Wingspan 13-18 mm; antenna brown-black (in male) or discontinuously yellow ventrally in female; frons entirely brown or with whitish coloration before eyes; labial palpus white, smooth; external transparent area large, oval (broader than high), of 3-4 cells in male and small, of 3 slightly elongated cells in female; apical area narrow in male and broad in female, dark; abdominal segments 2, 4, 6 with white (yellowish white) rings; abdomen in male entirely white ventrally, in female brown, with whitish scales.

GENITALIA.– Fig. 83.

BIOLOGY AND HABITAT.– Shrubby places, embankments, forest edges; hostplant: unknown (adults usually captured near various Lamiaceae such as *Calamintha nepeta* L., *Origanum vulgare* L., etc.); adult VI-VII.

DISTRIBUTION.– E Mediterranean: N Greece (captured by Blum and present authors); some additional specimens, which differ more or less from typical "*albiventris*", are known from Bulgaria and Greece.

84. Chamaesphecia osmiaeformis (Herrich-Schäffer, 1848)

(Pl. 8: 15)

Sesia osmiaeformis Herrich-Schäffer, 1848: pl. 9, fig. 52; 6(1852): 48. Type locality: Italy. Type material: lost.

DIAGNOSIS.– Wingspan 14-22 mm; antenna brown, darker apically, golden-yellow outside; frons brown, occasionally with indistinct yellow border before eyes; labial palpus ochreous, long hairy; ground coloration brown; external transparent area narrow, of 4-5 cells (Fig. 84a); longitudinal transparent area short in male and absent in female; discal spot of hindwing broad only to M2; hind tibia long hairy, ochreous in middle; 4th abdominal segment with narrow white ring and with triangular yellowish spot dorsally and laterally.

GENITALIA.– Fig. 84; basic morphology similar to that in the *C. doleriformis* group, but scopula absent as in *Chamaesphecia* s. str.

BIOLOGY AND HABITAT.– Dry grassy places, embankments, pastures, abandoned gardens; hostplants: *Salvia* spp., small herbaceous species (*S. verbenaca* L., *S. jaminiana* Noé, *S. phlomoides* Asso: Špatenka et al., 1999; recorded by Kallies and present authors); larva 1 year in root; adult V-VII.

DISTRIBUTION.– W Mediterranean: S Italy, Sicily, Sardinia (recorded by Garrevoet and Pühringer) and Corsica; one doubtful specimen in ZMHB labelled "Andalusia, Bartel".

85. Chamaesphecia ramburi (Staudinger, 1866)

(Pl. 8: 3)

Sesia ramburi Staudinger, 1866: 53. Type locality: Andalusia, Chiclana. Type material: lectotype ♂, ZMHB (des. by Špatenka & Laštůvka, 1988).

DIAGNOSIS.– Wingspan 15-25 mm; ground coloration grey or greyish brown; antenna yellow outside; frons black or brown with metallic sheen, occasionally with narrow yellow border before eyes; thorax whitish yellow laterally; external transparent area higher than broad, with straight border towards discal spot (Fig. 85a); ochreous spots distinct only in distal part of apical area; longitudinal transparent area in female very short or absent; discal spot of hindwing broad to M2 (Fig. 86b); abdominal segments 2, 4, 6 with white rings; hind tibia whitish or whitish ochreous in middle.

GENITALIA.– Fig. 85.

BIOLOGY AND HABITAT.– Dry stony or rocky places, roadsides, pastures; hostplants: *Phlomis lychnitis* L., *P. herba-venti* L. (reared by present authors, Bettag, Bläsius, Blum); larva in root; adult VI-VIII.

DISTRIBUTION.– W Mediterranean: Iberian Peninsula, SW France.

86. Chamaesphecia doleriformis (Herrich-Schäffer, 1846)

(Pl. 8: 1,2)

Sesia doleriformis Herrich-Schäffer, 1846: 69, part., pl. 9, fig. 49 (1848). Type locality: Croatia, Dalmatia. Type material: lost.

Sesia colpiformis Staudinger, 1856: 267. Type locality: Turkey, Brussa [Bursa] (restricted in the 1st edition). Type material: lost.

DIAGNOSIS.– Wingspan 12-22 mm; ground coloration brown, or ochreous-brown; antenna brown, golden yellow outside; frons brown, or metallic yellowish, or with yellow borders before eyes; thorax laterally below forewing yellow; external transparent area round or slightly broader than high, of 3-5 cells; apical area with ochreous or ochreous-orange spots (Fig. 86a); discal spot of hindwing broad to M2 (Fig. 86b); abdominal segments (2), 4, 6 with whitish rings and hind tibia in middle deep ochreous or ochreous-orange, strongly hairy (Fig. 86c) in nominotypical subspecies; in ssp. *colpiformis* abdominal segments 2, 4, 6 with white rings and hind tibia only slightly hairy (Fig. 86d).

GENITALIA.– Fig. 86; crista sacculi broadly covered with hairs; in female antrum long, slender, slightly broadened only distally; insignificant differences between subspecies have been found.

BIOLOGY AND HABITAT.– Dry grassy or stony places, embankments, pastures, field banks; hostplants: herbaceous *Salvia* spp. (*S. nemorosa* L., *S. pratensis* L., *S. sclarea* L., *S. verbenaca* L.: Wichra, 1966; Toševski, 1986a; Fiumi & Fabbri, 1996); larva 1-2 years in root; adult VI-VIII.

DISTRIBUTION.– E Mediterranean-Caspian: nominotypical subspecies in Italy, Dalmatia, S Balkan Peninsula; ssp. *colpiformis* in SE central Europe, N and E Balkan Peninsula, S Ukraine, S Russia.

REMARKS.– *Sesia doleriformis* and *S. colpiformis* do not show sympatric distributional pattern and the differences in both morphology and biology do not appear sufficient to distinguish the two taxa as separate species.

87. Chamaesphecia thracica Laštůvka, 1983

(Pl. 7: 14,15)

Chamaesphecia thracica Laštůvka, 1983a: 207. Type locality: Bulgaria, Michurin. Type material: holotype ♂, NMP.

DIAGNOSIS.– Wingspan 15-21 mm; distinguishable from other species of this group (*C. dumonti* to *C. staudingeri*) only with difficulty; group characters: antenna golden-yellow outside; eye with white border; labial palpus white basally and yellow distally; scapular spot on forewing base (occasionally absent); apical area without distinct spots; abdomen with narrow white rings on 2nd, 4th and 6th (7th in male) segments and with some orange rings before them; specific characters: external transparent area round, of (3)-5 cells in male and 3-(5) in female (Fig. 87a); orange abdominal rings on 2nd, 4th, 6th and (7th in male) segments, groups of orange scales on other segments.

GENITALIA.– Fig. 87; insignificant differences have been found in the species group; female genitalia distinguishable only with difficulty; antrum bulb-like and usually with more than 14 folds distally; lamella antevaginalis round the ostium with spinules up to 11 μ long (Laštůvka, 1983d).

BIOLOGY AND HABITAT.– Dry grassy places, pastures, shrubberies, embankments; hostplants: *Stachys thirkei* C. Koch (Laštůvka, 1983a), *Stachys germanica* L. (reared by Toševski and present authors); larva 1 year in root; adult VI-VII .

DISTRIBUTION.– E Mediterranean: S Italy, S and SE Balkans.

88. Chamaesphecia dumonti Le Cerf, 1922

(Pl. 7: 9,12)

Chamaesphecia dumonti Le Cerf, 1922: 35, pl. 540, figs 4537, 4538. Type locality: France, Valdeblore. Type material: lectotype ♂, MNHP (des. by Špatenka, 1992b).

Chamaesphecia similis Laštůvka, 1983a: 202. Type locality: Czech Republic, Klentnice. Type material: holotype ♂, NMP.

DIAGNOSIS.– Wingspan 13-21 mm; see *C. thracica*; borders before eyes often clear white; external transparent area large, round or oval, of 5 long cells in male and of 3-5 elongated cells in female (Fig. 88a); apical area narrow; forewing relatively broad distally; orange rings on all segments.

GENITALIA.– Fig. 88; see *C. thracica*; in female genitalia antrum bulb-like, with 6-12 longitudinal folds distally; spinules on lamella up to 15 μ long (Laštůvka, 1983d).

BIOLOGY AND HABITAT.– Dry "steppe" or rocky places, often on limestone; hostplants: *Stachys recta* L. (Schwarz, 1953; Laštůvka, 1983a), *S. plumosa* Griseb. (reared by Pinker), *S. iberica* Bieb. (Špatenka, pers. comm.), *S. antherocalyx* C. Koch, *S. thracica* Dav. (Špatenka et al., 1999); larva (1)2 years in root; adult VI-VIII.

DISTRIBUTION.– E Mediterranean-Caspian: from Germany (Herrmann & Bläsius, 1991; Stadie, 1998) and SE France across central, to SE and E Europe.

89. Chamaesphecia oxybeliformis (Herrich-Schäffer, 1846)

(Pl. 7: 6)

Sesia masariformis var. *oxybeliformis* Herrich-Schäffer, 1846: 69, pl. 7, fig. 36. Type locality: Russia. Type material: lost.

DIAGNOSIS.– Wingspan 18-23 mm; see *C. thracica*; external transparent area large, round in both sexes, of 5 cells (Fig. 89a); orange rings on all segments.

GENITALIA.– Fig. 89; see *C. thracica*; antrum in female genitalia not bulb-like and only with 3 longitudinal folds distally; spinules on lamella antevaginalis up to 22 μ long.

BIOLOGY AND HABITAT.– Dry grassy places, roadsides, pastures; hostplants: *Marrubium peregrinum* L. (Laštůvka, 1983a), *Phlomis pungens* Willd. (Gorbunov, pers. comm.); larva in root; adult VI-VII.

DISTRIBUTION.– E Mediterranean-Caspian: SE Balkans, Romania, S Ukraine and S Russia.

REMARKS.– An insufficiently known taxon.

90. *Chamaesphecia annellata* (Zeller, 1847)

(Pl. 7: 17,18)

Sesia annellata Zeller, 1847b: 415. Type locality: Turkey, Tlos. Type material: holotype ♀, BMNH.

DIAGNOSIS.– Wingspan 12-19 mm; see *C. thracica*; borders before eyes often yellowish; forewing narrower distally; external transparent area oval (broader than high), of 3-5 cells (Fig. 90a).

GENITALIA.– Fig. 90; see *C. thracica*; antrum in female genitalia not bulb-like, only with 0-3 folds distally; spinules on lamella antevaginalis up to 20 μ long.

BIOLOGY AND HABITAT.– Shrubby places, forest and field edges, embankments, ruderal habitats, village edges; hostplant: *Ballota nigra* L.; larva 1(2) years in root; pupa in root or in basal part of old stem; adult VI-VIII.

DISTRIBUTION.– E Mediterranean: SE central Europe, SE and E Europe.

91. *Chamaesphecia staudingeri* (Failla-Tedaldi, 1890)

(Pl. 7: 16)

Sesia staudingeri Failla-Tedaldi, 1890: 28. Type locality: Sicily, Castelbuono, Miliuni. Type material: not traced.

DIAGNOSIS.– An insufficiently known taxon; essential characters following original description: tegula with yellow border; thorax with yellow median line; forewing with two [male!] small transparent areas; external transparent area square or round; white spot on the forewing base [scapular spot ?]; abdomen with 3 yellow bands, on 1st segment [sic], in the middle and on the last segment; anal tuft black, yellow in the middle.

GENITALIA.– Not studied.

BIOLOGY AND HABITAT.– Unknown.

DISTRIBUTION.– Sicily, known from the type locality and from Ficuzza.

REMARKS.– According to the description, *Sesia staudingeri* apparently belongs to the "*annellata*" group. The presence of an endemic clearwing species in Sicily is improbable. Probably either *C. annellata* or *C. thracica* are involved, and both might be present. In spite of intensive investigations no recent authentic material of the "*annellata*" group has been obtained in Sicily.

92. Chamaesphecia proximata (Staudinger, 1891)

(Pl. 7: 4)

Sesia proximata Staudinger, 1891: 244. Type locality: Turkey, Hadjin [Saimbeyli]. Type material: lectotype ♂, ZMHB (des. by Špatenka & Laštůvka, 1988).

DIAGNOSIS.– Wingspan 15-26 mm; antenna brown or ochreous with brown apex dorsally; frons yellow; labial palpus strongly hairy in male, basally black, apically yellow in male and entirely yellow in female; scapular spot on forewing base; external transparent area in male usually large, of (3) 5 (6) cells; in female small, of 3 elongate and 1-2 small cells; hind tibia strongly hairy, yellow, with distinct black ring distally; abdominal segments 4, 6, 7 with broad and 2 (3, 5 in male) with narrow or indicated yellow rings.

GENITALIA.– Fig. 92.

BIOLOGY AND HABITAT.– Dry grassy, stony or rocky places, shrubby habitats, roadsides, pastures; hostplant: *Salvia sclarea* L.; larva 1 year in root; adult VI-VII (Laštůvka & Špatenka, 1984).

DISTRIBUTION.– E Mediterranean: S and SE Balkan Peninsula.

93. Chamaesphecia masariformis (Ochsenheimer, 1808)

(Pl. 7: 5)

Sesia masariformis Ochsenheimer, 1808: 173. Type locality: Austria, Vienna env. Type material: destroyed.

Sesia odyneriformis Herrich-Schäffer, 1846: 68, pl. 8, fig. 41. Type locality: Crete. Type material: holotype ♀, TMB.

Sesia allantiformis Eversmann, 1844: 103. Type locality: Russia, Ural Mts., Spask. Type material: lectotype ♂, ZISP (des. by Gorbunov, 1992).

Chamaesphecia djakonovi Popescu-Gorj & Căpuşe, 1966: 862. Type locality: Ukraine, Crimea, Karadag. Type material: holotype ♀, ZISP.

DIAGNOSIS.– Wingspan 17-27 mm; antenna yellow outside; frons black; labial palpus slightly hairy, orange ventrally; scapular spot on forewing base (Fig. 93a); external transparent area large, oval, of 5 cells in male; smaller, round, of 4-5 cells in female; abdomen with broad orange red rings on 2nd, 4th, 6th (7th in male) segments, orange spots on other segments; abdominal rings in specimens from Sicily and Crete yellow (known as *C. odyneriformis*); hind tibia strongly hairy, orange with black ring distally.

GENITALIA.– Fig. 93.

BIOLOGY AND HABITAT.– Dry, often stony or rocky places, embankments; hostplants: *Verbascum* spp., *Scrophularia canina* L. (reared by Toševski and present authors); larva 1-2 years in root; the old stem of the infested plant often falls off during winter; adult V-VII.

DISTRIBUTION.– E Mediterranean-Asiatic: SE central Europe, SE and E Europe.

REMARKS.– There is no reason for considering the yellow-banded specimens from Sicily and Crete, known as *C. odyneriformis*, to be a separate species.

94. Chamaesphecia nigrifrons (Le Cerf, 1911)

(Pl. 8: 18)

Sesia nigrifrons Le Cerf, 1911: 244. Type locality: France, Lardy. Type material: lectotype ♀, MNHP (des. by Špatenka, 1992b).

Chamaesphecia sevenari Lipthay, 1961: 213. Type locality: Hungary, Nográdszakál. Type material: holotype ♂, TMB.

DIAGNOSIS.– Wingspan 9-18 mm; antenna and frons black; labial palpus white basally and darker distally; forewing dark brown; discal spot broad; external transparent area small, round, of 3-4 cells; apical area without distinct spots; hind tibia black dorsally with conspicuous yellowish white distal end (Fig. 94a); abdominal segments 4, 6 (7 in male) with white or yellowish rings.

GENITALIA.– Fig. 94; *C. nigrifrons* is similar to the "*aerifrons*" group exteriorly, but has scopula absent and hairs in the distal part of valve are slightly smaller than in the proximal part as in *Chamaesphecia* s. str.

BIOLOGY AND HABITAT.– Forest edges and clearings, pastures; hostplant: *Hypericum perforatum* L. (reared by Špatenka; Bettag, 1991); larva 1 year in root; a distinct pile of reddish brown "sawdust" present at stem bases; the infested stem often falls off during winter; pupa in basal part of old stem; adult often emerges from end of remaining stem base; adult V-VII.

DISTRIBUTION.– Scattered in W, central and SE Europe (Le Cerf, 1911; Lipthay, 1961; Špatenka & Laštůvka, 1983; Steffny, 1990; Garrevoet & Laštůvka, 1998; Petersen & Bartsch, 1998; Bąkowski, Cungs, Toševski, pers. comm.).

95. Chamaesphecia bibioniformis (Esper, 1800)

(Pl. 7: 10,13)

Sphinx bibioniformis Esper, 1800: 30, pl. 44, figs 3, 4. Type locality: Hungary, Ofen [Budapest]. Type material: not traced.

Sesia tenthrediniformis var. *tengyraeformis* Boisduval, 1840: 42. Type locality: Spain, Andalusia, Granada. Type material: lectotype ♂, MNHP (des. by Leraut, 1985).

Chamaesphecia myrsinites Pinker, 1954: 182. Type locality: Macedonia, Ohrid. Type material: holotype ♂, NHMV.

DIAGNOSIS.– Wingspan 13-25 mm; antenna yellow outside; eye with white border; labial palpus shortly hairy, white or yellowish basally; scapular spot on forewing base; external transparent area large, round or oval, of 5 cells (Fig. 95a); in female smaller, round, of 3-5 cells; apical area with distinct ochreous spots between veins (Fig. 95a); discal spot of hindwing narrow, reaching M3 (only slightly narrower between M2 and M3)(Fig. 95b); abdominal segments 2, 4, 6 with white rings; abdomen more or less covered with yellow scales.

GENITALIA.– Fig. 95.

BIOLOGY AND HABITAT.– Scrub, roadsides, pastures; hostplants: *Euphorbia seguieriana* Neck., *E. myrsinites* L., *E. serrata* L., *E. nicaeensis* All., *E. characias* L., *E. macroclada* Boiss., *E. rigida* Bieb., etc. (Schwarz, 1953; Pinker, 1954; reared by Špatenka, present authors, etc.); larva (1)2 years in root; adult VI-VIII.

DISTRIBUTION.– W Palaearctic: SE central, S and E Europe.

REMARKS.– Populations from different parts of the range are partly different and they live in different *Euphorbia* species.– The SW European *Chamaesphecia tengyraeformis* may possibly represent a geographic subspecies.

96. Chamaesphecia anthraciformis (Rambur, 1832)

(Pl. 5: 8,9)

Sesia anthraciformis Rambur, 1832: 266. Type locality: [Corsica]. Type material: lectotype ♂, MNHP (des. by Špatenka, 1992b).

Sesia foeniformis Herrich-Schäffer, 1846: 78. Type locality: Sicily. Type material: lectotype ♀, ZMHB (des. by Špatenka & Laštůvka, 1988).

Sesia oryssiformis Herrich-Schäffer, 1846: 79, pl. 8, fig. 45. Type locality: Corsica. Type material: lost.

DIAGNOSIS.– Wingspan 15-28 mm; antenna, frons and labial palpus black; forewing, middle and hind tibia, abdomen dorsally and ventrally usually partly red; very variable species as to the extent of red coloration; occasionally entirely black, the form being known as *C. anthraciformis*; transparent areas (external and discal) present, or forewing entirely opaque.

GENITALIA.– Fig. 96.

BIOLOGY AND HABITAT.– Grassy habitats, rocks, pastures, roadsides, open forests; hostplants: *Euphorbia atlantis* Maire (Le Cerf, 1920), *E. nicaeensis* All. (Špatenka, pers. comm.), ? *E. myrsinites* L. (Rambur, 1832), *E. characias* L., *E. ceratocarpa* Ten. (reared by present authors); larva 1 year in root; adult V-VII.

DISTRIBUTION.– W Mediterranean: Corsica, Sardinia, Sicily, S and central Italy.

REMARKS.– The individual populations show different variation, e.g. the populations from NE Sicily (Peloritani) appear to be rather homogeneous ("*Chamaesphecia foeniformis*"), the populations from Madonia and NW Sicily (Palermo env.) are more variable ("*C. foeniformis*" and "*C. oryssiformis*"), the populations from SE Sicily (Mt. Iblei) are strongly variable (various forms of "*C. foeniformis*" and "*C. oryssiformis*"), and the largest number of described forms comes from North Africa, where the species is most variable.

97. Chamaesphecia palustris Kautz, 1927

(Pl. 7: 11)

Chamaesphecia palustris Kautz, 1927: 1. Type locality: Austria, Wilfleinsdorf bei Bruck a.d. Leitha. Type material: lectotype ♂, NHMV (des. by Špatenka, 1992b).

DIAGNOSIS.– Wingspan 22-31 mm; body light brown; antenna light brown with dark brown apex; frons brown; labial palpus long tufted; external transparent area nearly square, of 5 cells; longitudinal transparent area present in male, often absent in female; 4th abdominal segment with narrow, laterally broadened yellowish ring.

GENITALIA.– Fig. 97.

BIOLOGY AND HABITAT.– Marshes, ditches; hostplant: *Euphorbia palustris* L. (Kautz, 1927); larva 2 years in root and stem; pupa in a long tunnel (20-50 cm) in the old stem, affording escape from flood water; adult VI-VII.

DISTRIBUTION.– Scattered, W Palaearctic: SE central Europe (Kautz, 1927; Schwarz, 1953), Romania (Popescu-Gorj et al., 1958), NE Italy (Contarini & Fiumi, 1984), Ukraine, S Russia, N Anatolia (Kallies, Petersen, pers. comm.).

98. Chamaesphecia euceraeformis (Ochsenheimer, 1816)

(Pl. 7: 7)

Sesia euceraeformis Ochsenheimer, 1816: 171. Type locality: not mentioned. Type material: destroyed.

Sesia stelidiformis Freyer, 1836: 141, pl. 182, fig. 2. Type locality: not mentioned. Type material: lost.

DIAGNOSIS.– Wingspan 17-24 mm; ground coloration yellowish brown to brownish black; antenna with whitish subapical spot in female; frons brown; labial palpus long tufted in male; external transparent area round or higher than broad, of (4)5 cells (Fig. 98a); longitudinal transparent area present in male and usually absent in female; 4th abdominal segment with narrow yellow ring, distinctly broadened laterally; occasionally dorsal spot line present.

GENITALIA.– Fig. 98; see *C. amygdaloidis*.

BIOLOGY AND HABITAT.– Stony or rocky places, often on limestone, forest edges, scrub; hostplants: *Euphorbia epithymoides* L., *Euphorbia* sp.; larva 1-2 years in root; adult V-VII.

DISTRIBUTION.– W Palaearctic: scattered in W, central, S and E Europe.

99. Chamaesphecia amygdaloidis Schleppnik, 1933

(Pl. 7: 8)

Chamaesphecia stelidiformis f. *amygdaloidis* Schleppnik, 1933: 24. Type locality: Austria, NE Voralpen. Type material: lectotype ♂, NHMV (des. by Špatenka, 1992b).

DIAGNOSIS.– Wingspan 20-26 mm; ground coloration dark brown; antenna dark brown; frons dark brown; forewing margins broad and transparent areas small; external transparent area usually distinctly narrower than apical area (higher than broad)(Fig. 99a); 4th abdominal segment with yellowish ring broadened laterally; distinct dorsal spot line on abdomen.

GENITALIA.– Fig. 99; not significantly different from *C. euceraeformis*.

BIOLOGY AND HABITAT.– Subalpine stony or rocky places, meadows, pastures, forest edges, roadsides; hostplant: *Euphorbia austriaca* A. Kerner (Schleppnik, 1933, as *E. amygdaloides* L.); larva 2 years in root and/or basal part of stem; pupa in root or basal part of old stem; adult VII-VIII.

DISTRIBUTION.– NE limestone Alps: Upper Austria, Lower Austria and Styria (Schleppnik, 1933; Hamborg, 1994b).

REMARKS.– No clear differences from *C. euceraeformis* have been exactly stated; it is given here as a separate species following the prevailing view (cf. Malicky, 1968; Špatenka et al., 1993; Hamborg, 1994b). The two taxa have not yet been studied in their contact zone. More or less distinct external differences are also known to occur in other mountain populations of certain species.

100. Chamaesphecia crassicornis Bartel, 1912

(Pl. 8: 13)

Chamaesphecia crassicornis Bartel, 1912: 409, pl. 50m. Type locality: Kazakhstan, Uralsk. Type material: lectotype ♂, ZMHB (des. by Căpuşe, 1973).

DIAGNOSIS.– Wingspan 15-23 mm; antenna black, golden-yellow outside; frons black or brown, occasionally with ochreous scales before eyes; labial palpus slightly tufted in male; external transparent area broader than high, of 3 long cells (exceptionally of 1-2 further small cells)(Fig. 100a); apical area with ochreous spots between R3-Cu1; abdominal segments 2, 4, 6 with narrow white rings; dorsal spot line on abdomen (Fig. 100b).

GENITALIA.– Fig. 100; in female genitalia antrum distinctly sclerotised.

BIOLOGY AND HABITAT.– Dry grassy "steppe" habitats: grassy slopes, field banks, steppes; hostplant: *Euphorbia virgata* Waldst. & Kit. (Laštůvka, 1980); larva (1)2 years in root; the stem of the infested plant is often dry already in summer of the first year, but one-year-old larvae can hardly be reared to adulthood, or the adults obtained are distinctly smaller in size; adult VI-VIII.

DISTRIBUTION.– Caspian-Asiatic: E Austria (Sterzl, 1967), S Moravia (Laštůvka, 1988), S Slovakia (Laštůvka, 1980), Hungary (Laštůvka, 1990c), N Serbia (Toševski, pers. comm.), N Bulgaria (captured by Tkalců), E Romania (Špatenka, pers. comm.), S European Russia, W Kazakhstan.

101. Chamaesphecia leucopsiformis (Esper, 1800)

(Pl. 8: 14)

Sphinx leucopsiformis Esper, 1800: 25, pl. 41, figs 5, 6. Type locality: Hungary, Ofen [Budapest]. Type material: not traced.

DIAGNOSIS.– Wingspan 14-22 mm; antenna in dark specimens and in female often brown outside; frons brown; labial palpus nearly smooth, white ventrally; external transparent area broader than high, of 3 cells; apical area with ochreous spots between R5 and M3 (Fig. 101a); abdominal segment 4 with white ring; distinct dorsal line on abdomen (Fig. 101b).

GENITALIA.– Fig. 101; in female genitalia antrum weakly sclerotised.

BIOLOGY AND HABITAT.– Dry, sunny sandy and/or rocky (not limestone) or clayey habitats; hostplant: *Euphorbia cyparissias* L.; larva 1 year in root; growing larva at high summer usually causes the decay of the hostplant which becomes clearly visible; pupa in root or emergence tube of variable length from root to ground surface; adult VII (in SW Europe) VIII-IX (X).

DISTRIBUTION.– Scattered, European: central, parts of S Europe.

102. Chamaesphecia guriensis (Emich, 1872)

(Pl. 8: 9)

Sesia guriensis Emich, 1872: 63. Type locality: Georgia, Surebi. Type material: lectotype ♂, ZISP (des. by Gorbunov, 1992b).

DIAGNOSIS.– Wingspan 18-24 mm; ground coloration greyish black; antenna black, yellow outside; frons brownish black; labial palpus with long black scales in male; external transparent area nearly round or quadrate, large, of 4-5 cells; discal spot broad (square); longitudinal transparent area present in female; narrow white rings on 4th and 6th (7th) segments; 4th segment entirely and some other segments partly covered with whitish scales.

GENITALIA.– Fig. 102.

BIOLOGY AND HABITAT.– Subalpine habitats with *Rhododendron;* hostplant: *Euphorbia oblongifolia* C. Koch; larva 2 years in root; pupa in basal part of old stem; adult VI-VII (Gorbunov, 1991).

DISTRIBUTION.– Caucasus (Gorbunov, 1991).

103. Chamaesphecia hungarica (Tomala, 1901)

(Pl. 8: 10)

Sesia empiformis var. *hungarica* Tomala, 1901: 47. Type locality: Hungary, Budapest. Type material: holotype ♂, TMB.

Chamaesphecia deltaica Popescu-Gorj & Căpuşe, 1965: 341. Type locality: Romania, Periprava. Type material: holotype ♂, MGAB.

DIAGNOSIS.– Distinguishable from *C. tenthrediniformis* and *C. empiformis* with difficulty; group characters: antenna yellow outside; frons without white borders before eyes (Fig. 105a); labial palpus long tufted in male; apical area with distinct yellow spots; hind tibia with distinct black ring distally; abdomen with yellow scales or rings and narrow white rings on (2nd), 4th and 6th segments; specific characters: wingspan 16-23 mm; ground coloration dark; external transparent area round, of 5 cells; longitudinal transparent area present in female; light scales on abdomen greenish yellow rather than yellow, sharply delimited yellowish rings usually absent; white ring on 2nd segment occasionally absent.

GENITALIA.– Fig. 103; not significantly different from those of *C. empiformis* and *C. tenthrediniformis*.

BIOLOGY AND HABITAT.– Marshes, ditches; hostplants: *Euphorbia lucida* Waldst. & Kit., occasionally *E. palustris* L.; larva 1(2) years in root; pupa in basal part of old stem (max. 10-15 cm high); adult V-VI(VII).

DISTRIBUTION.– European with restricted range: E Poland (Bąkowski & Hołowiński, 1996), S Moravia (Marek, 1962), E Austria (Schleppnik, 1936), S Slovakia (Schwarz, 1953), Hungary (Tomala, 1901), N Serbia (Toševski, pers. comm.), and E Romania (Popescu-Gorj & Căpuşe, 1965, as *C. deltaica*).

REMARKS.– The species status has been confirmed by Issekutz (1950).

104. Chamaesphecia empiformis (Esper, 1783)

(Pl. 8: 11,12)

Sphinx empiformis Esper, 1783: 215, pl. 32, figs 1, 2. Type locality: Austria. Type material: lectotype ♀, SMW (des. by Naumann & Schroeder, 1980).

Chamaesphecia lastuvkai Špatenka, 1987: 12. Type locality: Bulgaria, Sliven. Type material: holotype ♀, TW.

DIAGNOSIS.– See *C. hungarica*; wingspan 11-22 mm; external transparent area usually round or broader than high (oval), usually of 4 cells (Fig. 104a); yellow spot between R3 and R4 inconspicuous or distinctly shorter than cell between R5 and M1 (Fig. 104a); apical area with distinct yellow spots; yellow rings (coloration) on abdomen usually more diffuse.

GENITALIA.– Fig. 104; not significantly different from *C. tenthrediniformis* and *C. hungarica.*

BIOLOGY AND HABITAT.– Dry grassy, sandy, stony or rocky places, embankments, pastures, forest edges, clearings (usually drier than in *C. tenthrediniformis*); hostplant: *Euphorbia cyparissias* L.; larva 1(2) years in root; adult V-IX.

DISTRIBUTION.– European: W, central, parts of S and E Europe.

REMARKS.– No significant specific differences have been found in the type material of *Chamaesphecia lastuvkai* (extremely yellow forms).

105. Chamaesphecia tenthrediniformis ([Denis & Schiffermüller], 1775)(Pl. 8: 7,8)

Sphinx tenthrediniformis [Denis & Schiffermüller], 1775: 44. Type locality: Austria, Vienna env., Moosbrunn. Type material: neotype ♂, NHMV (des. by Naumann & Schroeder, 1980).

Sesia taediiformis Freyer, 1836: 142, pl. 182, fig. 3. Type locality: not mentioned. Type material: lost.

DIAGNOSIS.– See *C. hungarica*; wingspan 12-21 mm; external transparent area usually higher than broad, of 4 cells; conspicuous yellow spot between R3 and R4 (Fig. 105b); longitudinal transparent area reduced or absent in female; abdominal segments 2, 4, 6 with narrow white rings and with more or less sharply defined yellow rings; yellow coloration often on other segments.

GENITALIA.– Fig. 105; not significantly different from *C. empiformis* and *C. hungarica.*

BIOLOGY AND HABITAT.– Mesophilous grassy places: roadsides, river banks, embankments, forest edges, pastures; hostplants: *Euphorbia esula* L. (Naumann & Schroeder, 1980), *E. salicifolia* Host (Laštůvka, 1990d), *Euphorbia* sp. (of the *E. esula* group, reared by Bläsius); larva 1 year in root; adult IV-VI(VIII).

DISTRIBUTION.– W Palaearctic: scattered in W, SW, central and SE Europe (Naumann & Schroeder, 1980; Laštůvka, 1986; Laštůvka & Laštůvka, 1988; Steffny, 1990; Bąkowski, 1995; Pühringer, 1996; Sobczyk & Rämisch, 1997; Bläsius, Kallies, Špatenka, Toševski, Wegner, pers. comm.).

106. Chamaesphecia astatiformis (Herrich-Schäffer, 1846)

(Pl. 8: 16,17)

Sesia astatiformis Herrich-Schäffer, 1846: 70, pl. 1, figs 5, 6. Type locality: not mentioned. Type material: lost.

DIAGNOSIS.– Wingspan 13-24 mm; antenna black, yellow outside; frons with whitish borders before eyes, or entirely yellow, or occasionally bronze (Fig. 106a); labial palpus nearly smooth, yellow (Fig. 106a); apical area of male broad, yellow, with dark veins and forewing broad distally; female forewing dark, longitudinal transparent area absent and apical area with indistinct or without light spots (Fig. 106b); hind tibia of male deep yellow without distinct dark ring distally, that of female orange with black distal ring; segments 2, 4, 6 with narrow white rings; abdomen more or less (in male often entirely) covered with yellow scales.

GENITALIA.– Fig. 106.

BIOLOGY AND HABITAT.– Dry, clayey or loess places or slopes, field banks, abandoned vineyards; hostplants: *Euphorbia esula* L., *E. salicifolia* Host, *Euphorbia* spp. (of this group); larva 1 year in root; infested plants dry off in autumn; usually sterile small plants are preferred (Laštůvka, 1983b); adult IV-VI(VII).

DISTRIBUTION.– Caspian-Asiatic: SE central, E and SE Europe.

Weismanniola Naumann, 1971

Weismanniola Naumann, 1971: 31.

Type species: *Sesia agdistiformis* Staudinger, 1866; by original designation [replacement name for *Weismannia* Spuler, 1910].

Weismannia Spuler, 1910: 317 [nec *Weismannia* Tutt, 1904].

Type species: *Sesia agdistiformis* Staudinger, 1866; by monotypy.

GENUS CHARACTERS.– Proboscis reduced; abdomen very slender; valve pointed, with specialised crest as in *Chamaesphecia*; crista sacculi with bifurcate hairs; scopula absent; aedeagus with two small cornuti apically. 1 species (monotypic).

107. Weismanniola agdistiformis (Staudinger, 1866)

(Pl. 2: 23)

Sesia agdistiformis Staudinger, 1866: 54. Type locality: Russia, Sarepta [Krasnoarmejskoe, Volgograd env.]. Type material: holotype ♂, ZMHB.

DIAGNOSIS.– Wingspan 15-24 mm; antenna greyish, very thin; frons metallic greyish; labial palpus smooth; forewing greyish brown; small external and discal transparent areas covered with yellowish scales; hindwing with broad greyish margin, entirely covered with translucent whitish scales; abdomen very slender (in male), without conspicuous rings; female unknown.

GENITALIA.– Fig. 107.

BIOLOGY AND HABITAT.– Steppes, bionomics unknown; adult V-VI.

DISTRIBUTION.– SE European Russia, W Kazakhstan.

PLATE 1.

Page

1. *Pennisetia hylaeiformis* (Laspeyres) ♂; Slovakia, Vršatec, ZL. 46
2. *P. bohemica* Králíček & Povolný ♂; Czech Republic, Prague, ZL. 47
3. *P. bohemica* Králíček & Povolný ♀; Czech Republic, Prague, ZL. 47
4. *Paranthrene tabaniformis tabaniformis* (Rottemburg) ♂; Czech Republic, Kersko, ZL. 50
5. *P. tabaniformis tabaniformis* f. *rhingiaeformis* (Hübner) ♀; Croatia, Trogir, ZL (Toševski leg.).
6. *P. tabaniformis synagriformis* (Rambur) ♂; Spain, Almansa, ZL.
7. *P. insolita polonica* (Schnaider) ♂; Slovakia, Tematín, ZL. 51
8. *P. insolita insolita* Le Cerf ♂; Turkey, Kadirli, ZL (K. Špatenka leg.).
9. *P. insolita hispanica* Špatenka & Laštůvka ♂; Spain, Albarracín, ZL.
10. *P. diaphana* Dalla Torre & Strand ♂; Macedonia, Skopje, ZL. 51
11. *Sesia pimplaeformis* Oberthür ♀; Azerbaidjan, Nakhichevan, ZL (O. Gorbunov leg.). 48
12. *S. melanocephala* Dalman ♂; Czech Republic, Velká Bíteš, ZL. 49
13. *S. apiformis* (Clerck) ♂; Czech Republic, Ostrovačice, ZL. 48
14. *S. apiformis* (Clerck) ♀ (black form); Czech Republic, Přibice, ZL.
15. *S. bembeciformis* (Hübner) ♂; Slovakia, Námestovo, ZL. 48
16. *S. bembeciformis* (Hübner) ♀; Czech Republic, Jestřebí, ZL.

1.8 × natural size

1
2
3
4
5
6
7
8
9
10
11
12
13
14
15
16

PLATE 2.

Page

1. *Tinthia tineiformis* (Esper) ♂; Spain, Requena, ZL. 44
2. *T. myrmosaeformis* (Herrich-Schäffer) ♂; Bulgaria, Michurin, ZL. 45
3. *Synanthedon melliniformis* (Laspeyres) ♂; Austria, Illmitz, ZL (D. Hamborg leg.). 57
4. *S. melliniformis* (Laspeyres) ♀; ditto.
5. *Tinthia brosiformis* (Hübner) ♂; Slovakia, Štúrovo, ZL. 45
6. *Synanthedon theryi* Le Cerf ♂; Spain, Cabo de Gata, ZL (M. Petersen leg.). 60
7. *S. codeti* (Oberthür) ♂; Spain, Ronda, ZL (R. Bläsius leg.). 59
8. *S. codeti* (Oberthür) ♀; Spain, Albarracín, ZL (K. Predota leg.).
9. *Osminia fenusaeformis* (Herrich-Schäffer) ♀; Turkey, Bürücek, ZL (O. Brodský leg.). 49
10. *Synanthedon conopiformis* (Esper) ♀; Slovakia, Štúrovo, ZL. 60
11. *S. tipuliformis* (Clerck) ♀; Czech Republic, Židlochovice, ZL. 61
12. *S. spuleri* (Fuchs) ♂; Slovakia, Gombasek, ZL. 61
13. *S. martjanovi* Sheljuzhko ♂; Russia, Zabajkalsk ZL). 58
14. *S. uralensis* (Bartel) ♂; Kazakhstan, Orenburg, KS (M. Nesterov leg.). 57
15. *S. soffneri* Špatenka ♂; Czech Republic, Kugen Hill, ZL (K. Špatenka leg.). ... 56
16. *S. flaviventris* (Staudinger) ♀; Czech Republic, Škrdlovice, ZL. 55
17. *S. loranthi* (Králíček) ♂; Slovakia, Malacky, ZL. 62
18. *S. loranthi* (Králíček) ♀; ditto.
19. *S. cephiformis* (Ochsenheimer) ♂; Czech Republic, Domašov u Brna, ZL. 62
20. *Bembecia fokidensis* Toševski ♂; Greece, Kato Luisi, ZL (K. Špatenka leg.). .. 70
21. *Synanthedon cephiformis* (Ochsenheimer) ♀; Slovakia, Zádiel, ZL. 62
22. *Bembecia himmighoffeni* (Staudinger) ♀; Spain, Albarracín, ZL (K. Predota leg.). 73
23. *Weismanniola agdistiformis* (Staudinger) ♂; Russia, Sarepta, ZL. 97
24. *Bembecia iberica* Špatenka ♂; Spain, Sra de Gúdar, ZL. 69
25. *B. iberica* Špatenka ♀; ditto.
26. *B. himmighoffeni* (Staudinger) ♂; France, Pertuis-Bonde, ZL. 73
27. *B. volgensis* Gorbunov ♂; Russia, Vyazovka, AK (A. & V. Isajeva leg.). 65
28. *B. hymenopteriformis* (Bellier) ♂; Spain, Ronda, ZL. 63
29. *B. hymenopteriformis* (Bellier) ♀; ditto.
30. *Chamaesphecia albiventris* (Lederer) ♂; Mesopotamia, ZL. 85
31. *Bembecia volgensis* Gorbunov ♀; Russia, Ryabina, AK (V. Zolotuhin leg.). 65

1.8 × natural size

1
2
3
4
5
6
7
8
9
10
11
12
13
14
15
16
17
18
19
20
21
22
23
24
25
26
27
28
29
30
31

PLATE 3.

Page

1. *Synanthedon scoliaeformis* (Borkhausen) ♀; Slovakia, Horná Súča, ZL. 52
2. *S. mesiaeformis* (Herrich-Schäffer) ♀; Hungary, Darány, ZL (D.Hamborg leg.).. 52
3. *S. polaris* (Staudinger) ♂; Sweden, Hr, Mittädalen, ZL (N. Ryrholm leg.). 55
4. *S. spheciformis* ([Denis & Schiffermüller]) ♀; Czech Republic, Adamov, ZL. 53
5. *S. vespiformis* (Linnaeus) ♀; Czech Republic, Maršovice, ZL. 59
6. *S. vespiformis* (Linnaeus) ♂; ditto.
7. *S. stomoxiformis riefenstahli* Špatenka ♂; Spain, El Molinillo, ZL. 53
8. *S. stomoxiformis stomoxiformis* (Hübner) ♀; Slovakia, Kamenica n/Hronom, ZL.
9. *S. formicaeformis* (Esper) ♂; Czech Republic, Pouzdřany, ZL. 54
10. *S. culiciformis* (Linnaeus) ♀; Turkey, Ödemis-Salihli, ZL (E. Hüttinger leg.). ... 54
11. *S. culiciformis* (Linnaeus) ♂; Czech Republic, Černá Hora, ZL.
12. *S. culiciformis* (Linnaeus) ♂; Czech Republic, Bylnice-Sidonie (aberration), ZL.
13. *S. andrenaeformis* (Laspeyres) ♀; Czech Republic, Pavlovské vrchy Hills, ZL. . 56
14. *S. andrenaeformis* (Laspeyres) ♂; Austria, Grazer Feld (aberration, cf. Hamborg, 1994a), ZL (D. Hamborg leg.).
15. *S. andrenaeformis* (Laspeyres) ♂; Slovakia, Trenčianske Teplice, ZL.
16. *S. myopaeformis myopaeformis* (Borkhausen) ♀; Spain, Ademuz, ZL. 58
17. *S. myopaeformis typhiaeformis* (Borkhausen) ♀; Italy, Rome, ZL (C. Prola leg.).
18. *S. myopaeformis cruentata* (Mann) ♂; Sicily, Palermo, ZL (F. Wagner leg.).

1.8 × natural size

1
2
3
4
5
6
7
8
9
10
11
12
13
14
15
16
17
18

PLATE 4.

Page

1. *Bembecia ichneumoniformis* ([Denis & Schiffermüller]) ♀; Slovakia, Zádiel, ZL. 65
2. *B. ichneumoniformis* ([Denis & Schiffermüller]) ♂; France, Pertuis-Bonde, ZL.
3. *B. pavicevici pavicevici* Toševski ♂; Macedonia, Pletvar, ZL (I. Toševski leg.). .. 67
4. *B. albanensis albanensis* (Rebel) ♂; Czech Republic, Bulhary, ZL. 66
5. *B. albanensis albanensis* (Rebel) ♀; Slovakia, Silická Jablonica, ZL.
6. *B. albanensis kalavrytana* (Sheljuzhko) ♂; Greece, Kalavrita, ZL (E. Schwabe & J. Gelbrecht leg.).
7. *B. scopigera* (Scopoli) ♂; Spain, Sra de Gúdar, ZL. 68
8. *B. scopigera* (Scopoli) ♀; Slovakia, Štúrovo, ZL.
9. *B. uroceriformis* (Treitschke) ♂; Spain, Uleila del Campo, ZL. 73
10. *B. lomatiaeformis* (Lederer) ♂; Greece, Taygetos, ZL. 64
11. *B. abromeiti* Kallies & Riefenstahl ♂, paratype; Spain, Mallorca, Cala Canyamel, KS (A. Kallies & A. Musolff leg.). 65
12. *B. uroceriformis* (Treitschke) ♀; Slovakia, Plešivec, ZL. 73
13. *B. sareptana* (Bartel) ♀, holotype; Russia, Sarepta, ZMHB (H. Rangnow leg.)... 64
14. *B. fibigeri* Laštůvka & Laštůvka ♂; Spain, Vivel del Rio Martín, ZL. 67
15. *B. fibigeri* Laštůvka & Laštůvka ♀; ditto.
16. *B. sanguinolenta* (Herrich-Schäffer) ♀; Turkey, Göreme, ZL (M. Petersen leg.). . 72
17. *B. sirphiformis* (Lucas) ♂; France, Pertuis-Bonde, ZL. 71
18. *B. sirphiformis* (Lucas) ♀; ditto.

1.8 × natural size

1
2
3
4
5
6
7
8
9
10
11
12
13
14
15
16
17
18

Plate 5.

Page

1. *Bembecia priesneri* Kallies, Petersen & Riefenstahl ♂, paratype; Turkey, Kappadocia, KS. 68
2. *B. priesneri* Kallies, Petersen & Riefenstahl ♀, paratype; Turkey, Kappadocia, AK (K. Špatenka leg.).
3. *Bembecia blanka* ♂; Crete, Fourfouras, ZL (K. Špatenka leg.). 69
4. *B. megillaeformis* (Hübner) ♂; Croatia, Bunica, ZL (D. Stadie leg.). 70
5. *B. megillaeformis* (Hübner) ♀; Slovakia, Zádiel, ZL.
6. *B. puella* Laštůvka ♀, paratype; Slovakia, Plešivec, ZL. 71
7. *B. puella* Laštůvka ♂, paratype; Slovakia, Plešivec, ZL.
8. *Chamaesphecia anthraciformis* (Rambur) ♂; Sicily, Piana degli Albanesi, ZL (M. Petersen leg.). 92
9. *C. anthraciformis* (Rambur) ♂; Sicily, S. Fratello, ZL (M. Petersen leg.).
10. *Pyropteron chrysidiformis chrysidiformis* (Esper) ♂; Spain, Albarracín, ZL. 74
11. *P. chrysidiformis chrysidiformis* (Esper) ♂; France, Pertuis-Bonde, ZL.
12. *P. chrysidiformis chrysidiformis* (Esper) ♂; Spain, Albarracín, ZL.
13. *P. chrysidiformis sicula* Le Cerf ♂; Italy, Campoforogna, ZL (C. Prola leg.).
14. *P. minianiformis* (Freyer) ♂; Bulgaria, Michurin, ZL. 74
15. *P. minianiformis* (Freyer) ♀; Bulgaria, Primorsko, ZL.
16. *P. doryliformis doryliformis* (Ochsenheimer) ♂; Spain, Riópar, ZL. 75
17. *P. doryliformis doryliformis* (Ochsenheimer) ♀; Spain, Riópar, ZL.
18. *P. doryliformis icteropus* (Zeller) ♂; Sicily, S. Fratello, ZL (M. Petersen leg.).

1.8 × natural size

1
2
3
4
5
6
7
8
9
10
11
12
13
14
15
16
17
18

PLATE 6.

Page

1. *Pyropteron aistleitneri* Špatenka ♀, holotype; Spain, Sra de Guillimona, TW (E. Aistleitner leg.). 78
2. *P. hispanica* Kallies ♂; Spain, Ronda, ZL. 76
3. *P. hispanica* Kallies ♀; Spain, Carratraca, ZL.
4. *P. triannuliformis* (Freyer) ♂; Czech Republic, Pustiměř, ZL. 75
5. *P. triannuliformis* (Freyer) ♀; Yugoslavia, Budva, ZL.
6. *P. meriaeformis* (Boisduval) ♂; Spain, Anglés, ZL. 76
7. *P. koschwitzi* Špatenka ♂, holotype; Spain, Aranjuez, TW (U. Koschwitz leg.) 77
8. *P. kautzi* (Reisser) ♂; Spain, Sierra Nevada, ZL (F. Pühringer leg.). 78
9. *P. mannii* (Lederer) ♂; Bulgaria, Primorsko, ZL. 81
10. *P. muscaeformis* (Esper) ♂; Czech Republic, Popice u Znojma, ZL. 77
11. *P. muscaeformis* (Esper) ♀; Slovakia, Borský Jur, ZL.
12. *P. mannii* (Lederer) ♀; Bulgaria, Michurin, ZL. 81
13. *P. affinis* (Staudinger) ♂; Slovakia, Jablonov n/Turnou, ZL. 79
14. *P. leucomelaena* (Zeller) ♂; France, Pertuis, ZL. 79
15. *P. leucomelaena* (Zeller) ♀; Spain, Requena, ZL.
16. *Dipchasphecia lanipes* (Lederer) ♀; Turkey, ZL. 81
17. *Chamaesphecia chalciformis* (Esper) ♂; Bulgaria, Primorsko, ZL. 83
18. *C. schmidtiiformis* (Freyer) ♀; Italy, Lago di Garda, ZL (W. Ramme leg.). 83

1.8 × natural size

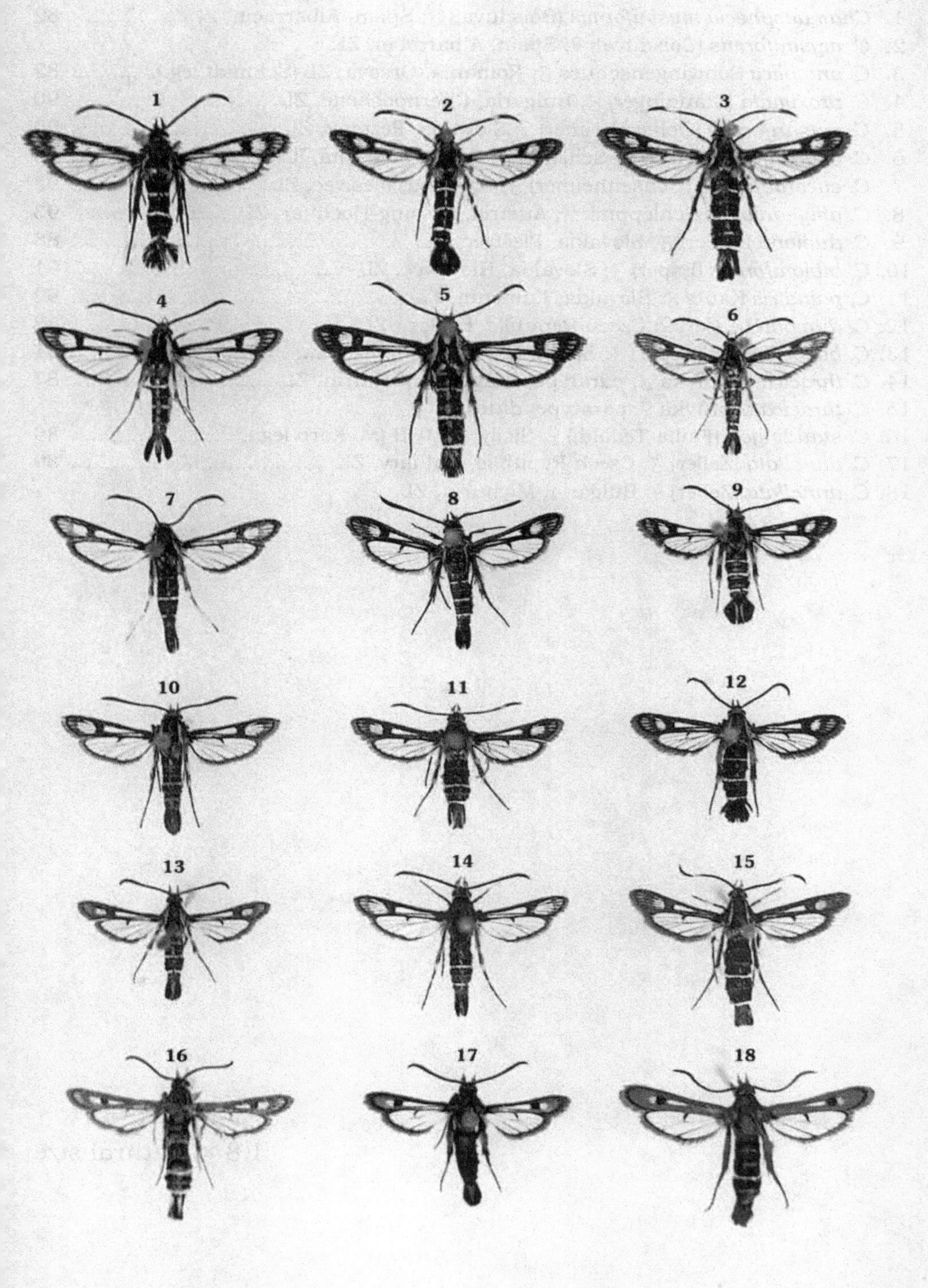
1
2
3
4
5
6
7
8
9
10
11
12
13
14
15
16
17
18

PLATE 7.

Page

1. *Chamaesphecia mysiniformis* (Boisduval) ♂; Spain, Albarracín, ZL. 82
2. *C. mysiniformis* (Boisduval) ♀; Spain, Albarracín, ZL.
3. *C. anatolica* Schwingenschuss ♂; Romania, Orşova, ZL (Schmidt leg.). 82
4. *C. proximata* (Staudinger) ♂; Bulgaria, Chernoochene, ZL. 90
5. *C. masariformis* (Ochsenheimer) ♀; Slovakia, Pezinok, ZL. 90
6. *C. oxybeliformis* (Herrich-Schäffer) ♂; Russia, Sarepta, KS. 88
7. *C. euceraeformis* (Ochsenheimer) ♂; Slovakia, Plešivec, ZL. 93
8. *C. amygdaloidis* Schleppnik ♂; Austria, Lassing-Hochkar, ZL. 93
9. *C. dumonti* Le Cerf ♂; Slovakia, Plešivec, ZL. 88
10. *C. bibioniformis* (Esper) ♂; Slovakia, Hlohovec, ZL. 91
11. *C. palustris* Kautz ♂; Slovakia, Kamenín, ZL. 92
12. *C. dumonti* Le Cerf ♀; Czech Republic, Pavlov, ZL. 88
13. *C. bibioniformis* (Esper) ♂; Spain, El Molinillo, ZL. 91
14. *C. thracica* Laštůvka ♂, paratype; Bulgaria, Michurin, ZL. 87
15. *C. thracica* Laštůvka ♀, paratype; ditto.
16. *C. staudingeri* (Failla-Tedaldi) ♀; Sicily, ZMHB (M. Korb leg.). 89
17. *C. annellata* (Zeller) ♂; Czech Republic, Bulhary, ZL. 89
18. *C. annellata* (Zeller) ♀; Bulgaria, Michurin, ZL.

1.8 × natural size

1
2
3
4
5
6
7
8
9
10
11
12
13
14
15
16
17
18

PLATE 8.

Page

1. *Chamaesphecia doleriformis colpiformis* (Staudinger) ♂; Czech Republic, Dolní Dunajovice, ZL. 87
2. *C. doleriformis doleriformis* (Herrich-Schäffer) ♂; Italy, Metaponto, ZL.
3. *C. ramburi* (Staudinger) ♂; Spain, Diezma, ZL. 86
4. *C. anthrax* Le Cerf ♂; Spain, Albarracín, ZL. 84
5. *C. aerifrons* (Zeller) ♂; France, Pertuis, ZL. 85
6. *C. alysoniformis* (Herrich-Schäffer) ♀; Bulgaria, Michurin, ZL. 84
7. *C. tenthrediniformis* ([Denis & Schiffermüller]) ♂; Czech Republic, Velké Bílovice, ZL. 96
8. *C. tenthrediniformis* ([Denis & Schiffermüller]) ♀; ditto.
9. *C. guriensis* (Emich) ♀; Georgia, ZMHB. 94
10. *C. hungarica* (Tomala) ♂; Slovakia, Iža, ZL. 95
11. *C. empiformis* (Esper) ♂; Slovakia, Hrádok n/Váhom, ZL. 95
12. *C. empiformis* (Esper) ♀; Czech Republic, Bojkovice, ZL.
13. *C. crassicornis* Bartel ♂; Czech Republic, Bulhary, ZL. 94
14. *C. leucopsiformis* (Esper) ♂; Slovakia, Kravany n/Dunajom, ZL. 94
15. *C. osmiaeformis* (Herrich-Schäffer) ♂; Morocco, Ifrane, ZL (K. Špatenka leg.). . 86
16. *C. astatiformis* (Herrich-Schäffer) ♂; Czech Republic, Velké Bílovice, ZL. 96
17. *C. astatiformis* (Herrich-Schäffer) ♀; ditto.
18. *C. nigrifrons* (Le Cerf) ♂; Slovakia, Silica, ZL. 91

1.8 × natural size

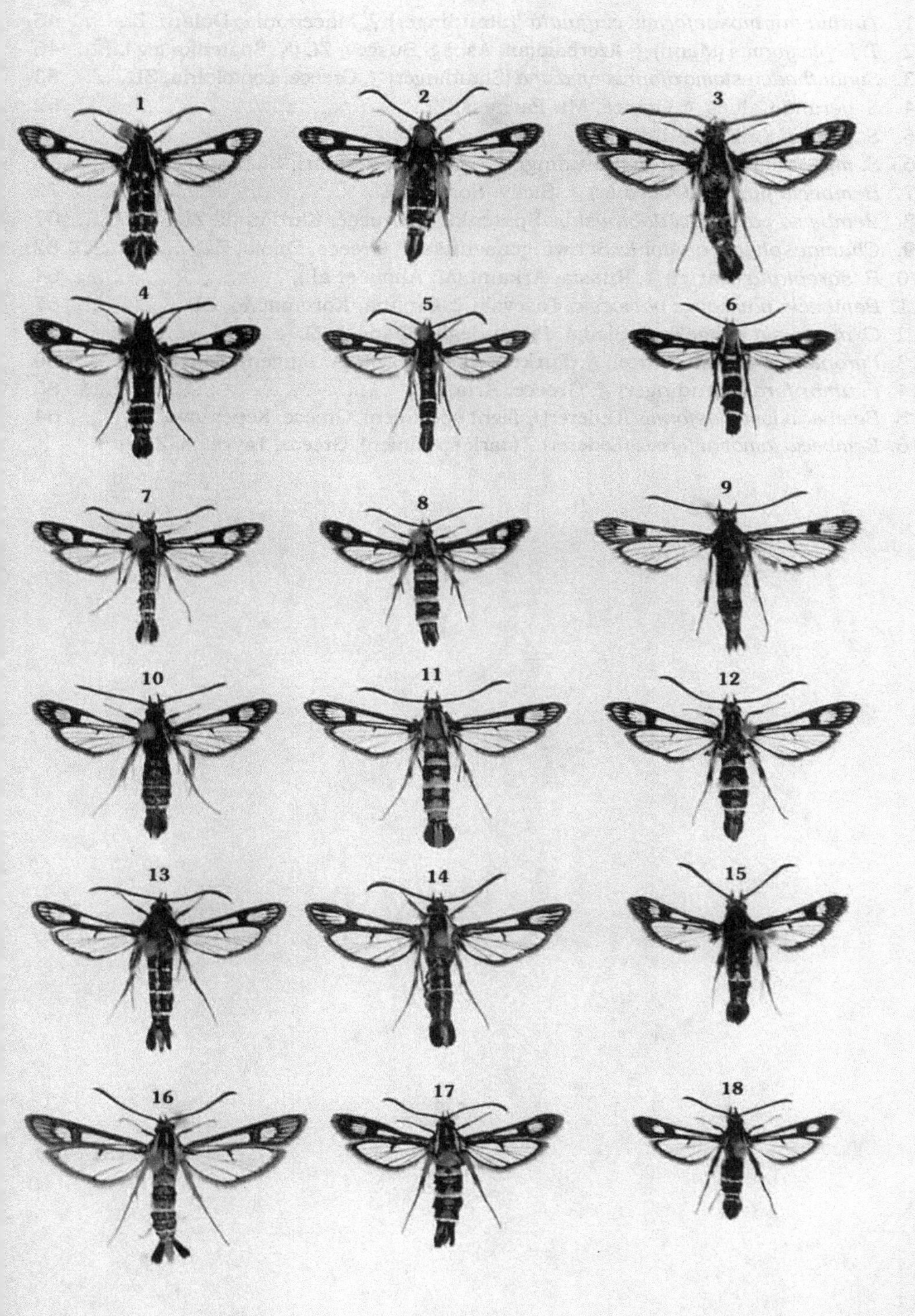
1
2
3
4
5
6
7
8
9
10
11
12
13
14
15
16
17
18

PLATE 9.

Page

1. *Tinthia myrmosaeformis cingulata* (Staudinger) ♀; Macedonia, Dolani, ZL. 45
2. *T. hoplisiformis* (Mann) ♂; Azerbaidjan, Ashagi Busgov, ZL (K. Špatenka leg.)........ 46
3. *Synanthedon stomoxiformis amasina* (Staudinger) ♂; Greece, Leptokaria, ZL......... 53
4. *S. geranii* Kallies ♂; Greece, Mt. Parnass, ZL. 62
5. *S. geranii* Kallies ♀; ditto.
6. *S. myopaeformis graeca* (Staudinger) ♂; Greece, Exohori, ZL. 58
7. *Bembecia flavida* (Oberthür) ♂; Sicily, Sortino, ZL. 72
8. *Bembecia pavicevici dobrovskyi* Špatenka ♂; Greece, Kardamili, ZL. 67
9. *Chamaesphecia anatolica* Schwingenschuss ♂; Greece, Dilofo, ZL. 82
10. *B. sareptana* (Bartel) ♂; Russia, Arkaim (M. Ahola et al.). 64
11. *Bembecia pavicevici pavicevici* Toševski ♀; Croatia, Koromačno, ZL. 67
12. *C. maurusia* Püngeler ♀; Sicilia, Piana degli Albanesi, ZL. 84
13. *Pyropteron cirgisa* (Bartel) ♂; Turkey, Aksaray, ZL (A. Lingenhöle leg.). 80
14. *P. umbrifera* (Staudinger) ♂; Greece, Árta, ZL. 80
15. *Bembecia lomatiaeformis* (Lederer) ♀ (light specimen); Greece, Kepessovo, ZL........ 64
16. *Bembecia lomatiaeformis* (Lederer) ♀ (dark specimen); Greece, Taygetos, ZL.

1.8 × natural size

1
2
3
4
5
6
7
8
9
10
11
12
13
14
15
16

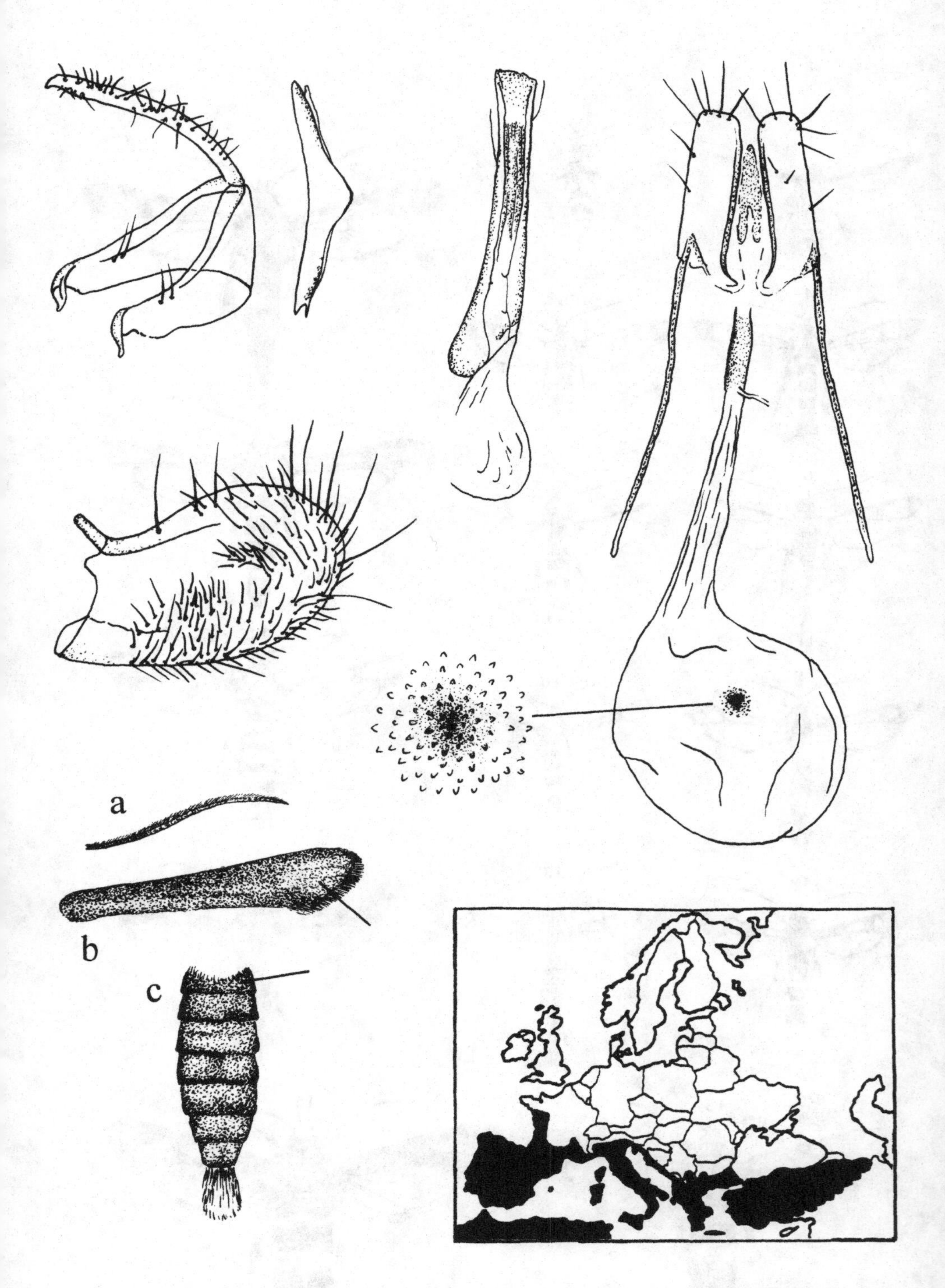

1. *Tinthia tineiformis* (Esper) – Bulgaria, a: antenna, b: forewing, c: abdomen (ZL).

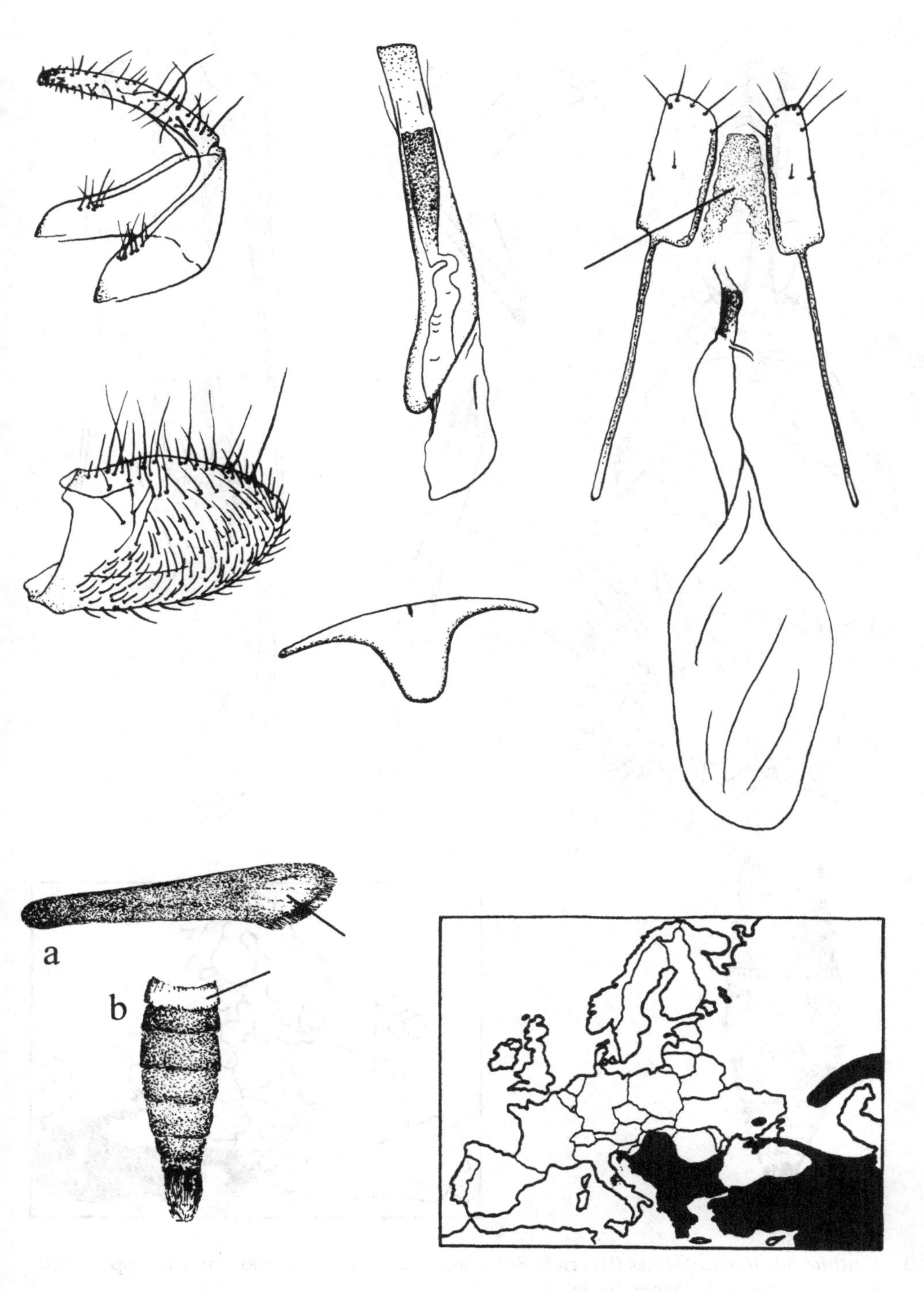

2. *Tinthia brosiformis* (Hübner) – Slovakia, a: forewing, b: abdomen (ZL).

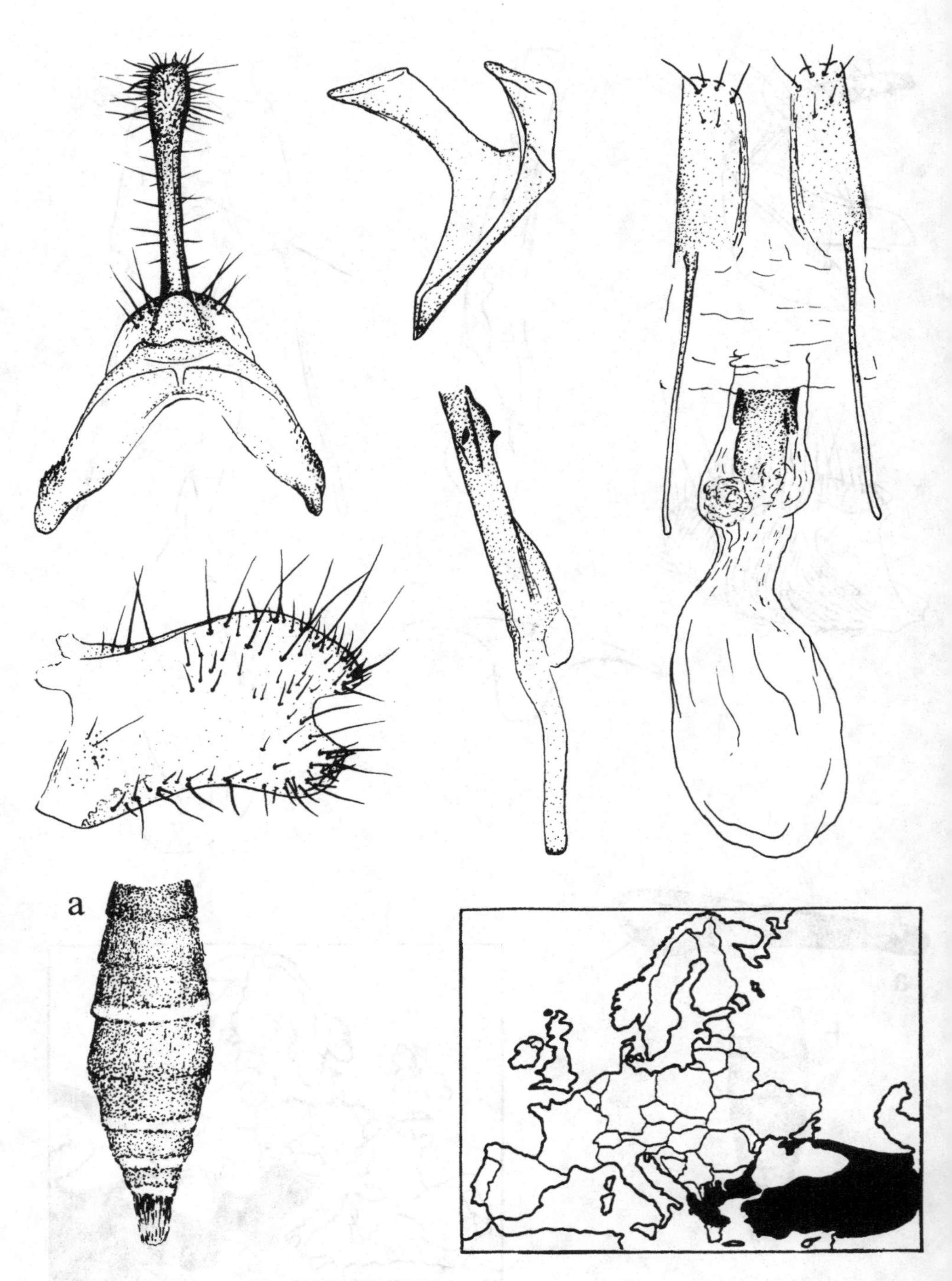

3. *Tinthia myrmosaeformis* (Herrich-Schäffer) – Bulgaria, a: abdomen of ssp. *cingulata* (Staudinger) – Macedonia (ZL).

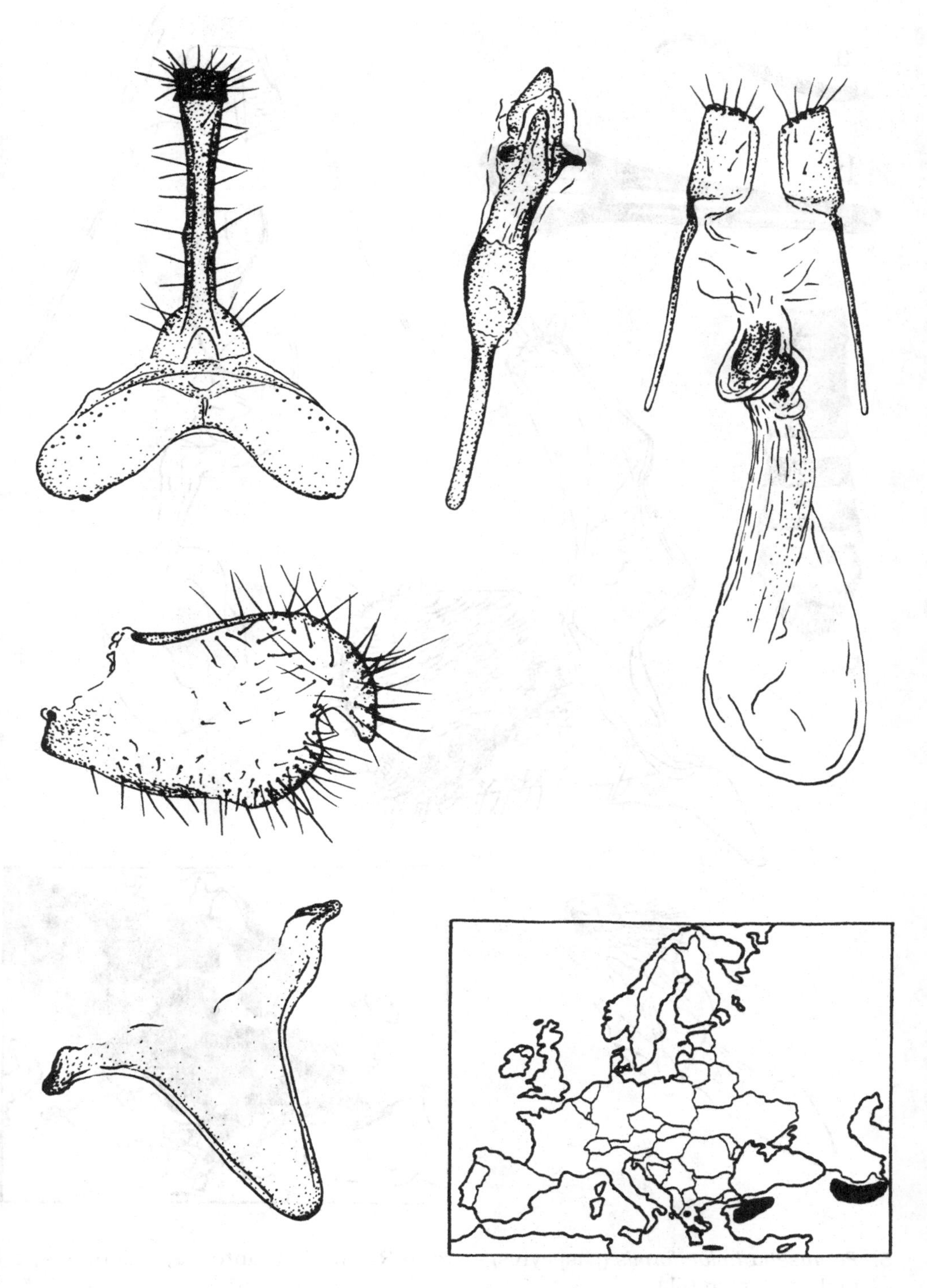

4. *Tinthia hoplisiformis* (Mann) – Azerbaidjan (ZL, ex KS).

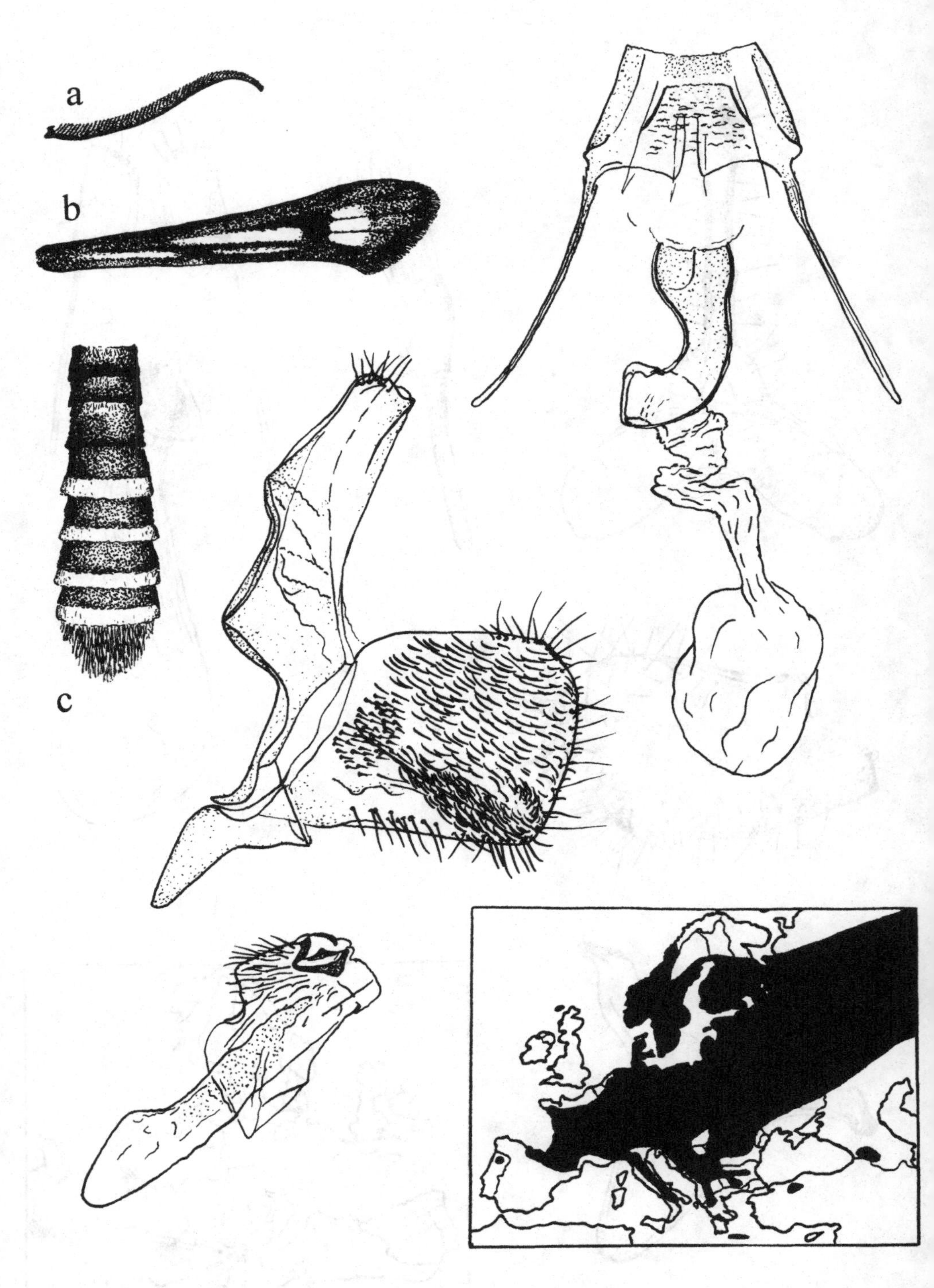

5. *Pennisetia hylaeiformis* (Laspeyres) – Czech Republic, a: antenna, b: forewing, c: male abdomen (ZL).

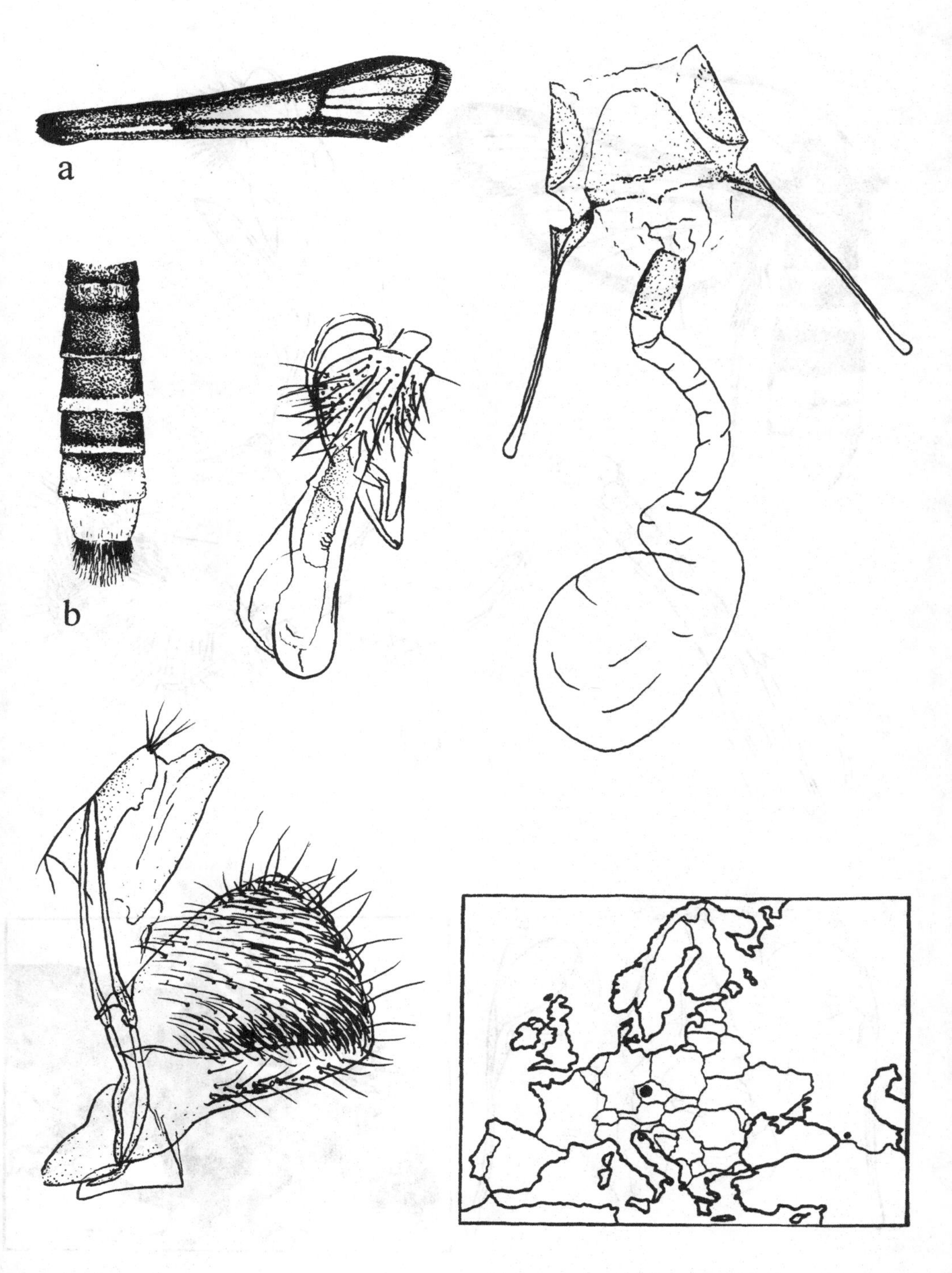

6. *Pennisetia bohemica* Králíček & Povolný – Czech Republic, a: forewing, b: male abdomen (ZL).

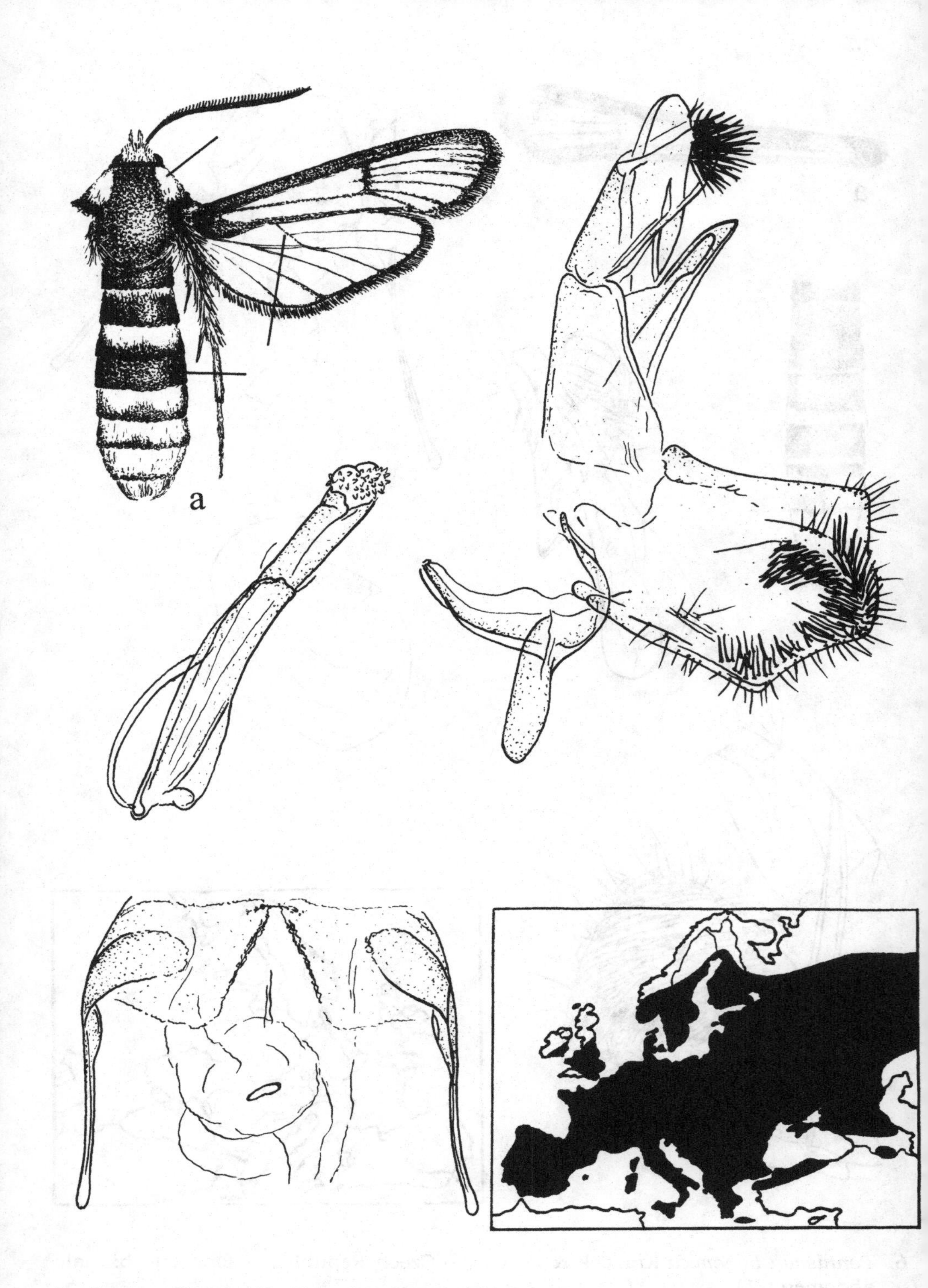

7. *Sesia apiformis* (Clerck) – Czech Republic, a: ♂ (ZL).

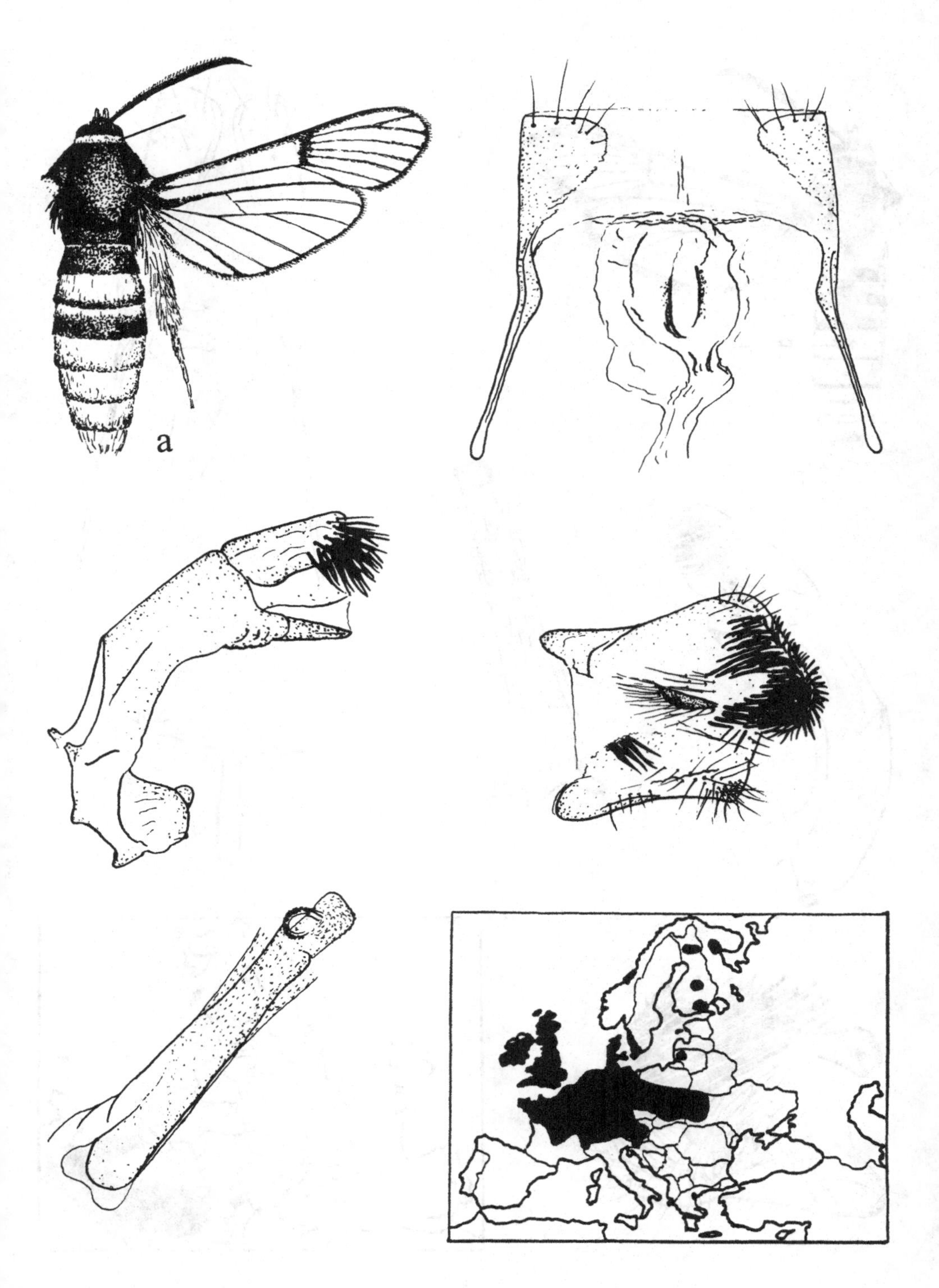

8. *Sesia bembeciformis* (Hübner) – Czech Republic, a: ♂ (ZL).

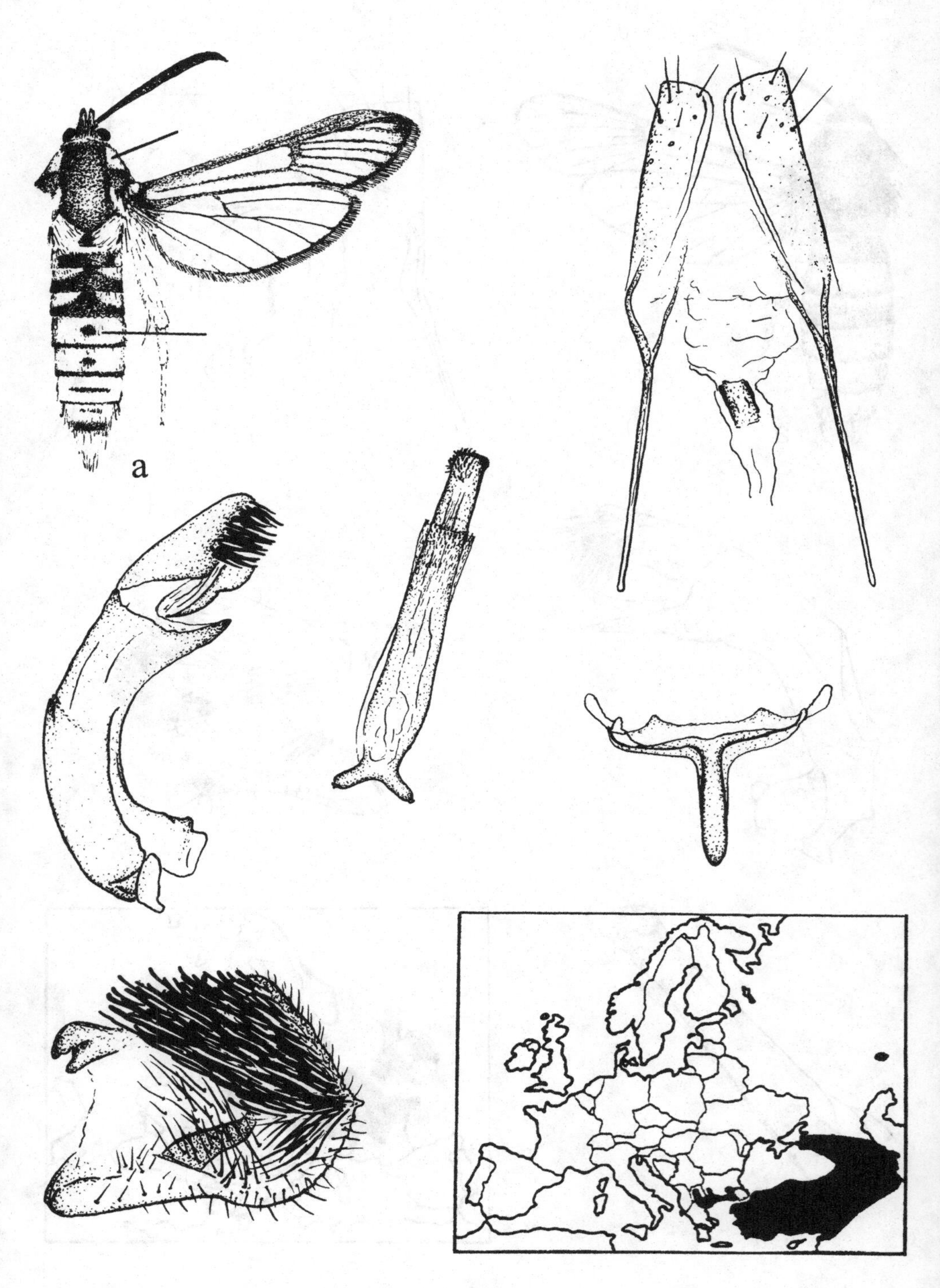

9. *Sesia pimplaeformis* Oberthür – Azerbaidjan, a: ♀ (ZL, ex OG).

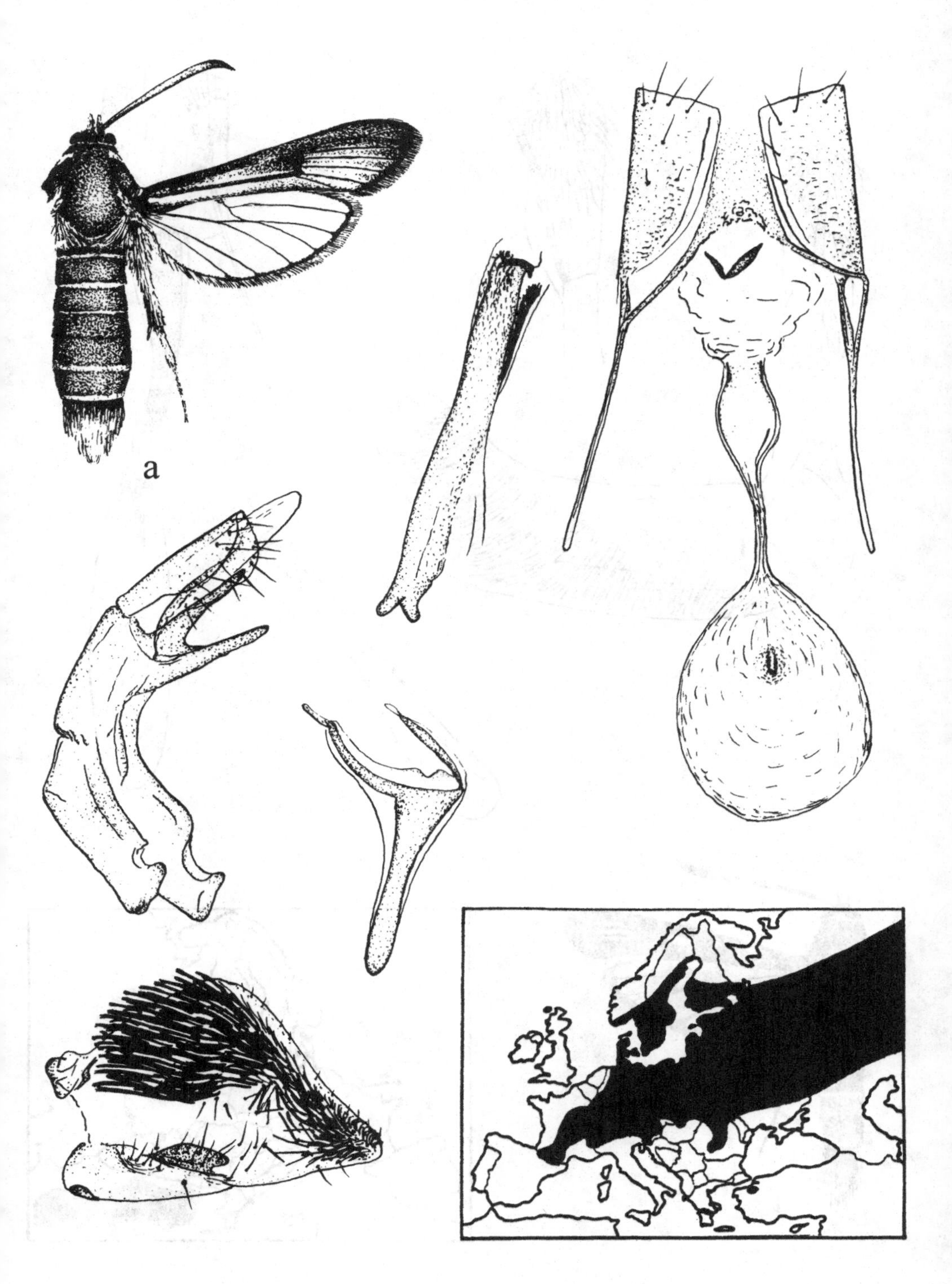

10. *Sesia melanocephala* Dalman – Czech Republic, a: ♀ (ZL).

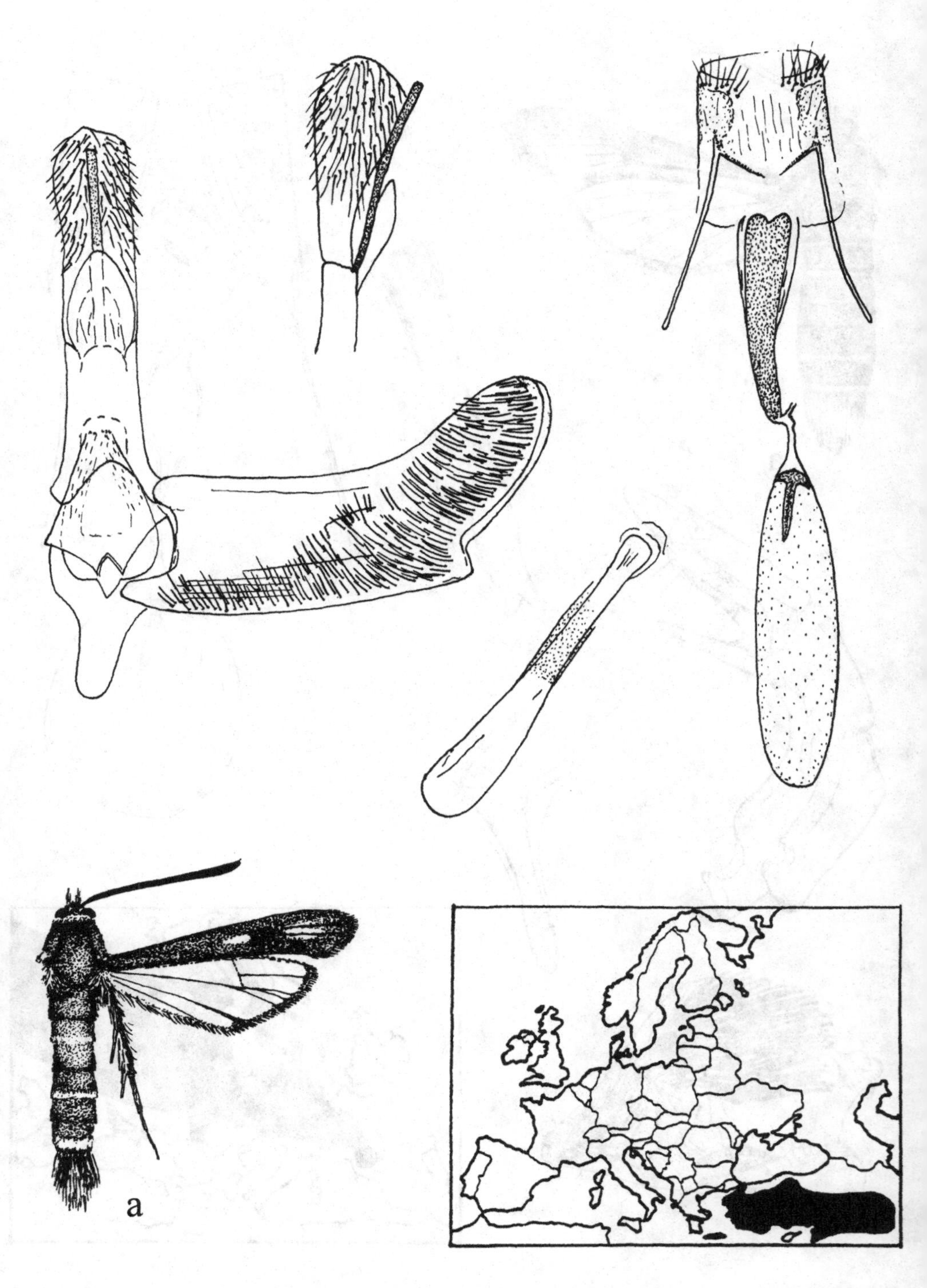

11. *Osminia fenusaeformis* (Herrich-Schäffer) – Turkey, a: ♀ (ZL, ex O. Brodský).

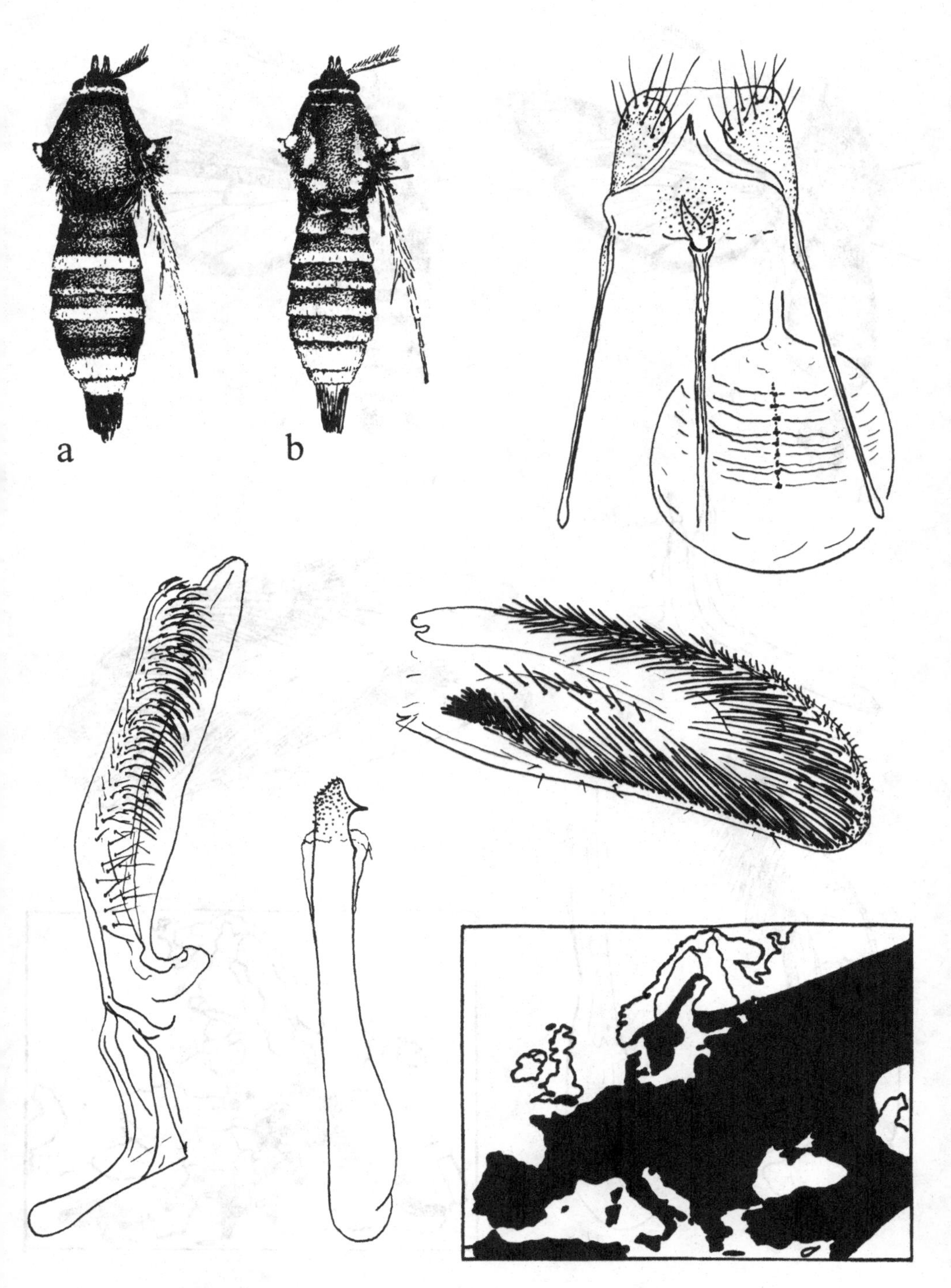

12. *Paranthrene tabaniformis* (Rottemburg) – Slovakia, a: ssp. *tabaniformis* – Slovakia, b: ssp. *synagriformis* (Rambur) – Spain (ZL).

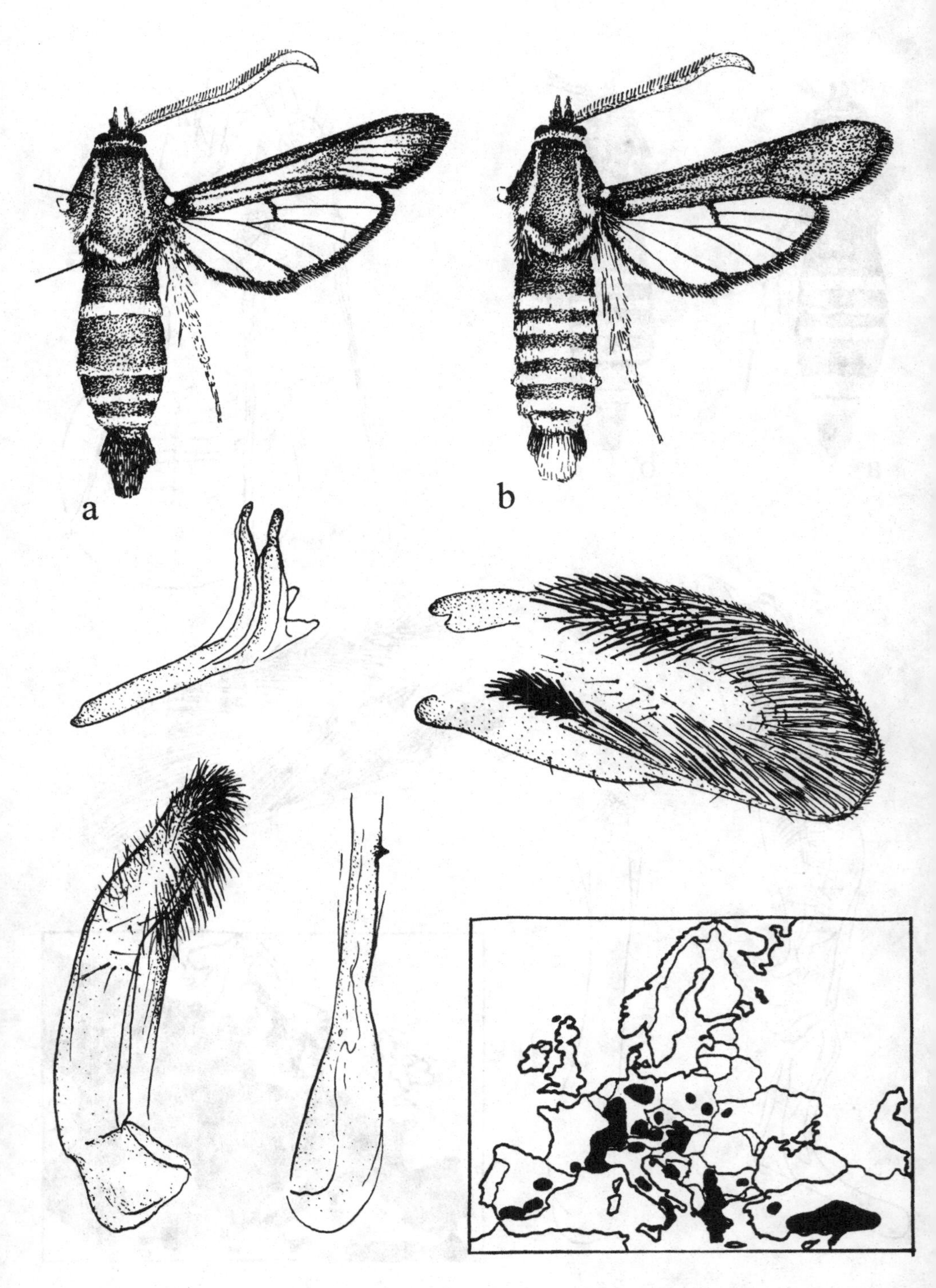

13. *Paranthrene insolita* Le Cerf – Slovakia, a: ssp. *polonica* Schnaider ♂ – Slovakia, b: ssp. *hispanica* Špatenka & Laštůvka ♂ – Spain (ZL).

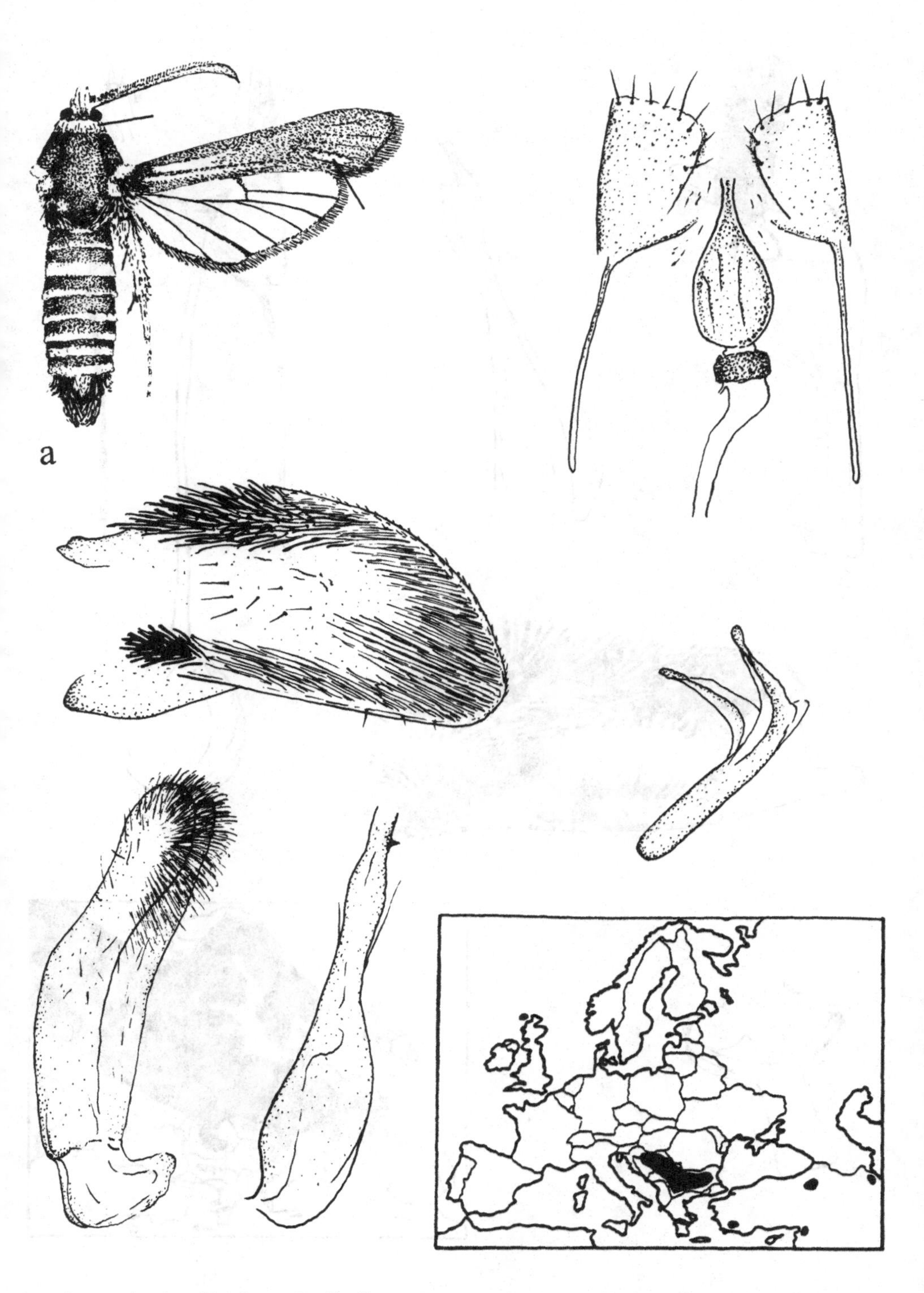

14. *Paranthrene diaphana* Dalla Torre & Strand – Yugoslavia, a: ♂ (ZL, ex IT).

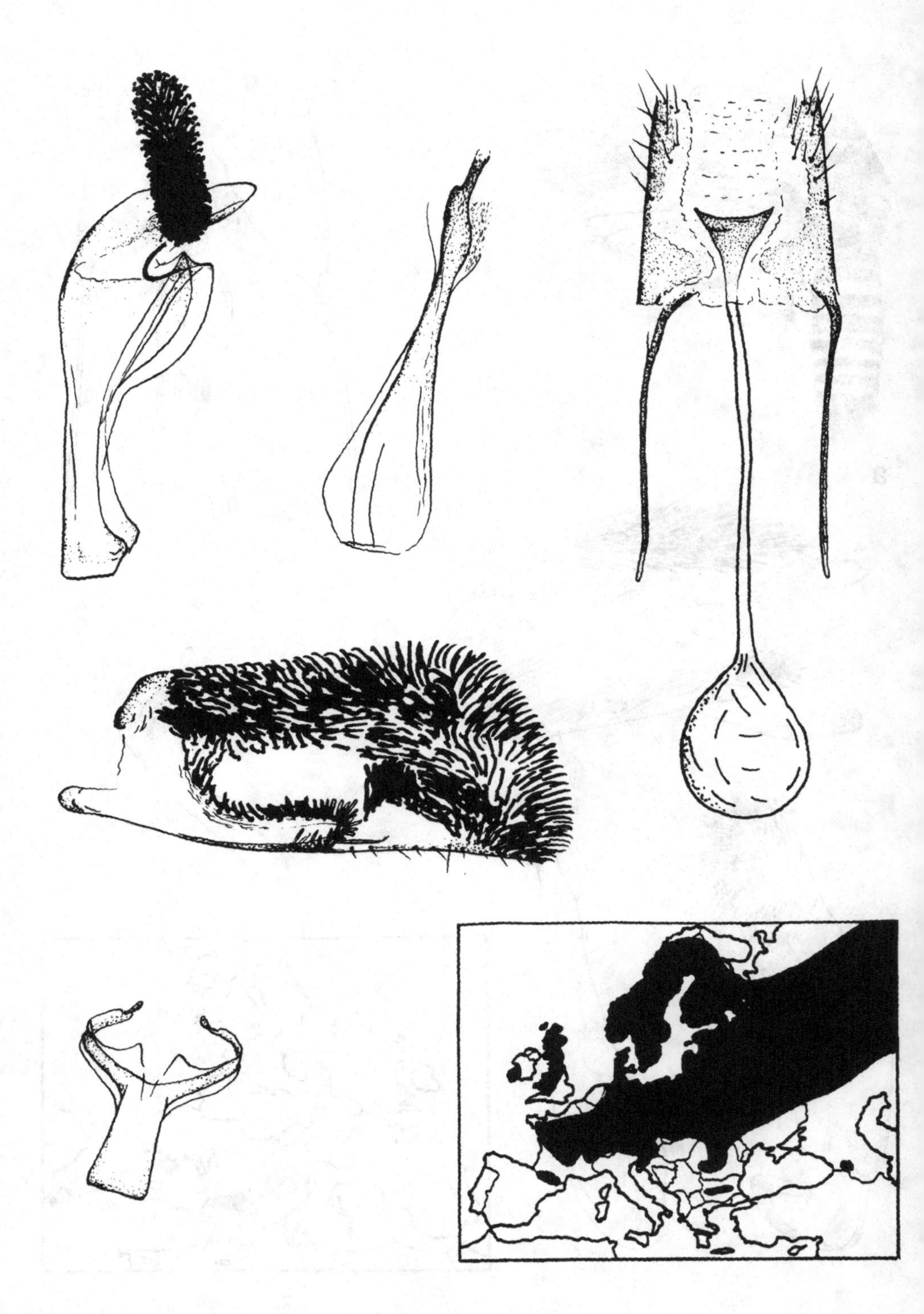

15. *Synanthedon scoliaeformis* (Borkhausen) – Czech Republic (ZL).

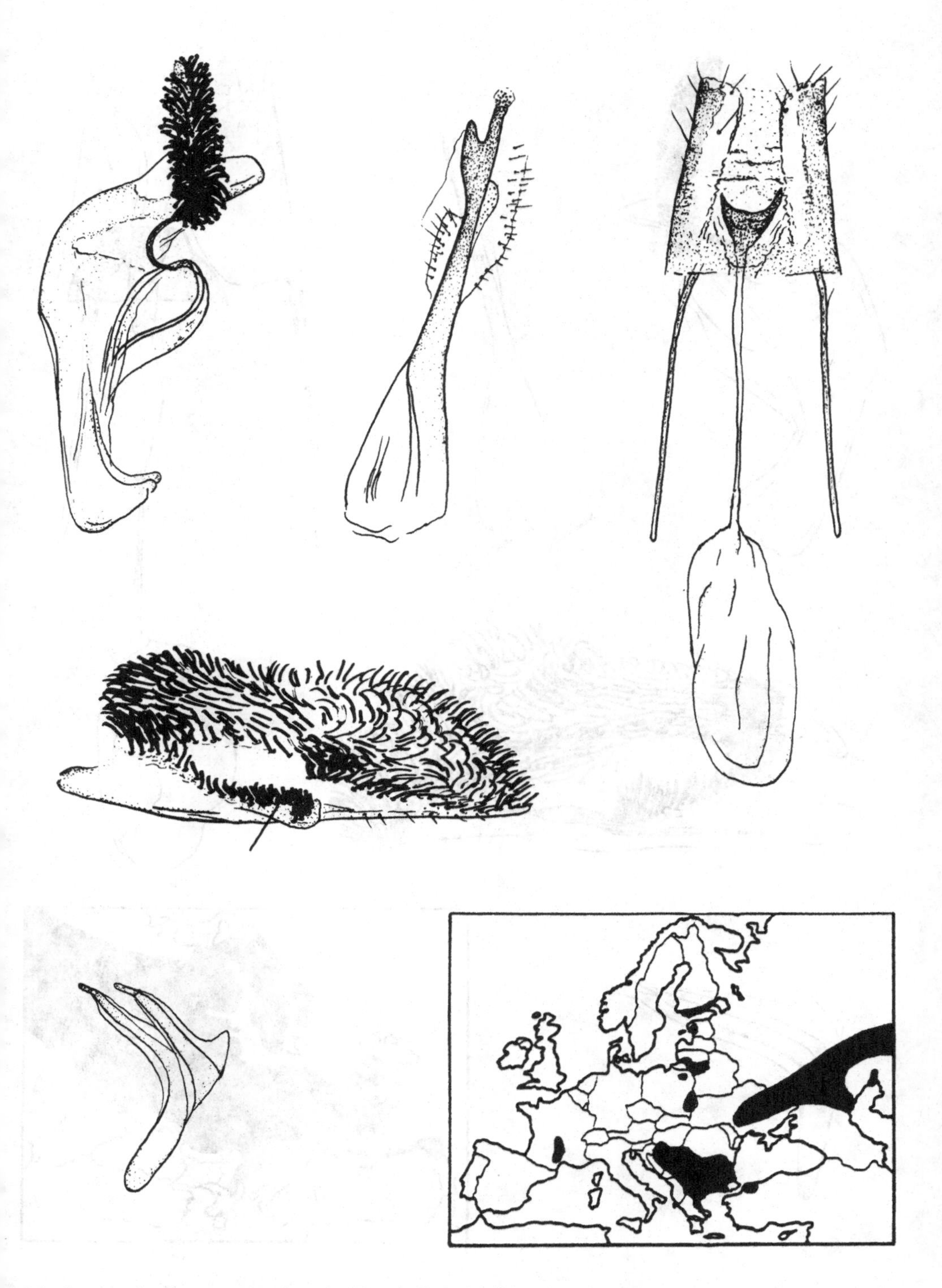

16. *Synanthedon mesiaeformis* (Herrich-Schäffer) – Bulgaria (ZL).

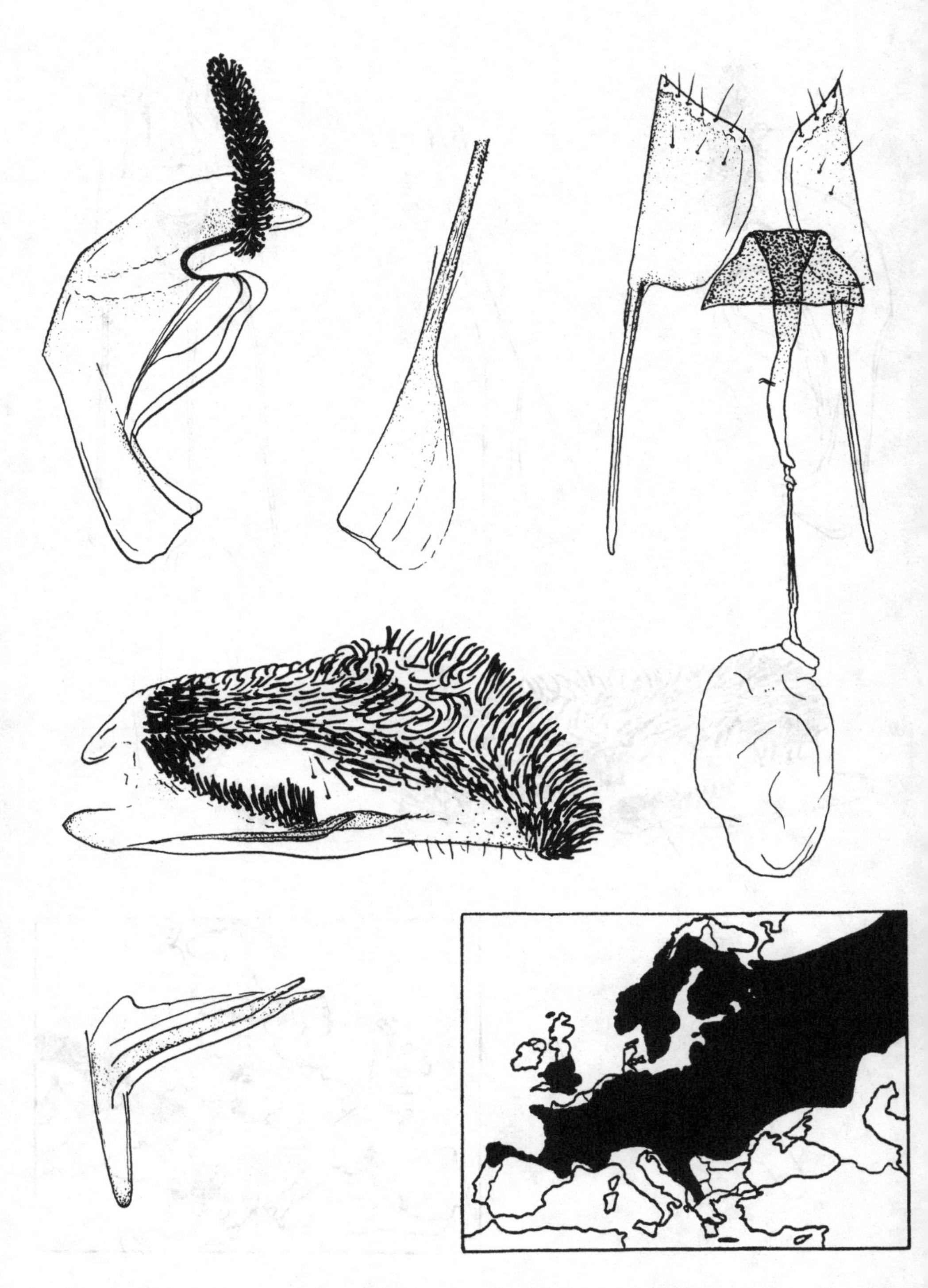

17. *Synanthedon spheciformis* ([Denis & Schiffermüller]) – Czech Republic (ZL).

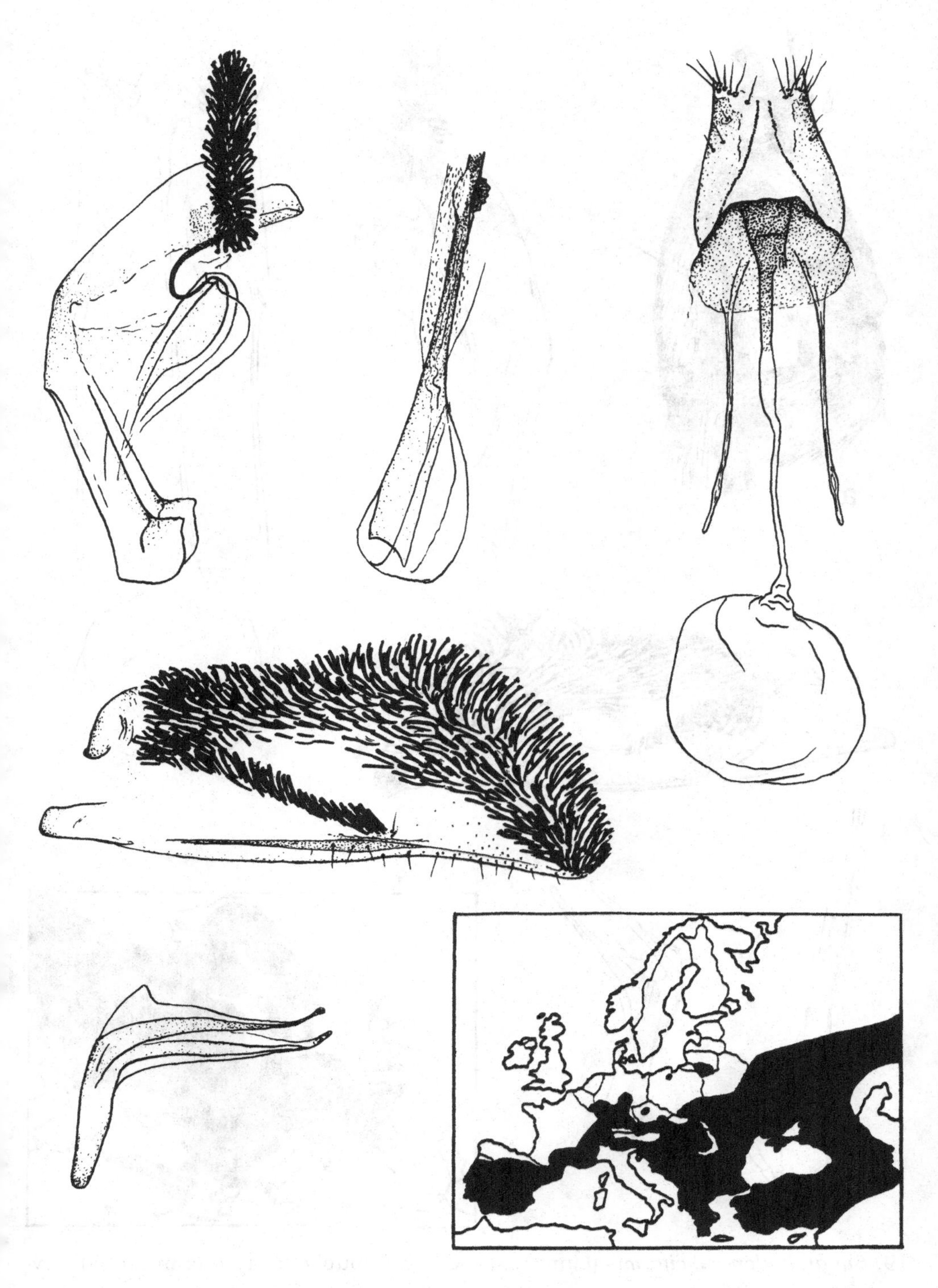

18. *Synanthedon stomoxiformis* (Hübner) – Czech Republic (ZL).

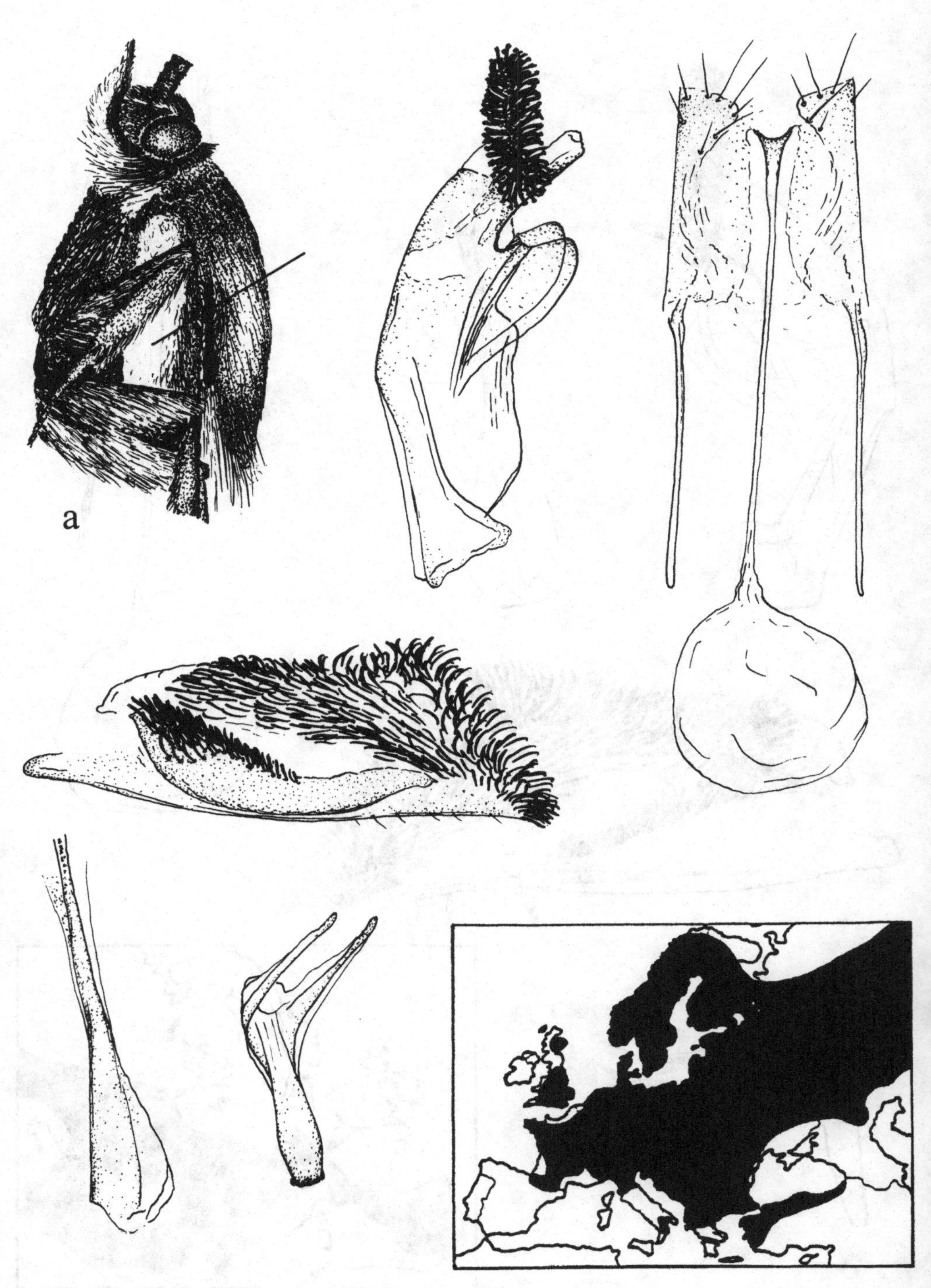

19. *Synanthedon culiciformis* (Linnaeus) – Czech Republic, a: thorax in lateral view (ZL).

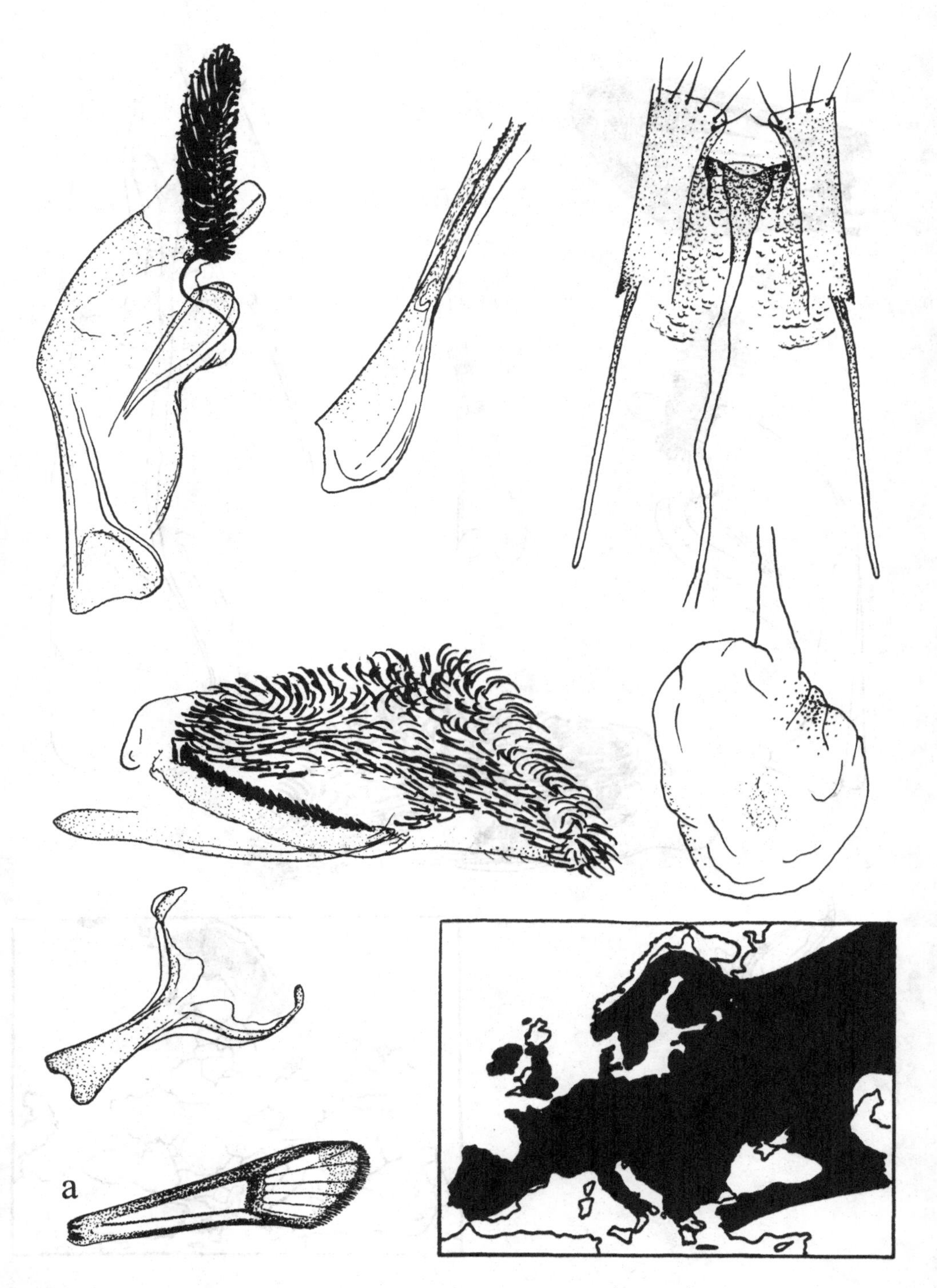

20. *Synanthedon formicaeformis* (Esper) – Czech Republic (ZL), a: forewing of ssp. *serica* Alphéraky – Russia (AK).

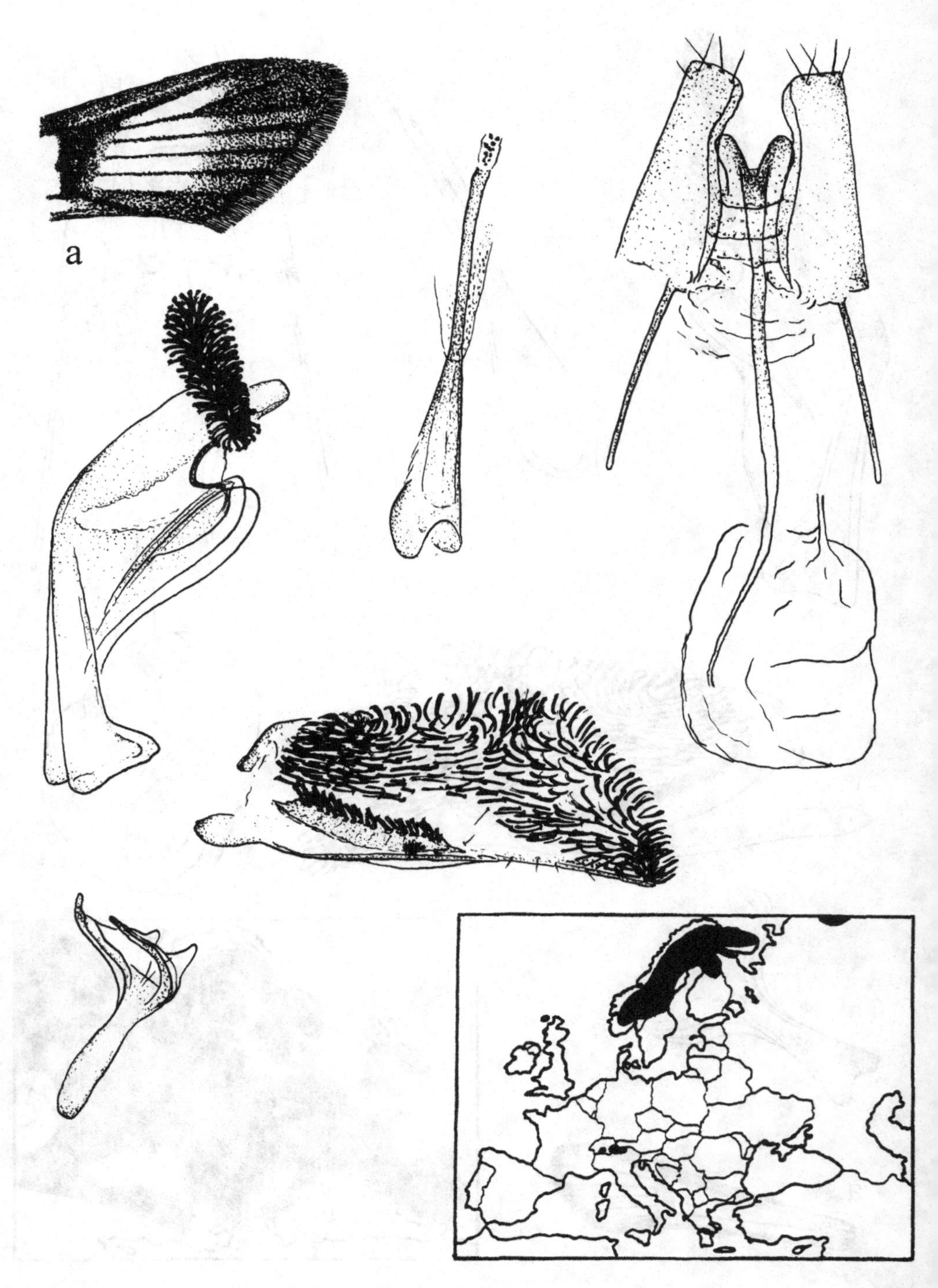

21. *Synanthedon polaris* (Staudinger) – Sweden, a: external transparent area (ZL, ex N. Ryrholm).

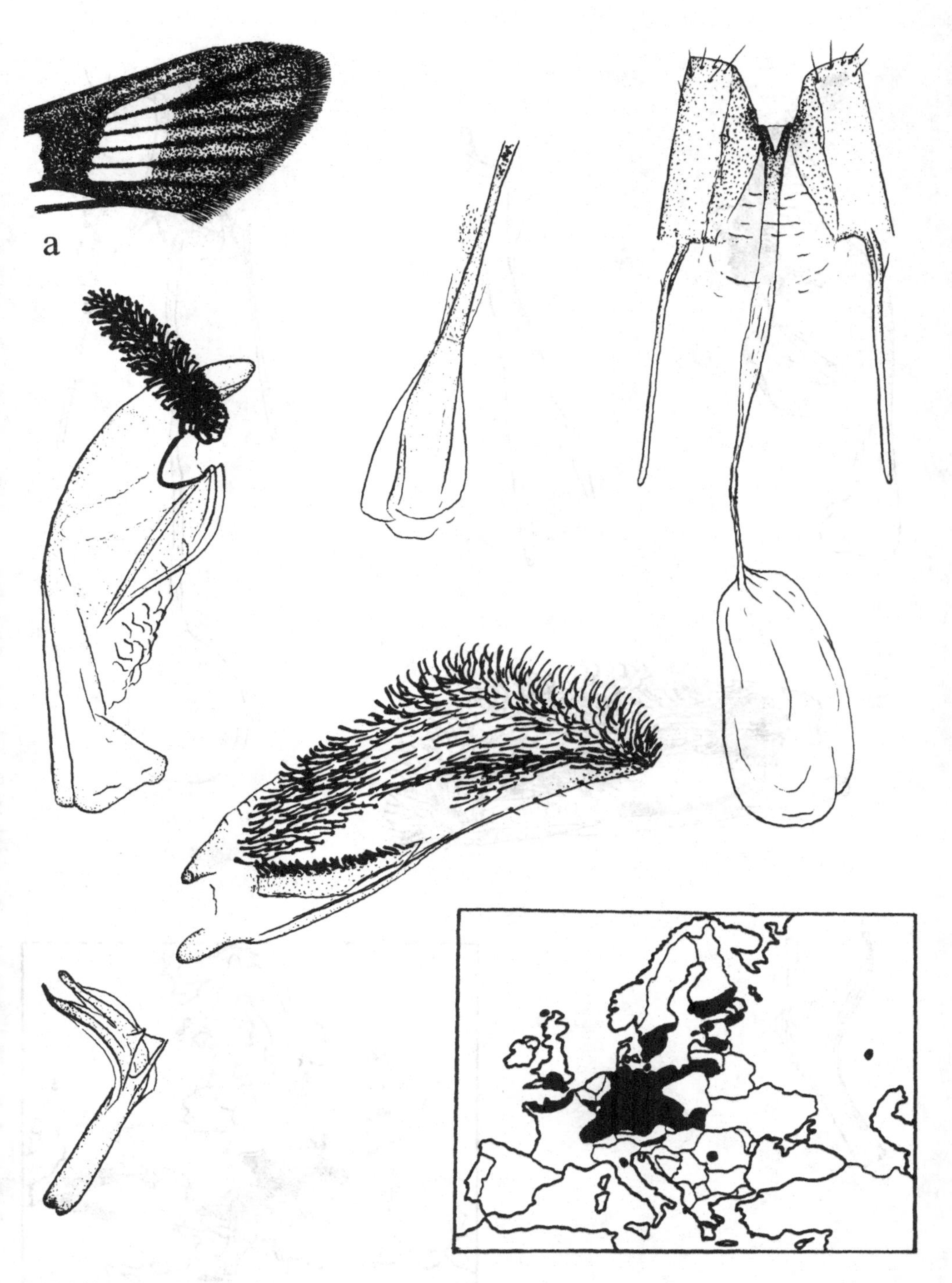

22. *Synanthedon flaviventris* (Staudinger) – Czech Republic, a: external transparent area (ZL).

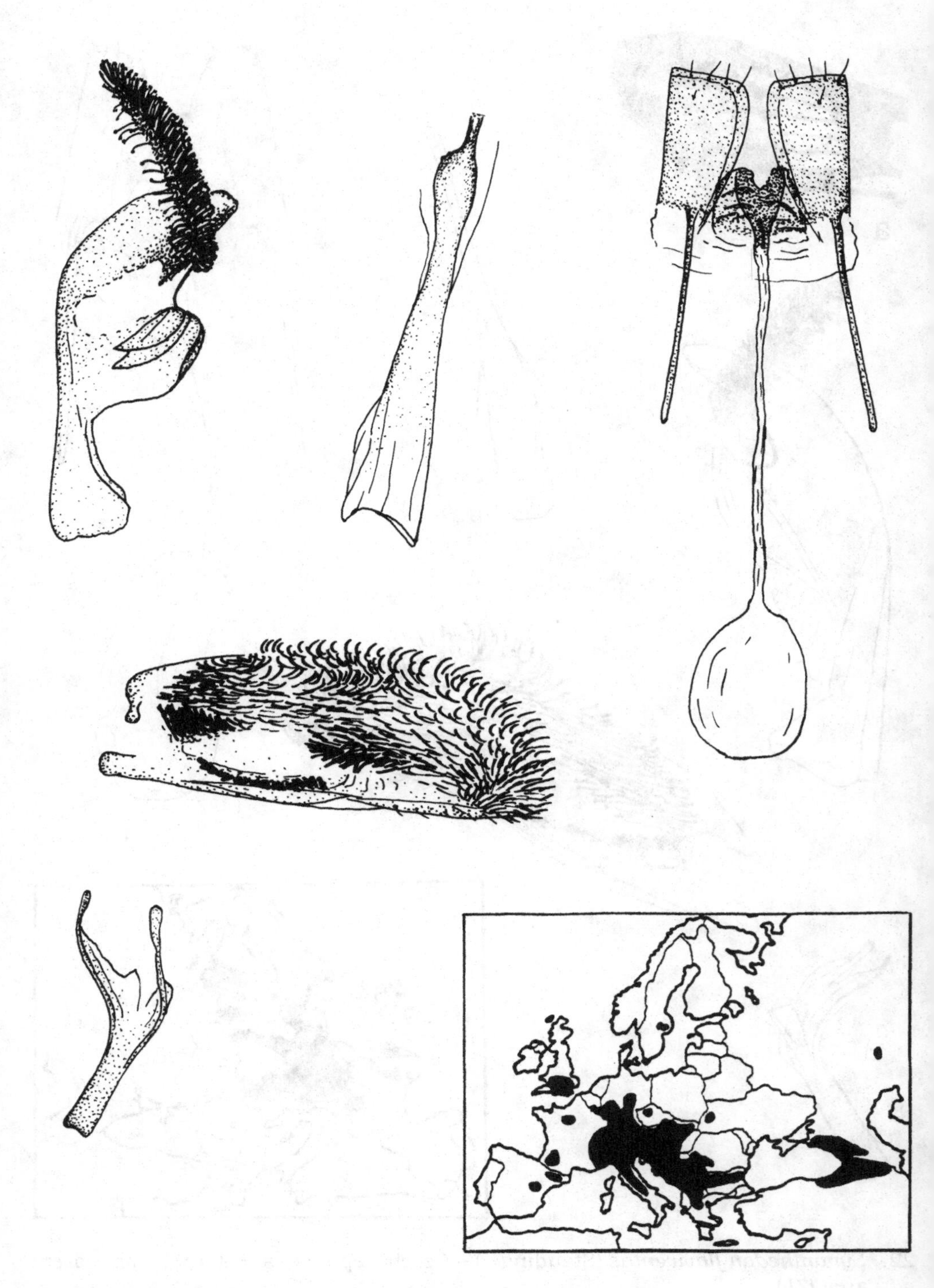

23. *Synanthedon andrenaeformis* (Laspeyres) – Slovakia (ZL).

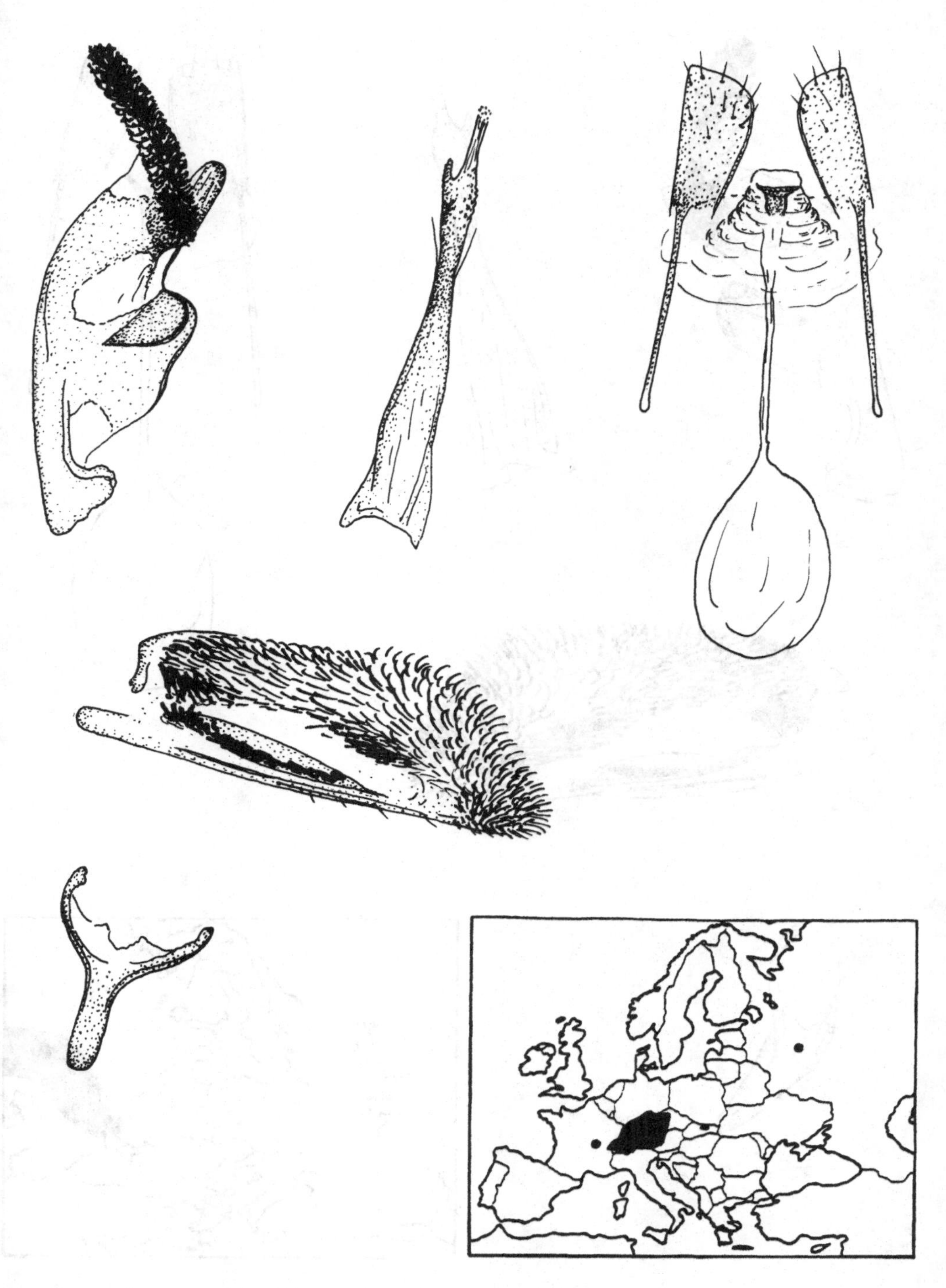

24. *Synanthedon soffneri* Spatenka – Czech Republic (ZL).

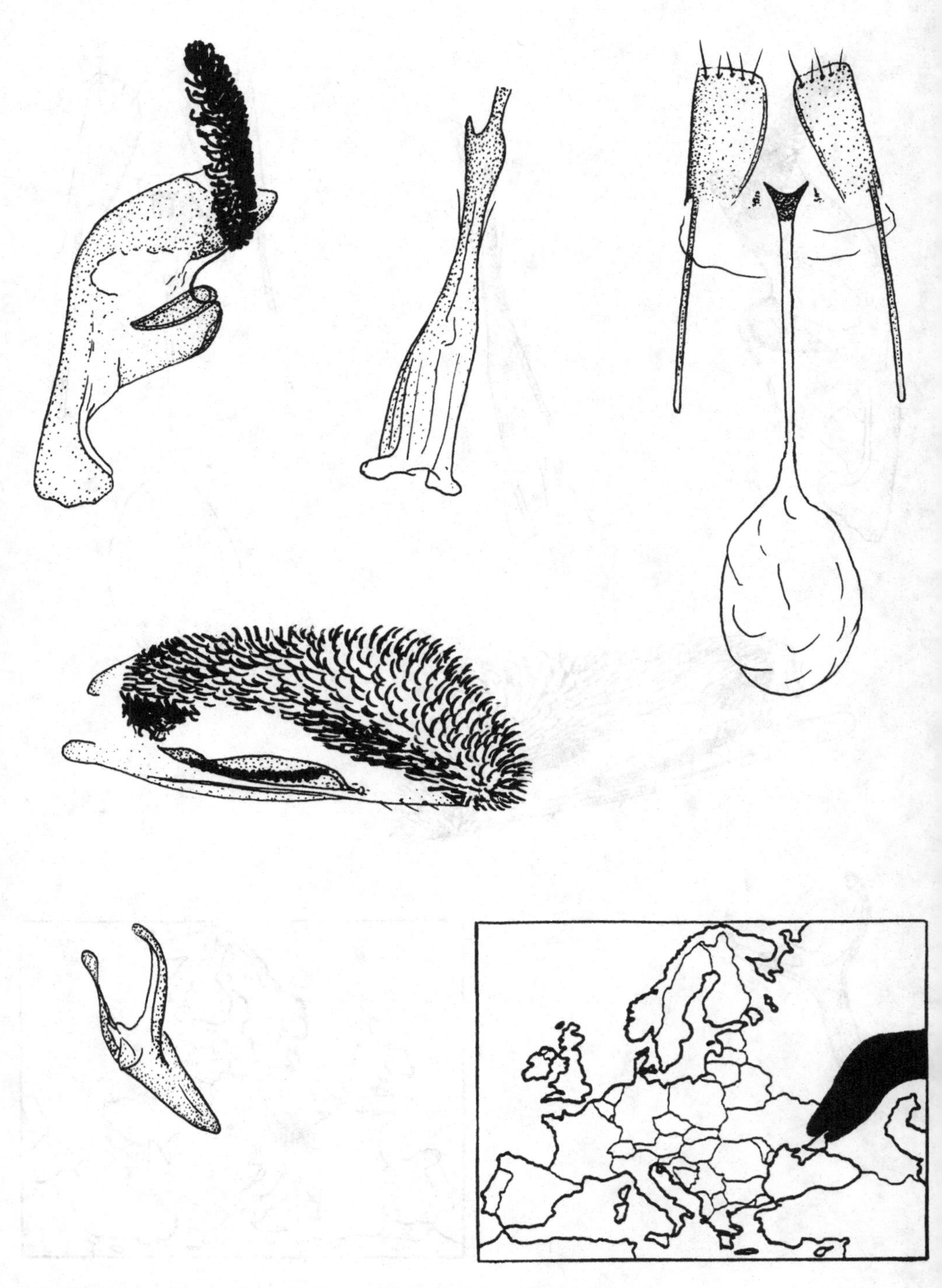

25. *Synanthedon uralensis* (Bartel) – Kazakhstan (ZL, ex OG).

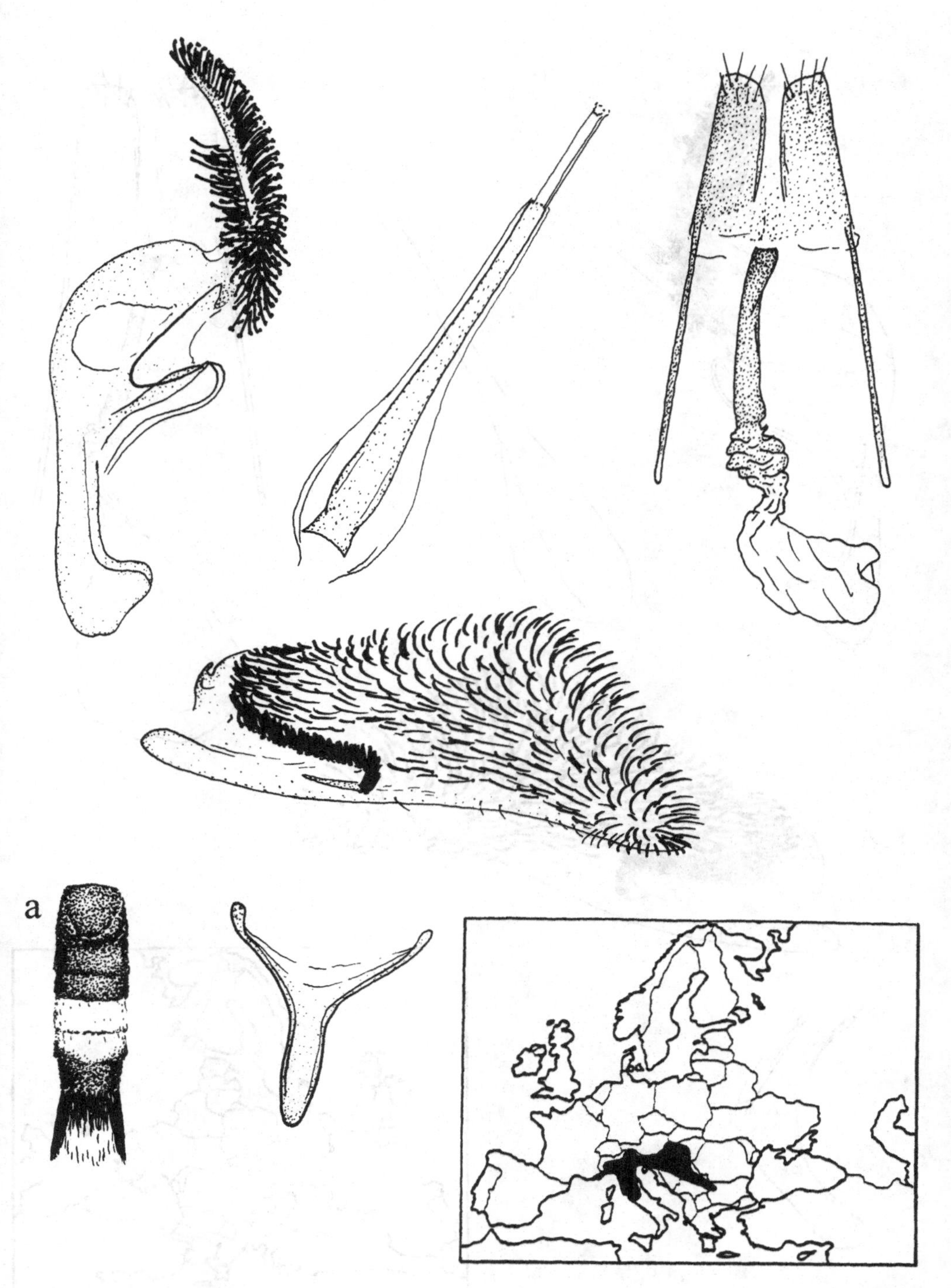

26. *Synanthedon melliniformis* (Laspeyres) – Austria, a: female abdomen in ventral view (ZL, ex D. Hamborg).

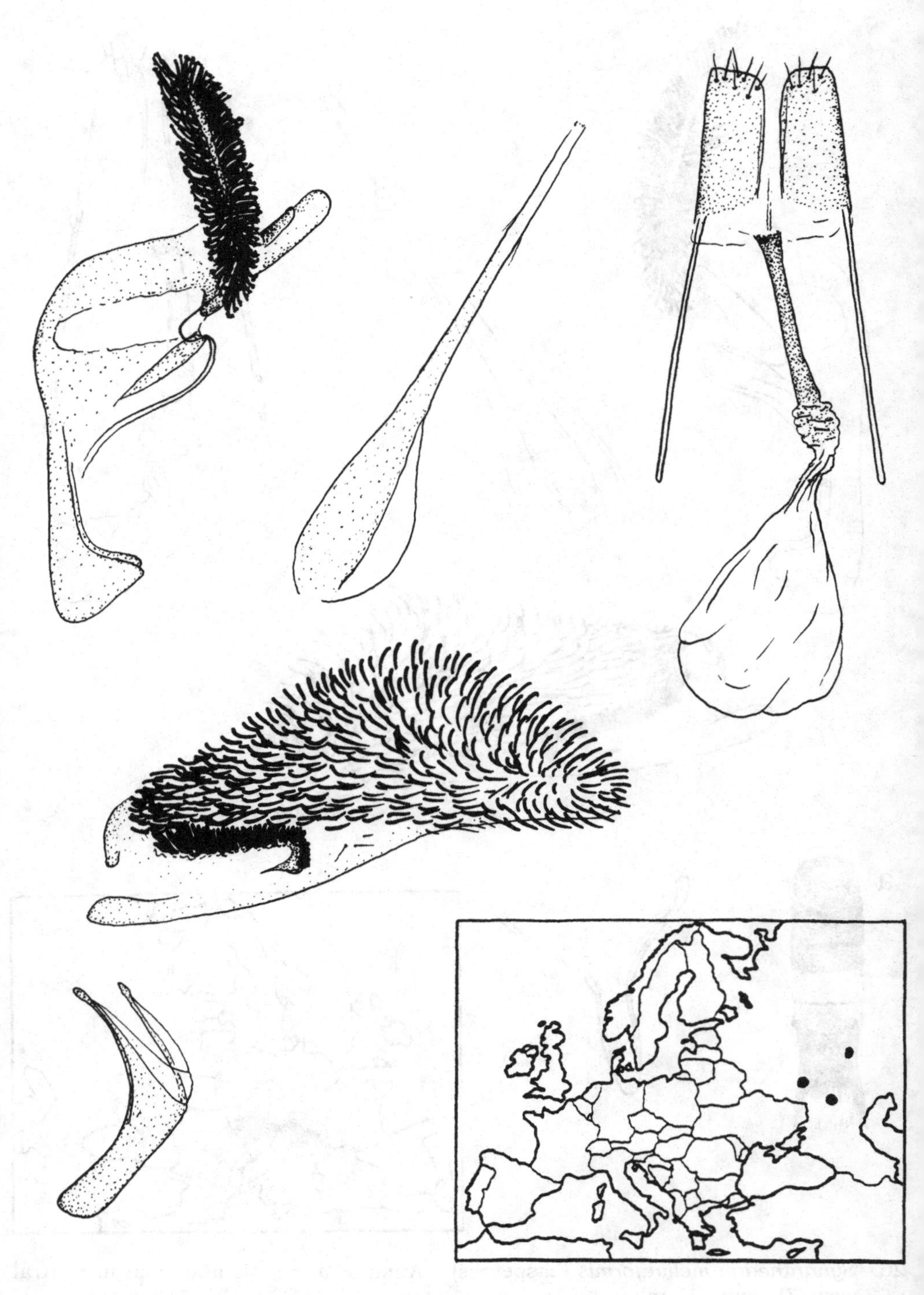

27. *Synanthedon martjanovi* Sheljuzhko – Russia (OG).

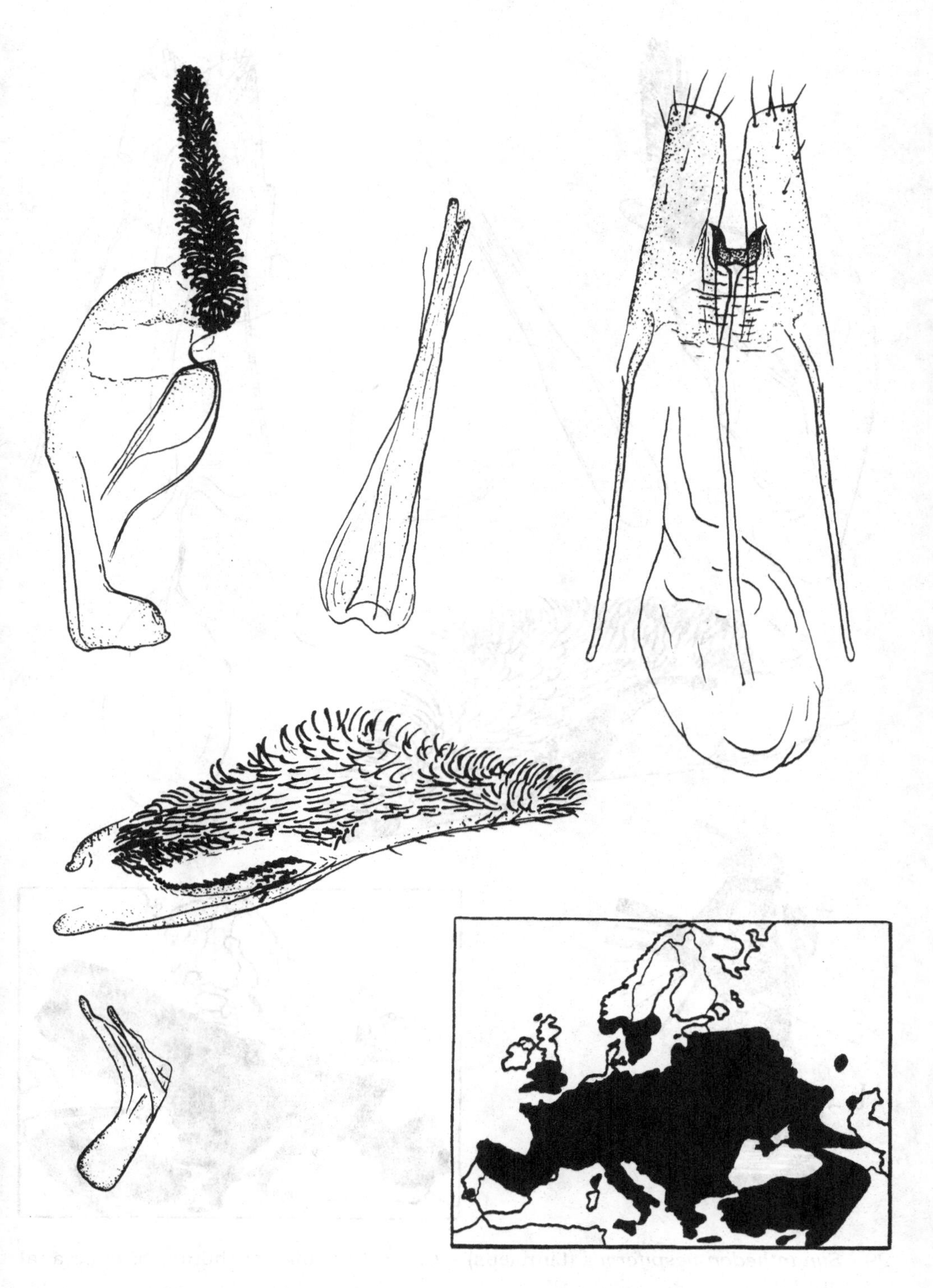

28. *Synanthedon myopaeformis* (Borkhausen) – Czech Republic (ZL).

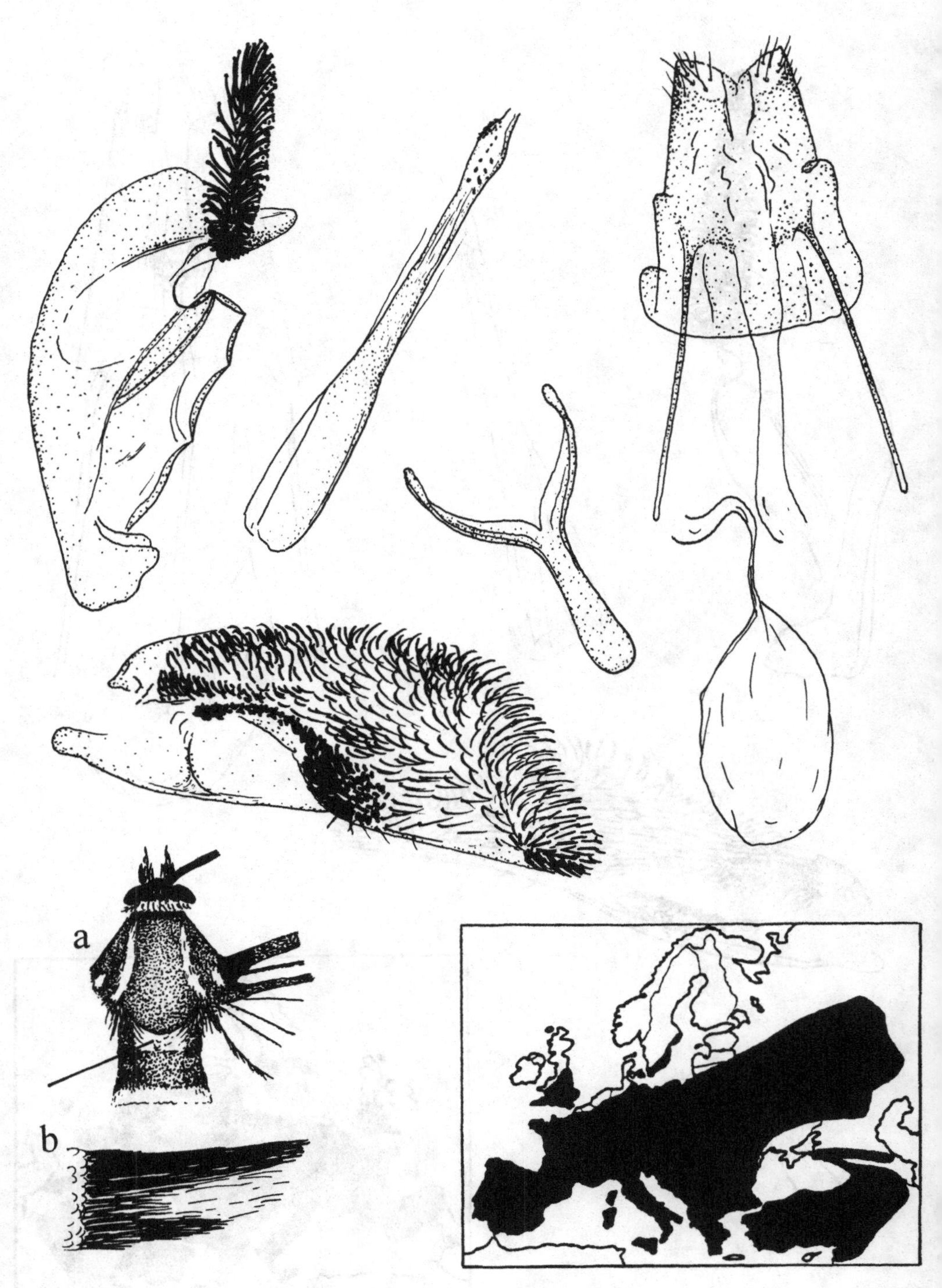

29. *Synanthedon vespiformis* (Linnaeus) – Czech Republic, a: thorax, b: male anal tuft (ZL).

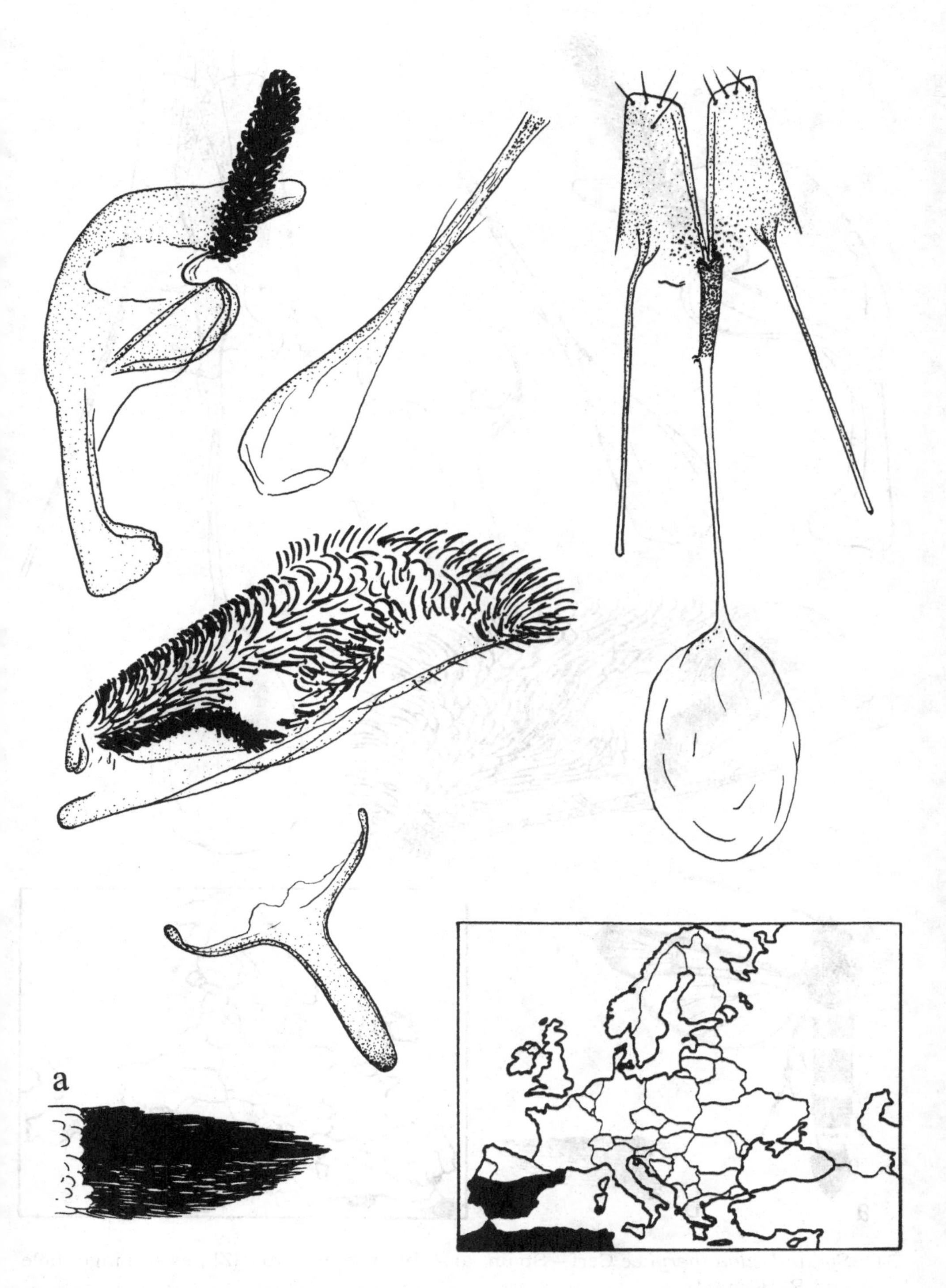

30. *Synanthedon codeti* (Oberthür) – Spain, a: male anal tuft (ZL, ex R. Bläsius).

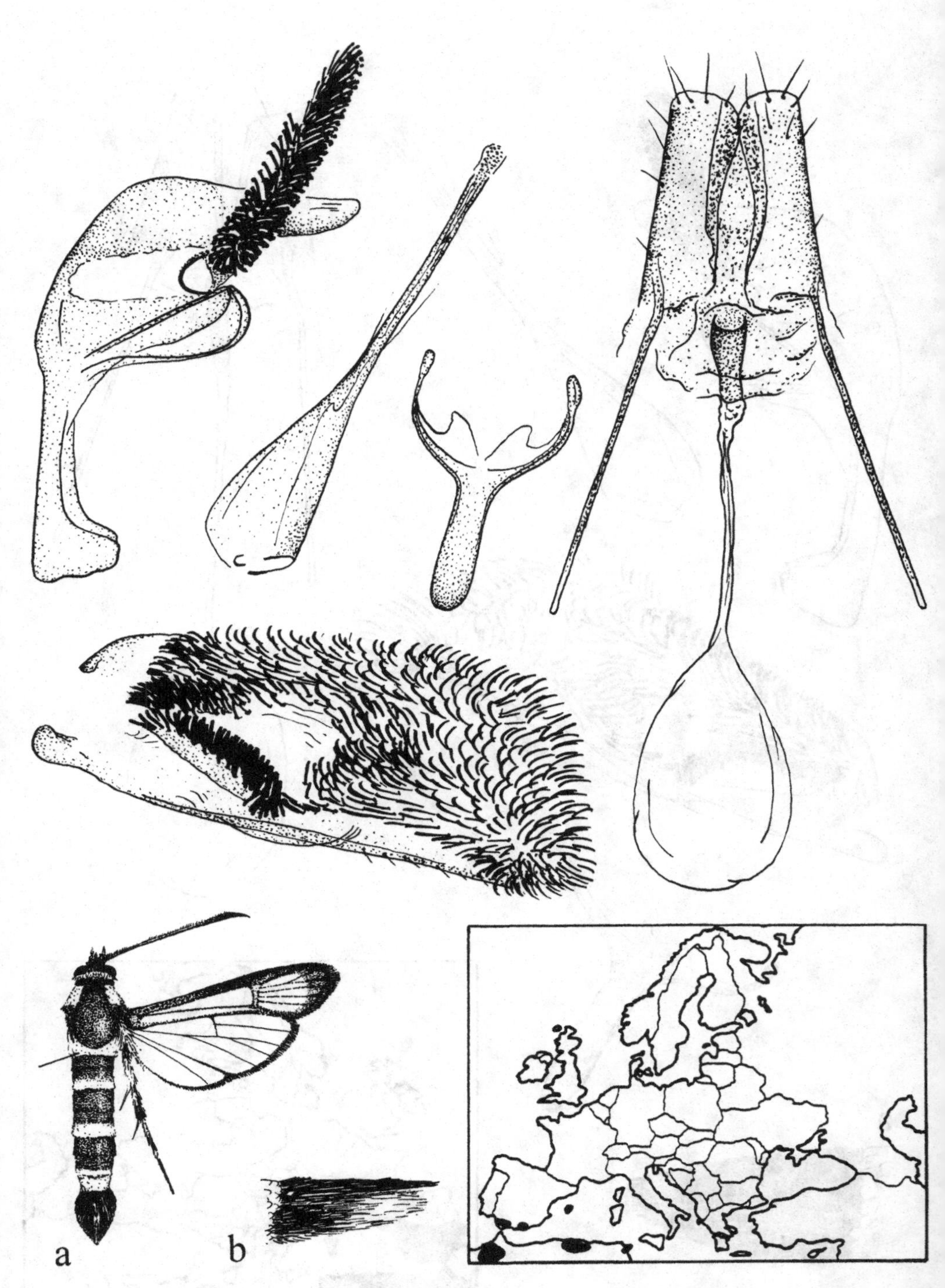

31. *Synanthedon theryi* Le Cerf – Spain, a: ♂, b: male anal tuft (ZL, ex A. Lingenhöle and R. Bläsius).

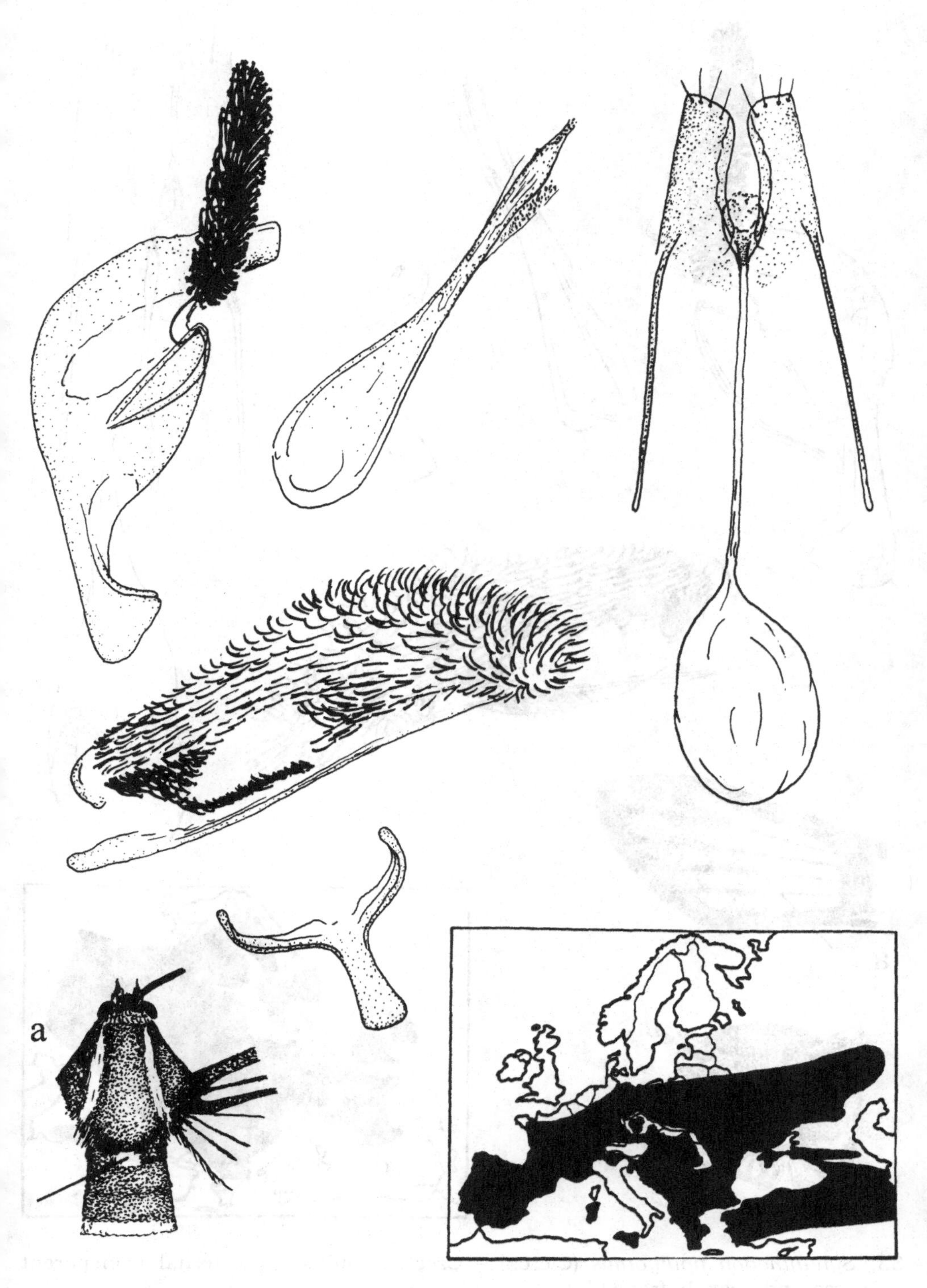

32. *Synanthedon conopiformis* (Esper) – Czech Republic, a: thorax (ZL).

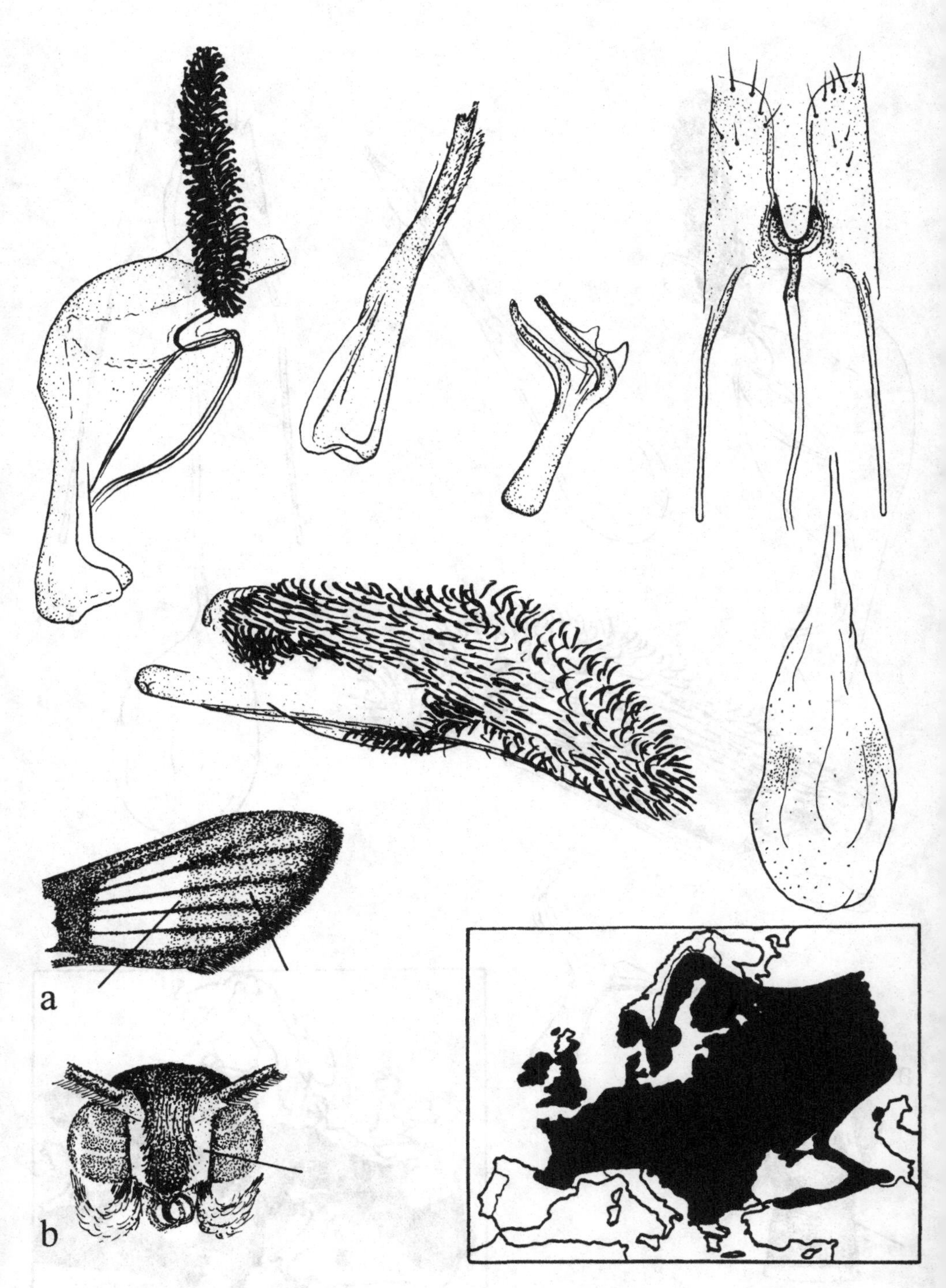

33. *Synanthedon tipuliformis* (Clerck) – Czech Republic, a: external transparent area and apex, b: frons (ZL).

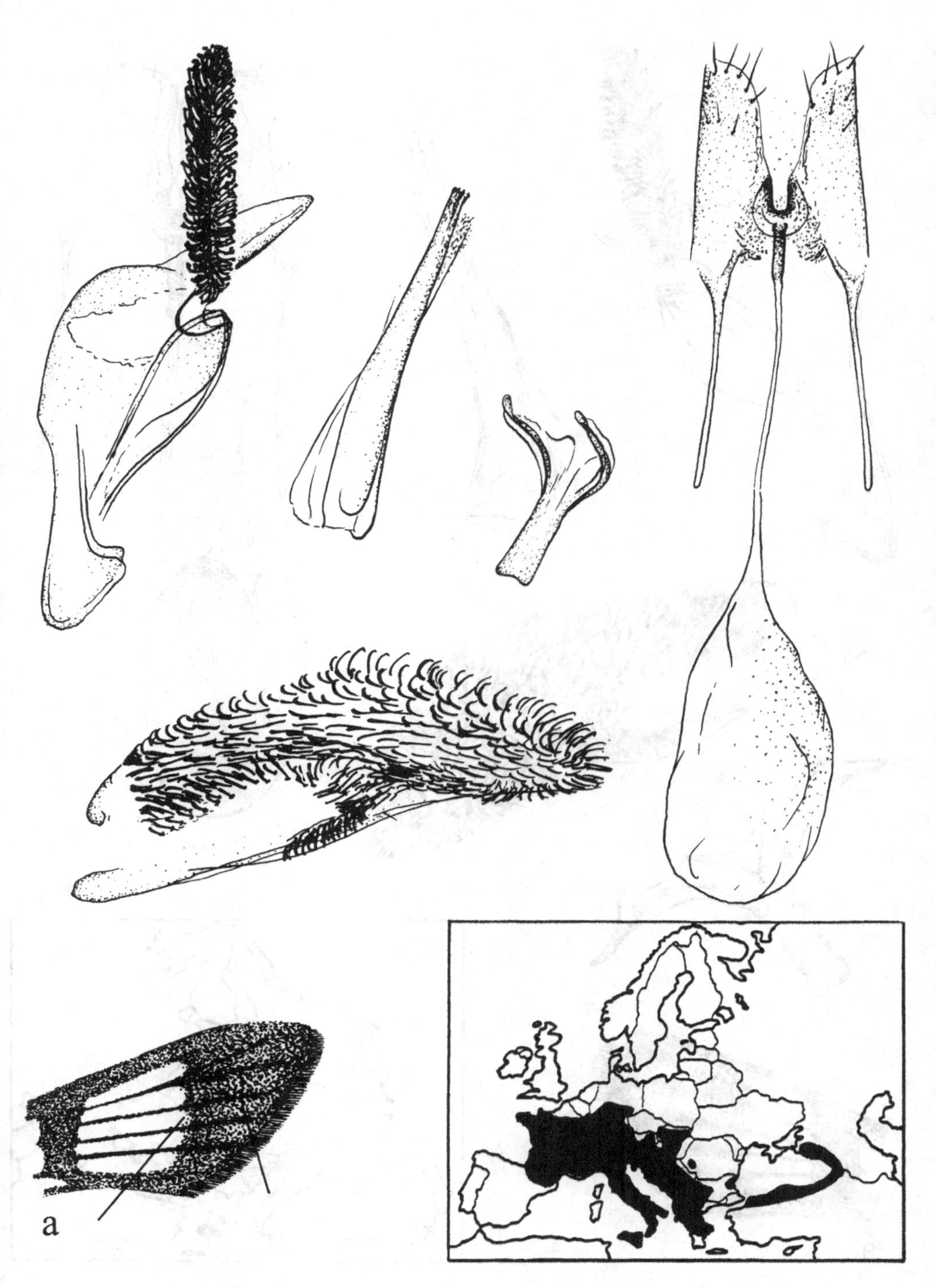

34. *Synanthedon spuleri* (Fuchs) – Slovakia, a: external transparent area and apex (ZL).

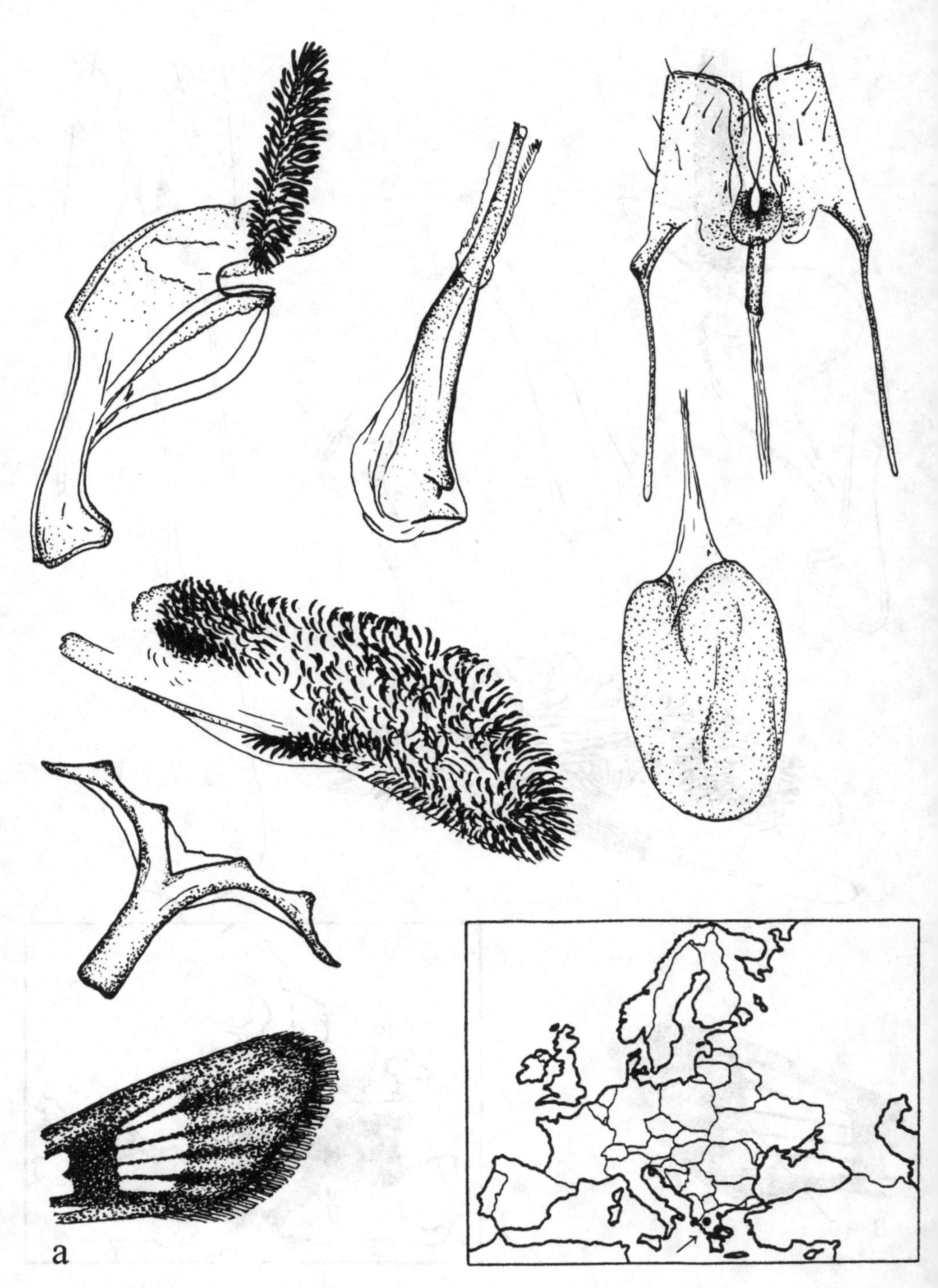

35. *Synanthedon geranii* Kallies – Greece, a: external transparent area and apex (ZL).

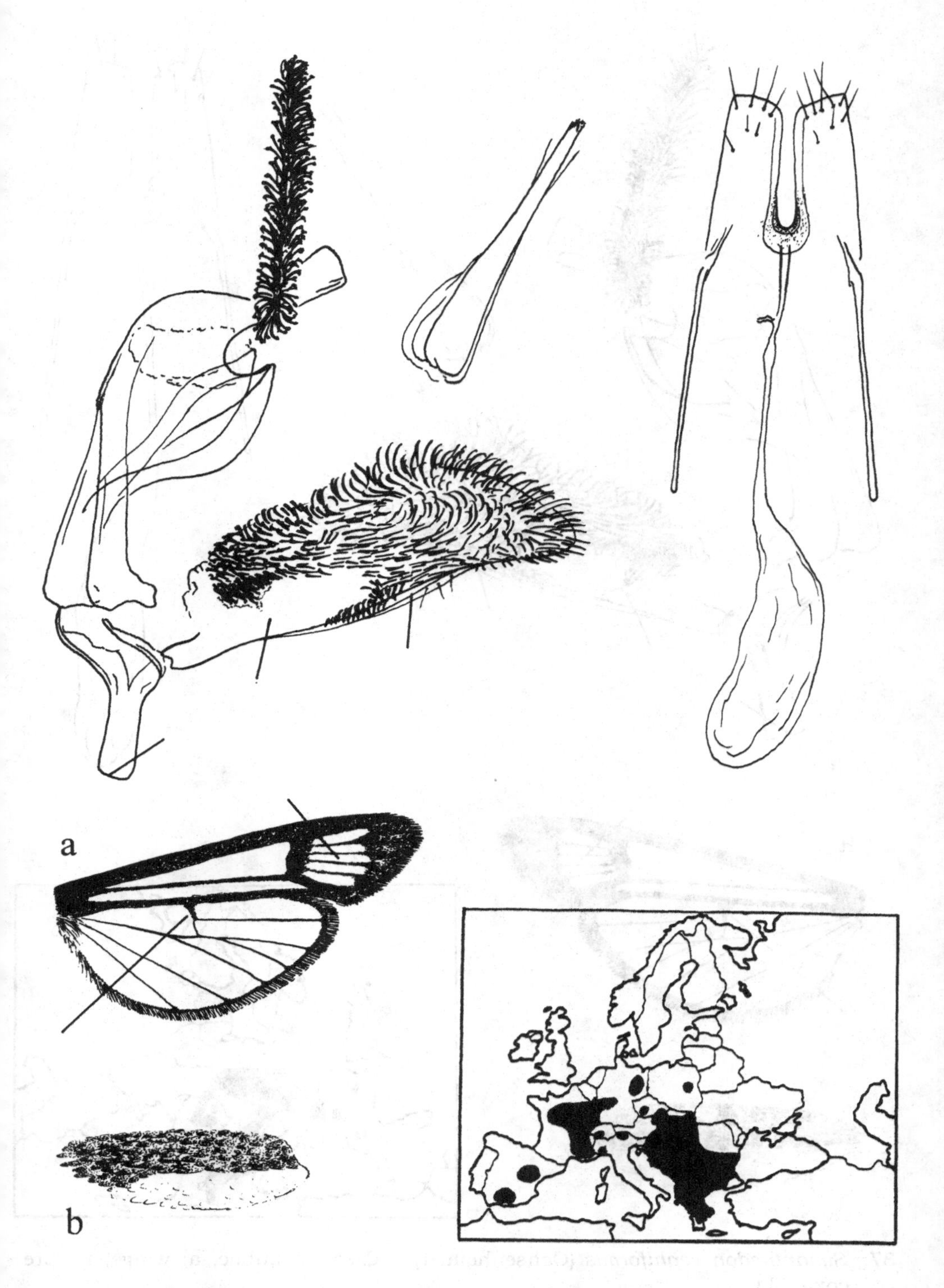

36. *Synanthedon loranthi* (Králíček) – Czech Republic, a: wings, b: fore coxa (ZL).

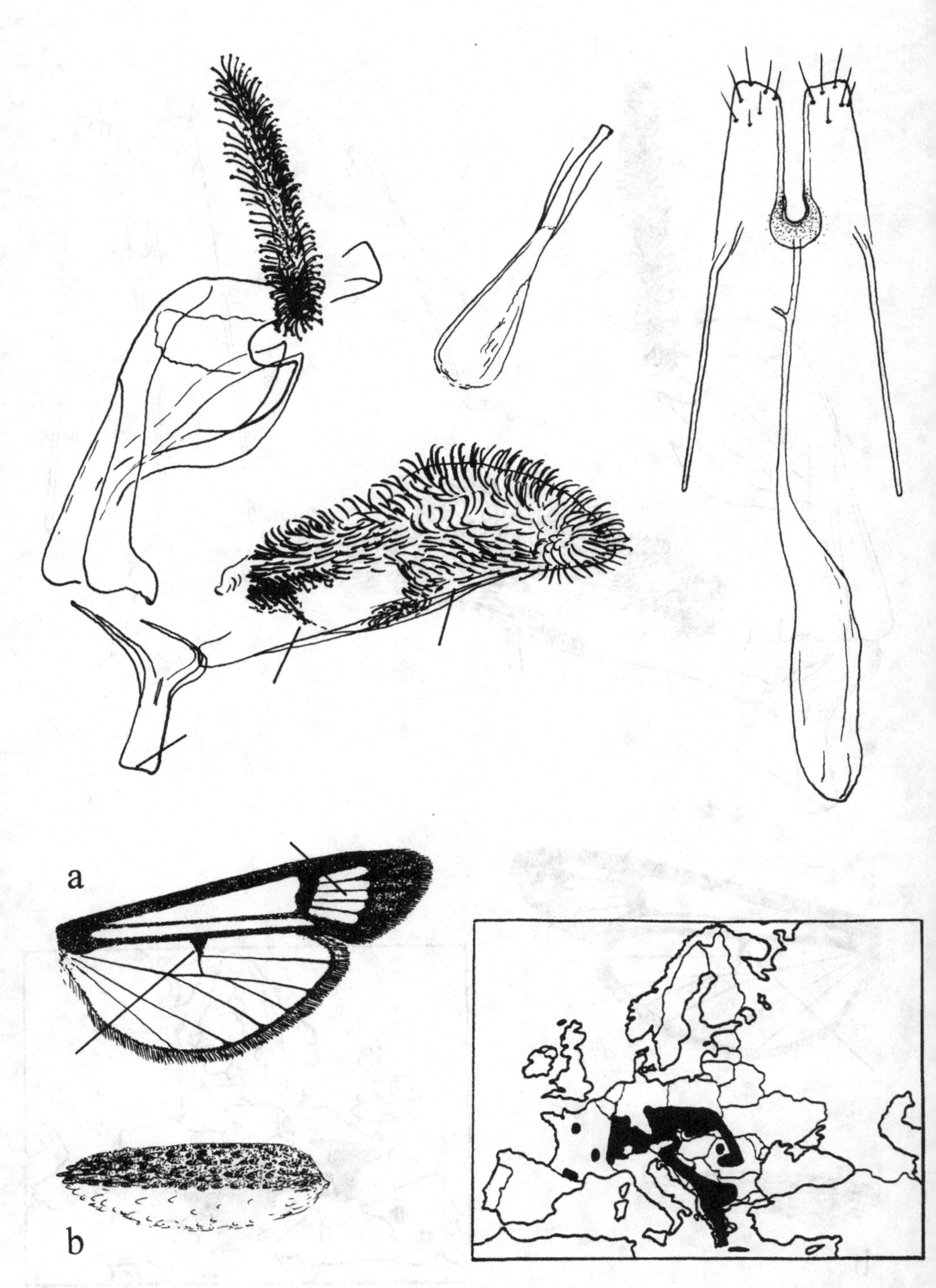

37. *Synanthedon cephiformis* (Ochsenheimer) – Czech Republic, a: wings, b: fore coxa (ZL).

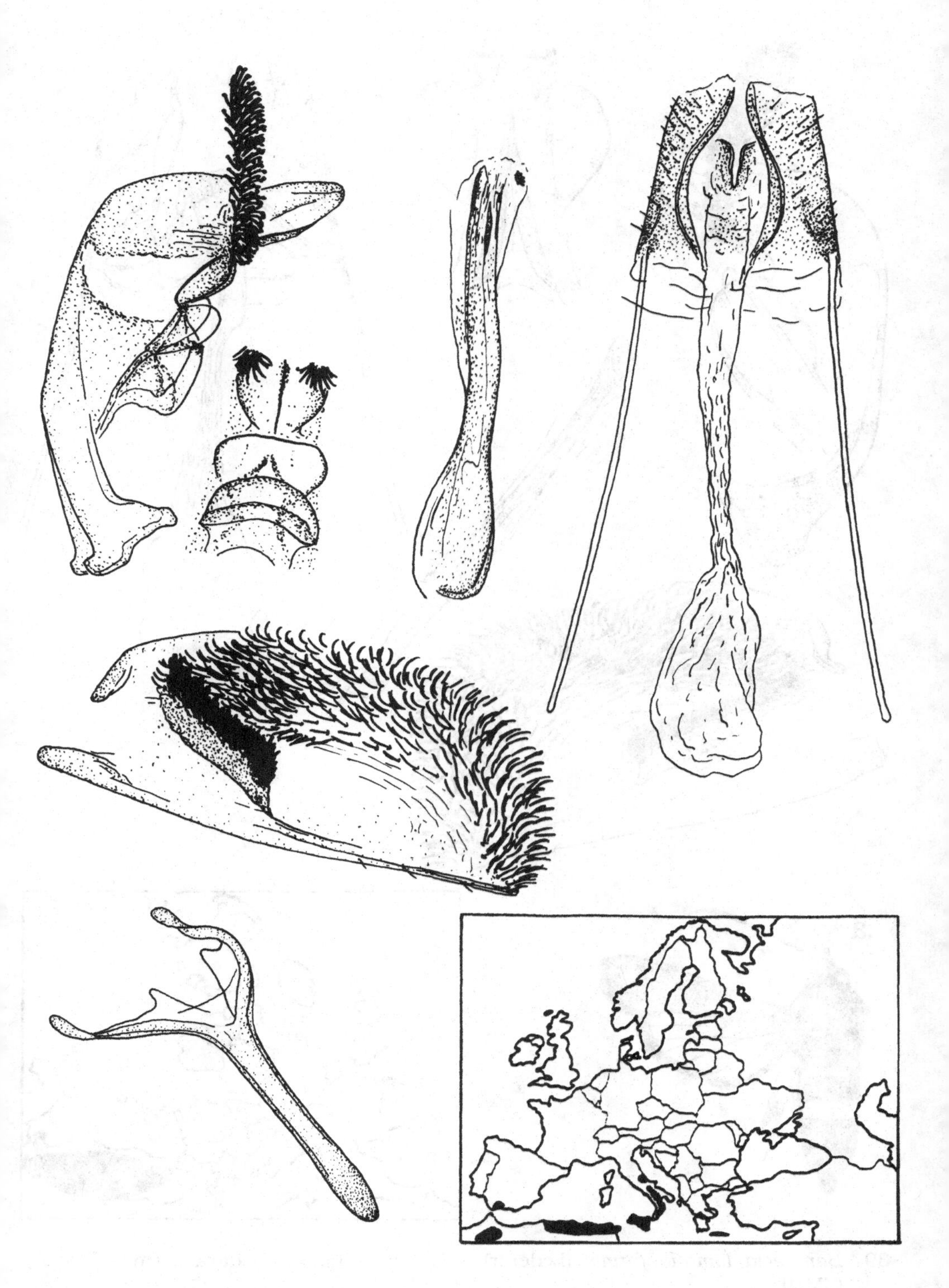

38. *Bembecia hymenopteriformis* (Bellier) – Spain (ZL).

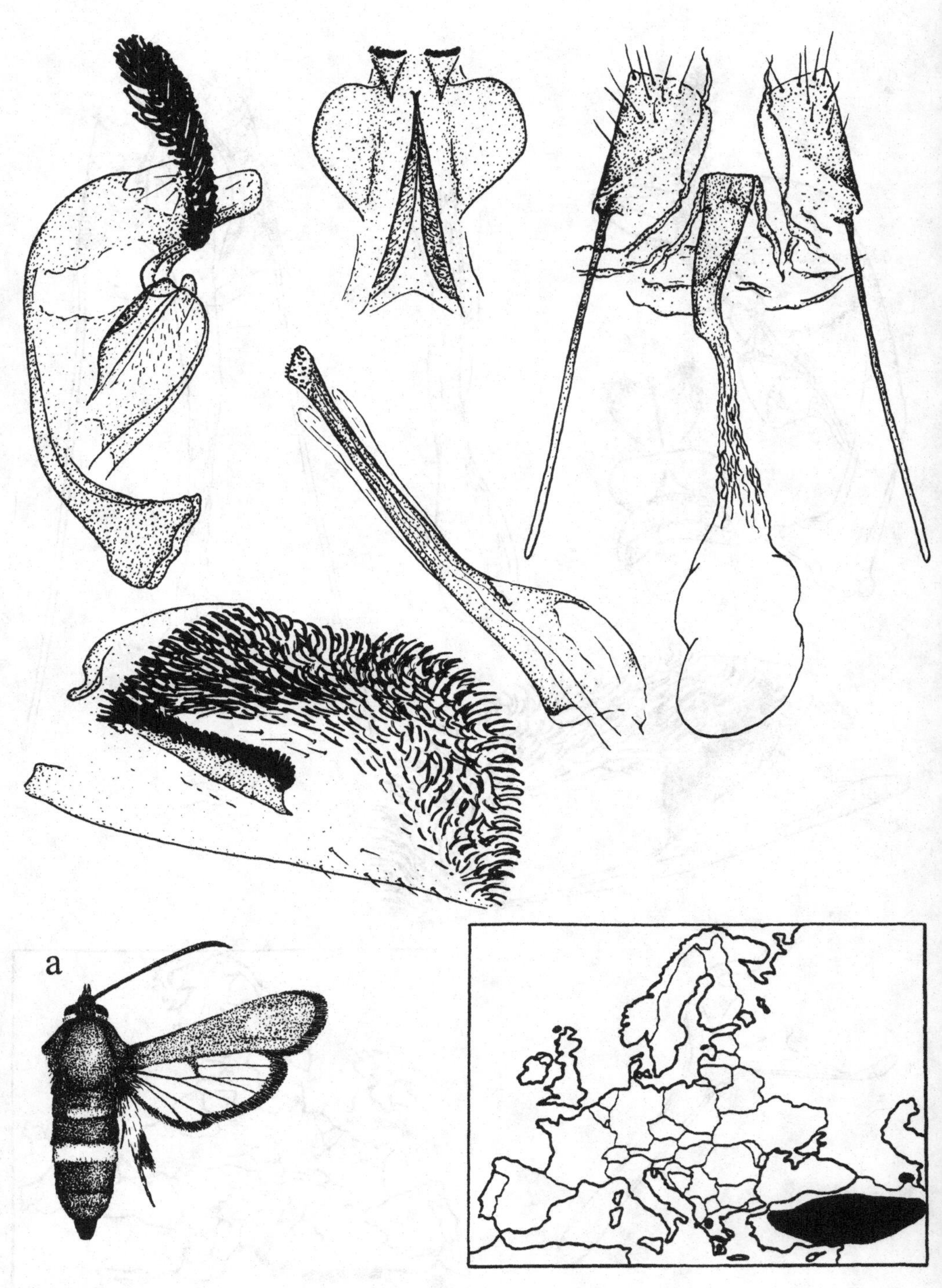

39. *Bembecia lomatiaeformis* (Lederer) – Greece (ZL), a: ♀, dark form – Turkey (ZMHB).

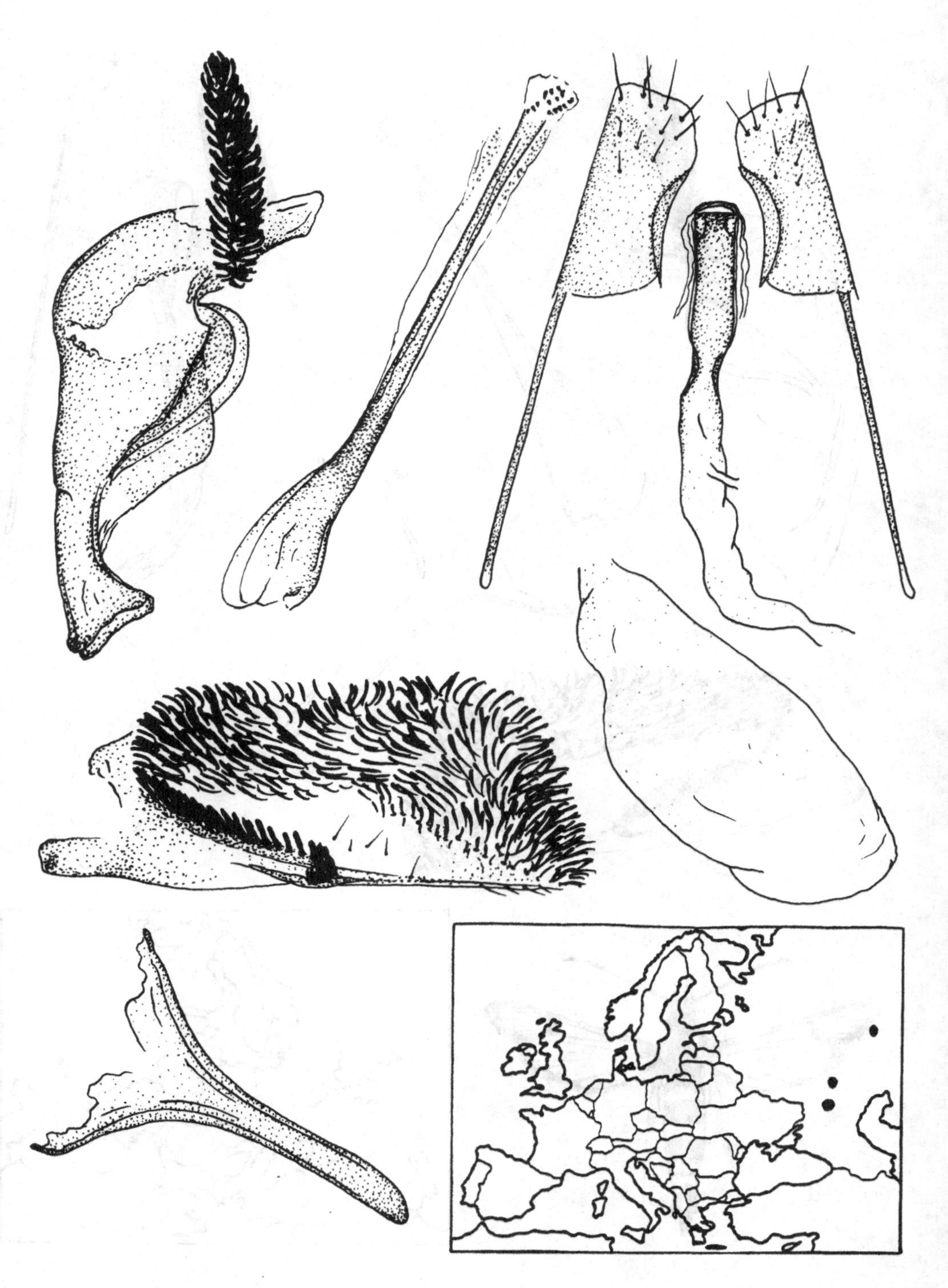

40. *Bembecia sareptana* (Bartel) – Russia (♂, M. Ahola et al., ♀, holotype, ZMHB).

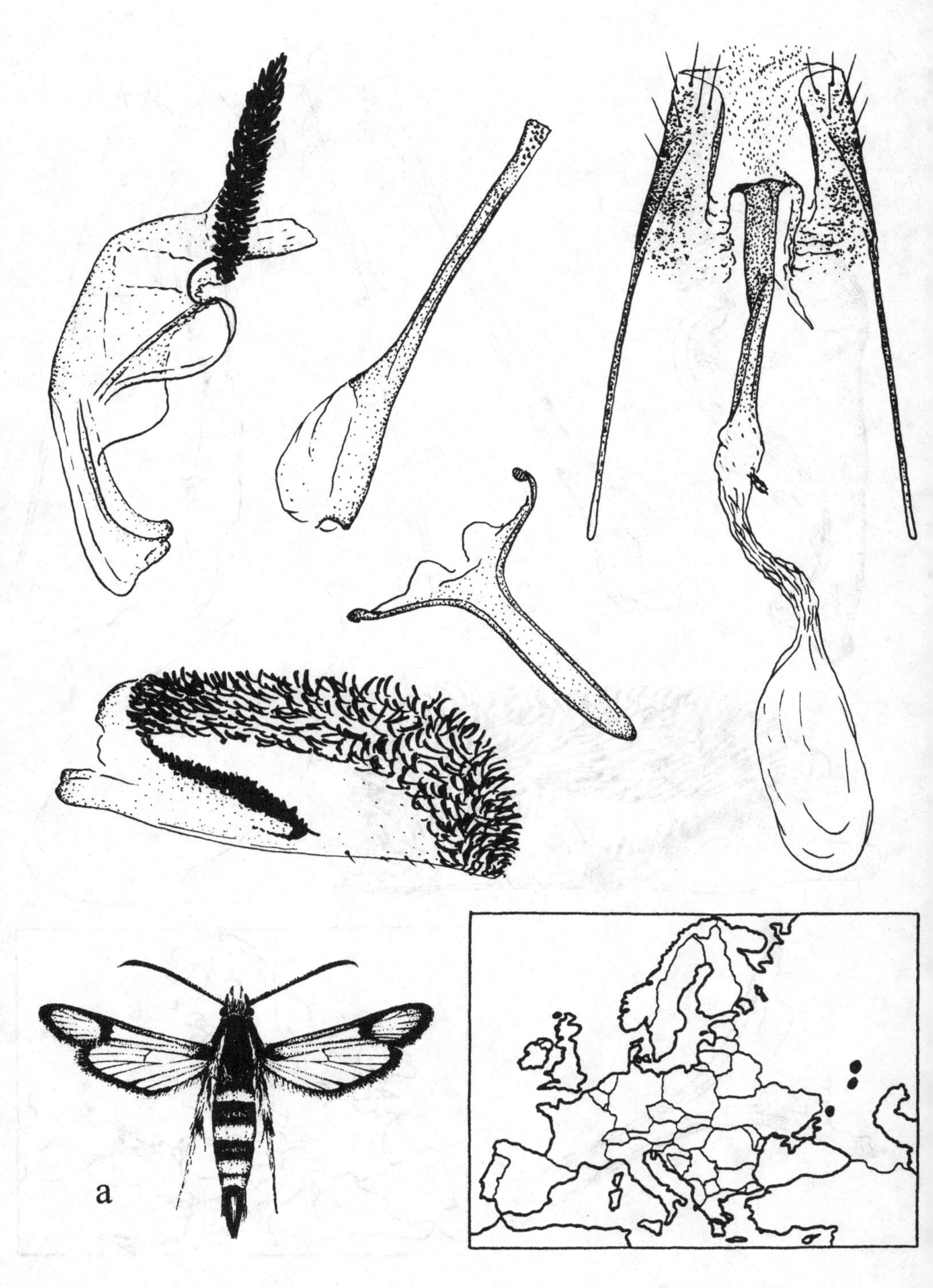

41. *Bembecia volgensis* Gorbunov – Russia (♂ OG, ♀ AK), a: ♂, holotype (ZISP).

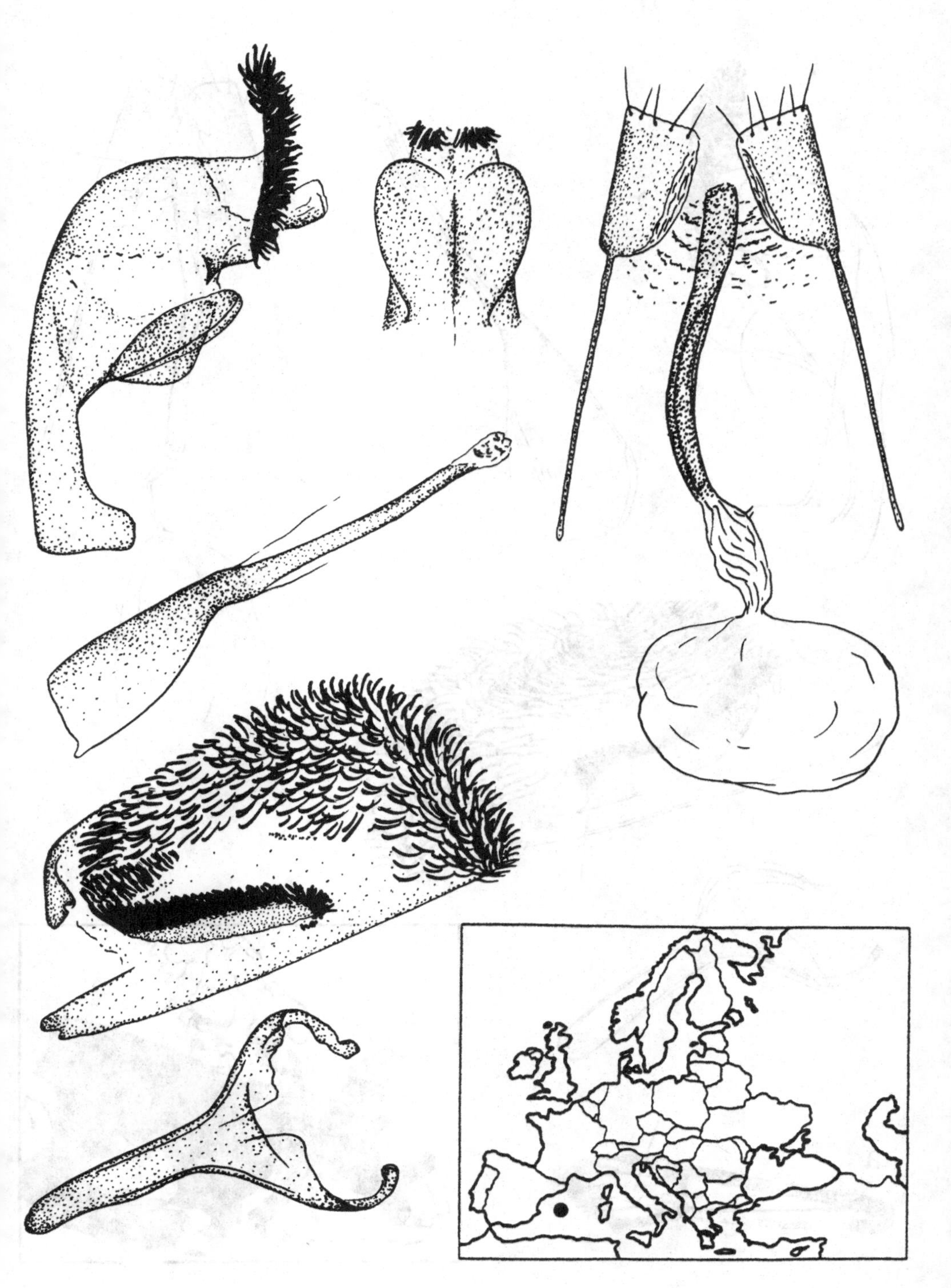

42. *Bembecia abromeiti* Kallies & Riefenstahl – Mallorca, after Sobczyk in Kallies & Riefenstahl (2000), a: ♂, paratype (AK).

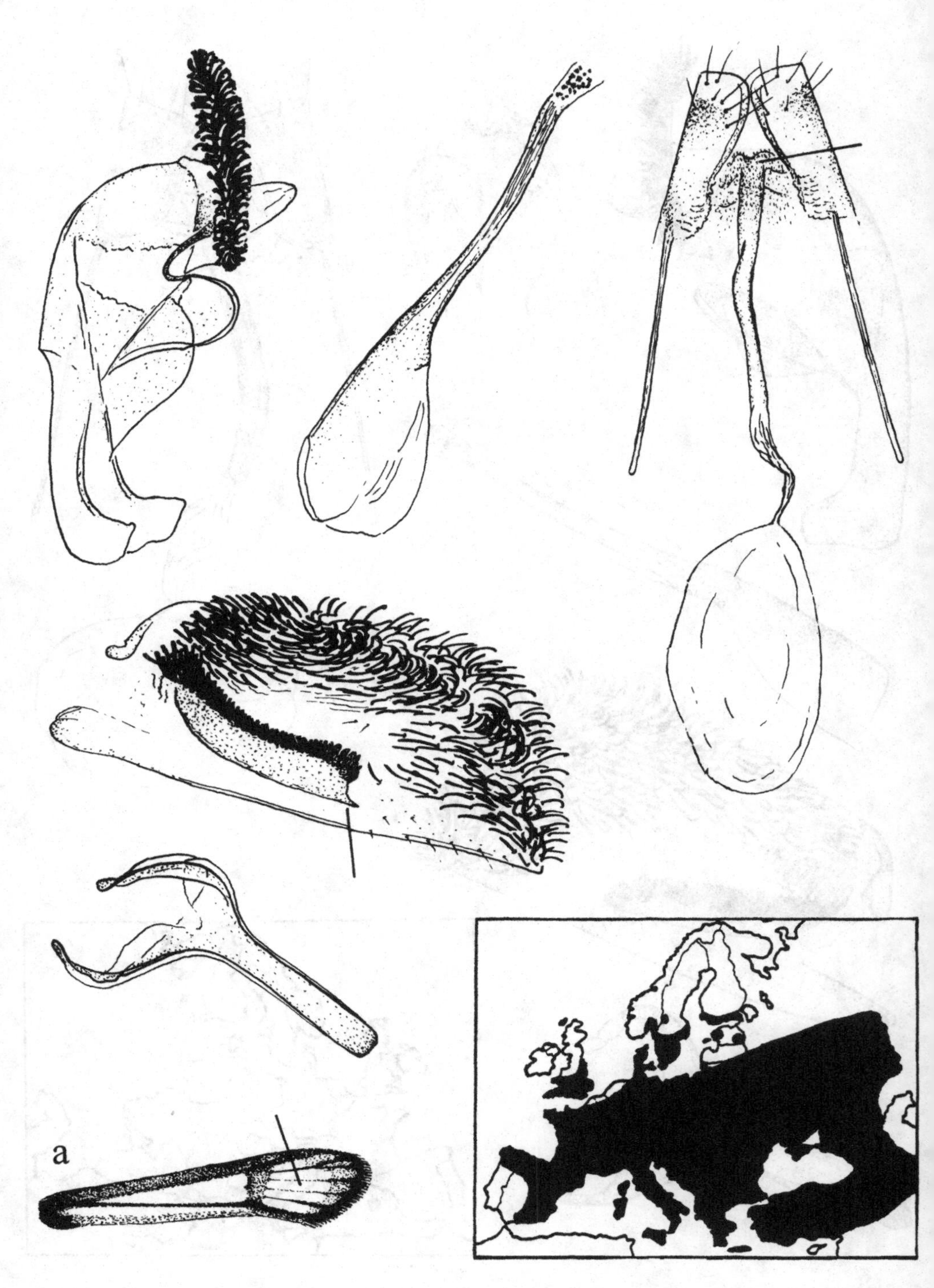

43. *Bembecia ichneumoniformis* ([Denis & Schiffermüller]) – Czech Republic, a: male forewing (ZL).

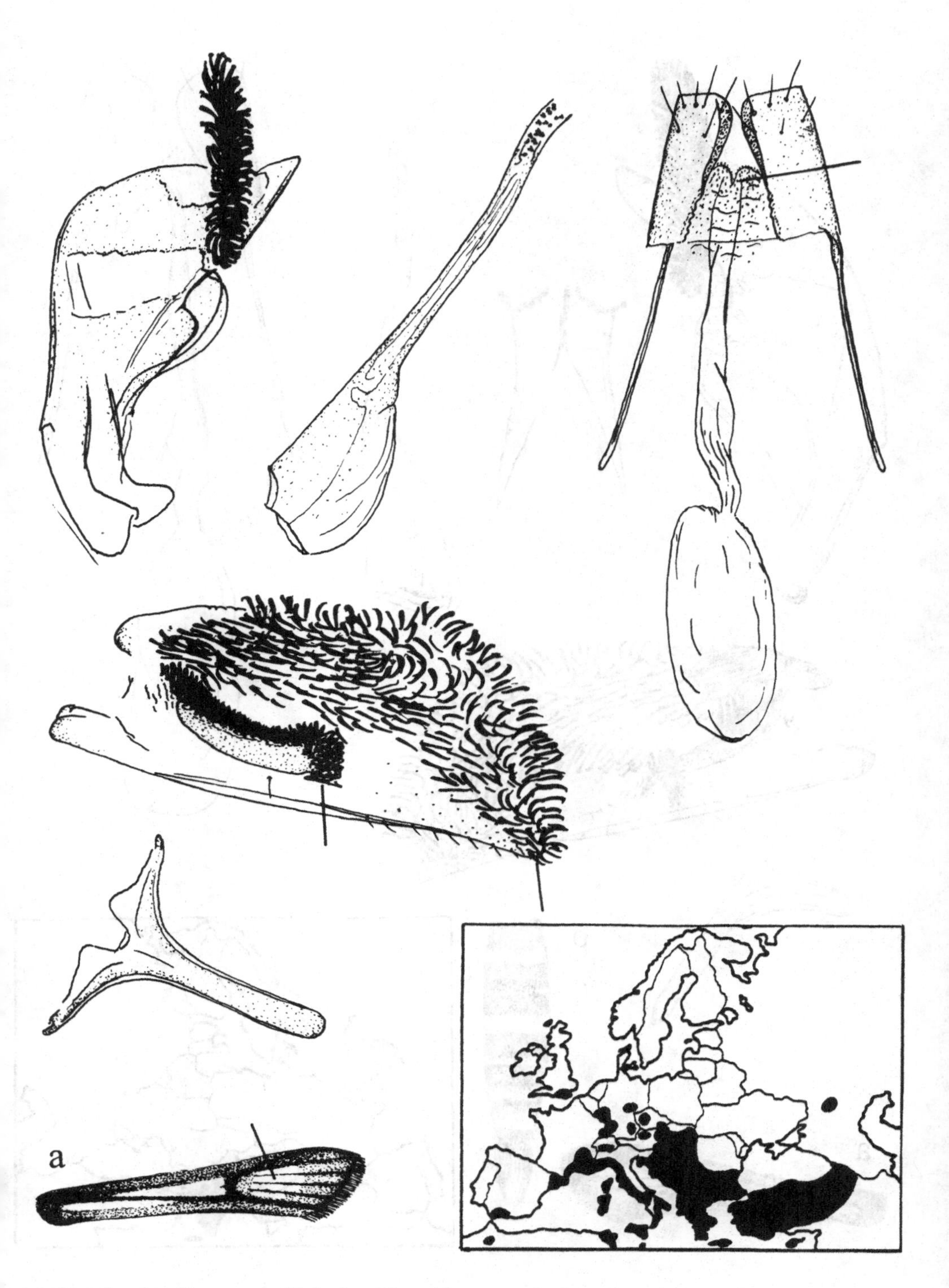

44. *Bembecia albanensis* (Rebel) – Slovakia, a: male forewing (ZL).

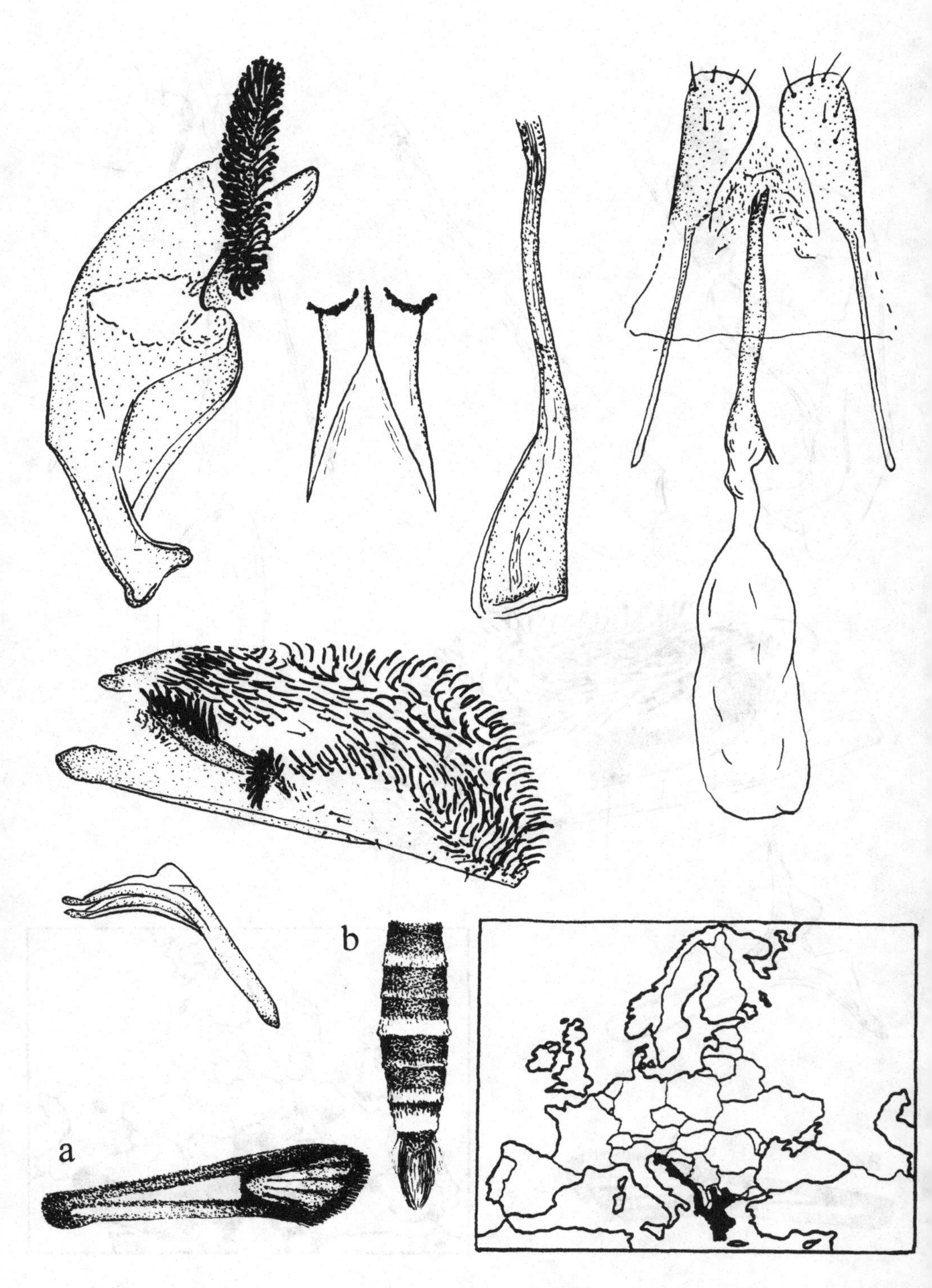

45. *Bembecia pavicevici* Toševski – Croatia, a: male forewing, b: male abdomen (ZL).

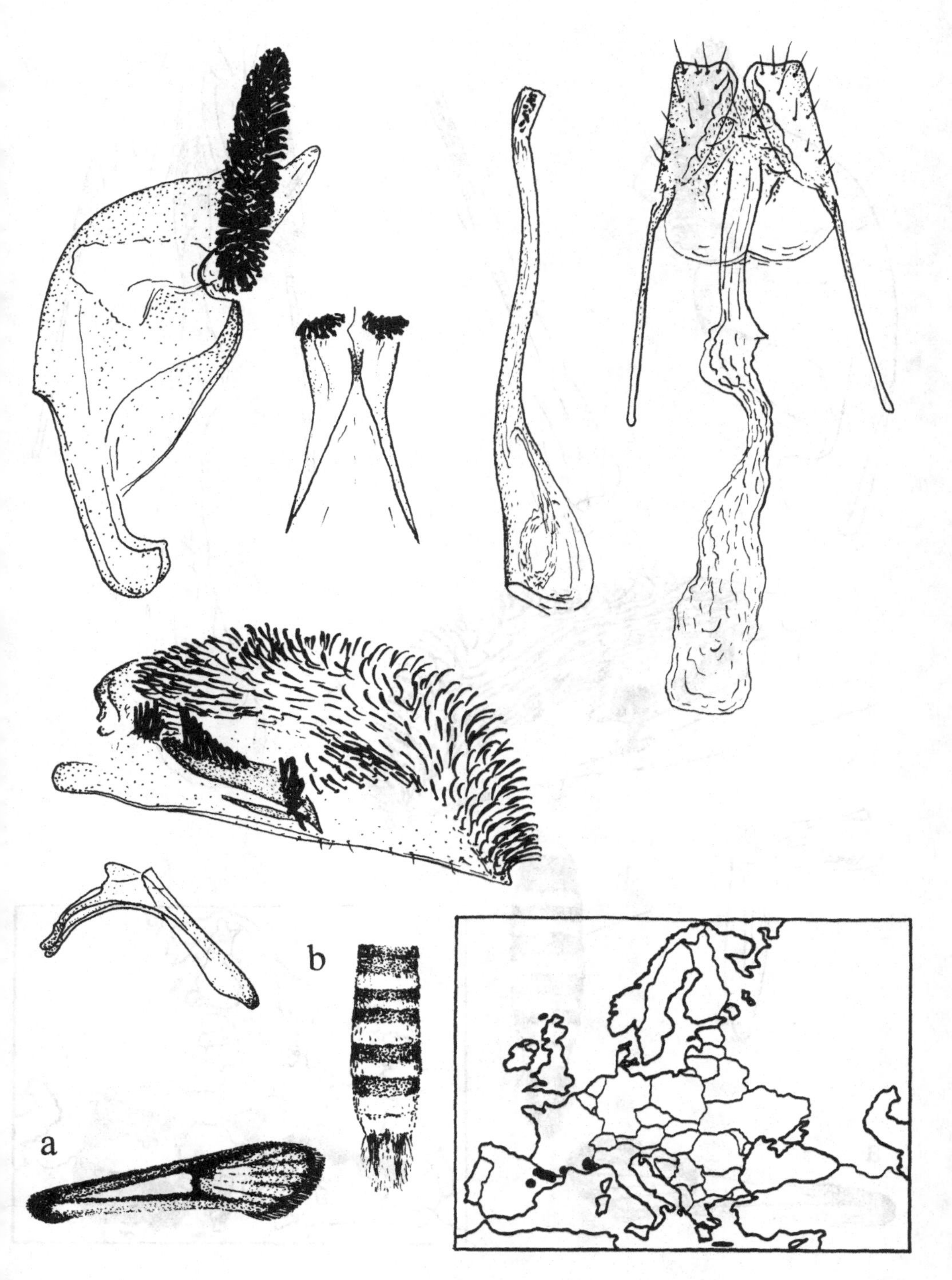

46. *Bembecia fibigeri* Laštůvka & Laštůvka – Spain, a: male forewing, b: male abdomen (ZL).

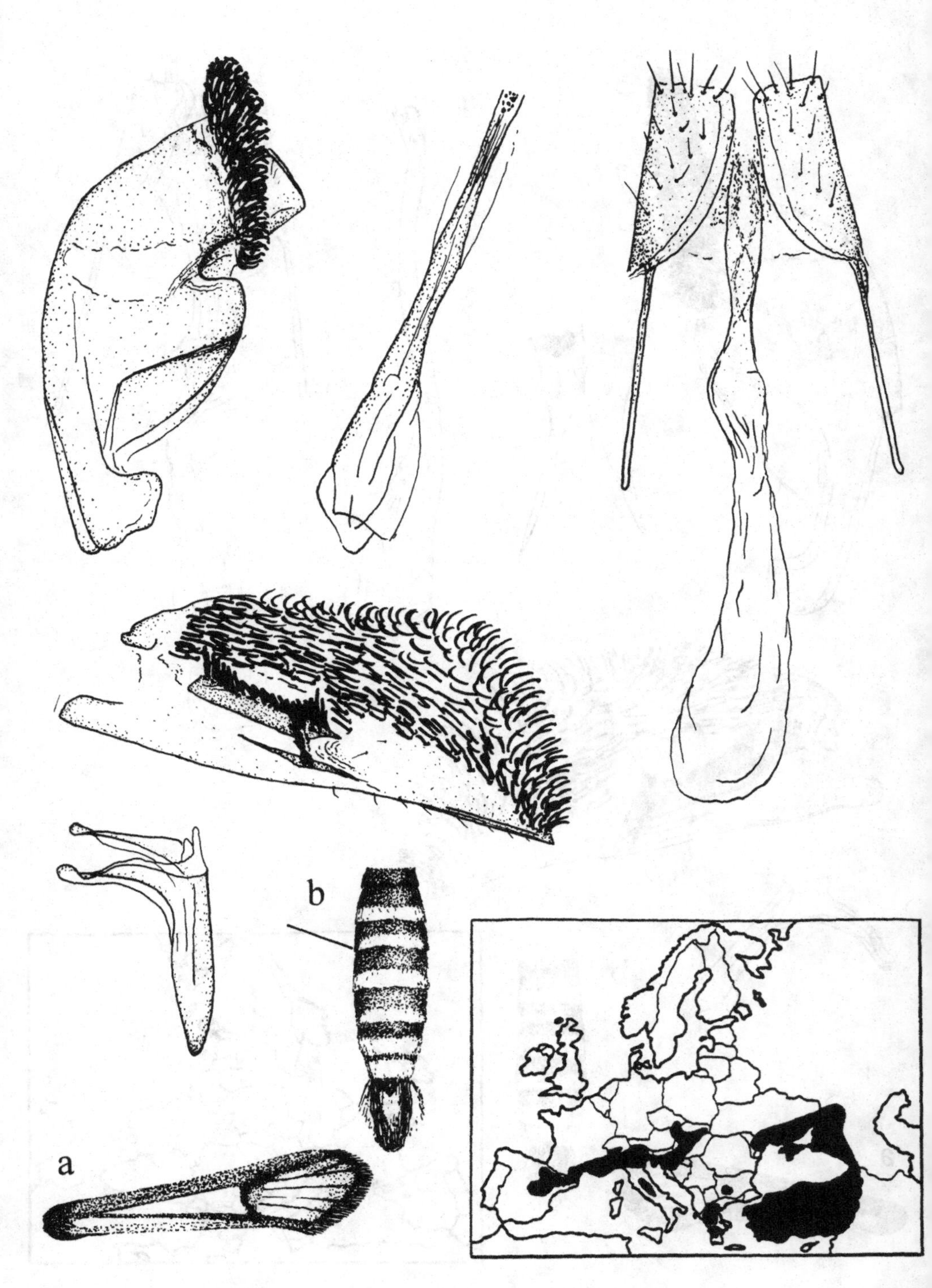

47. *Bembecia scopigera* (Scopoli) – Czech Republic, a: male forewing, b: male abdomen (ZL).

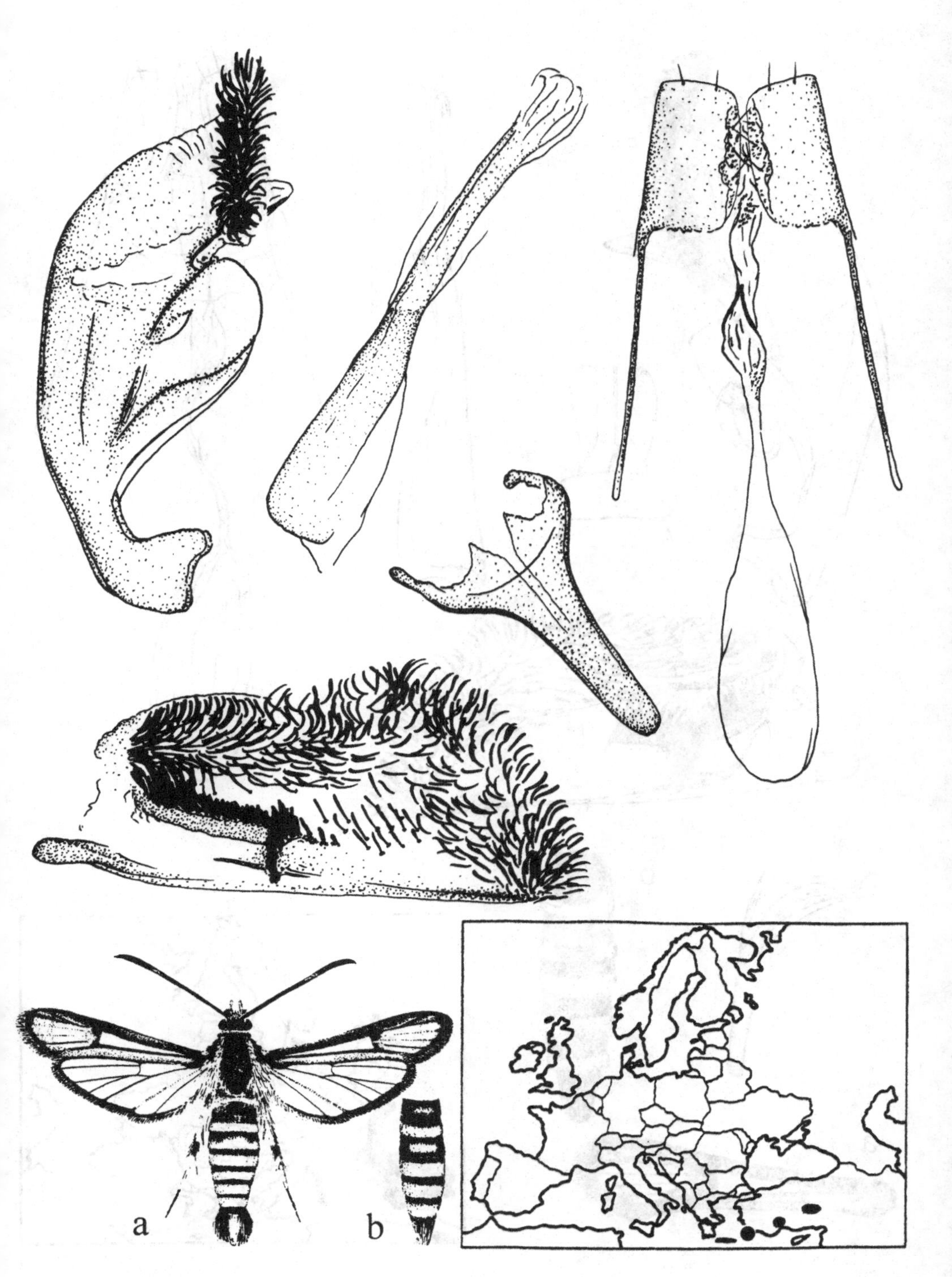

48. *Bembecia priesneri* Kallies, Petersen & Riefenstahl – Turkey, a: ♂, paratype, b: female abdomen (KS).

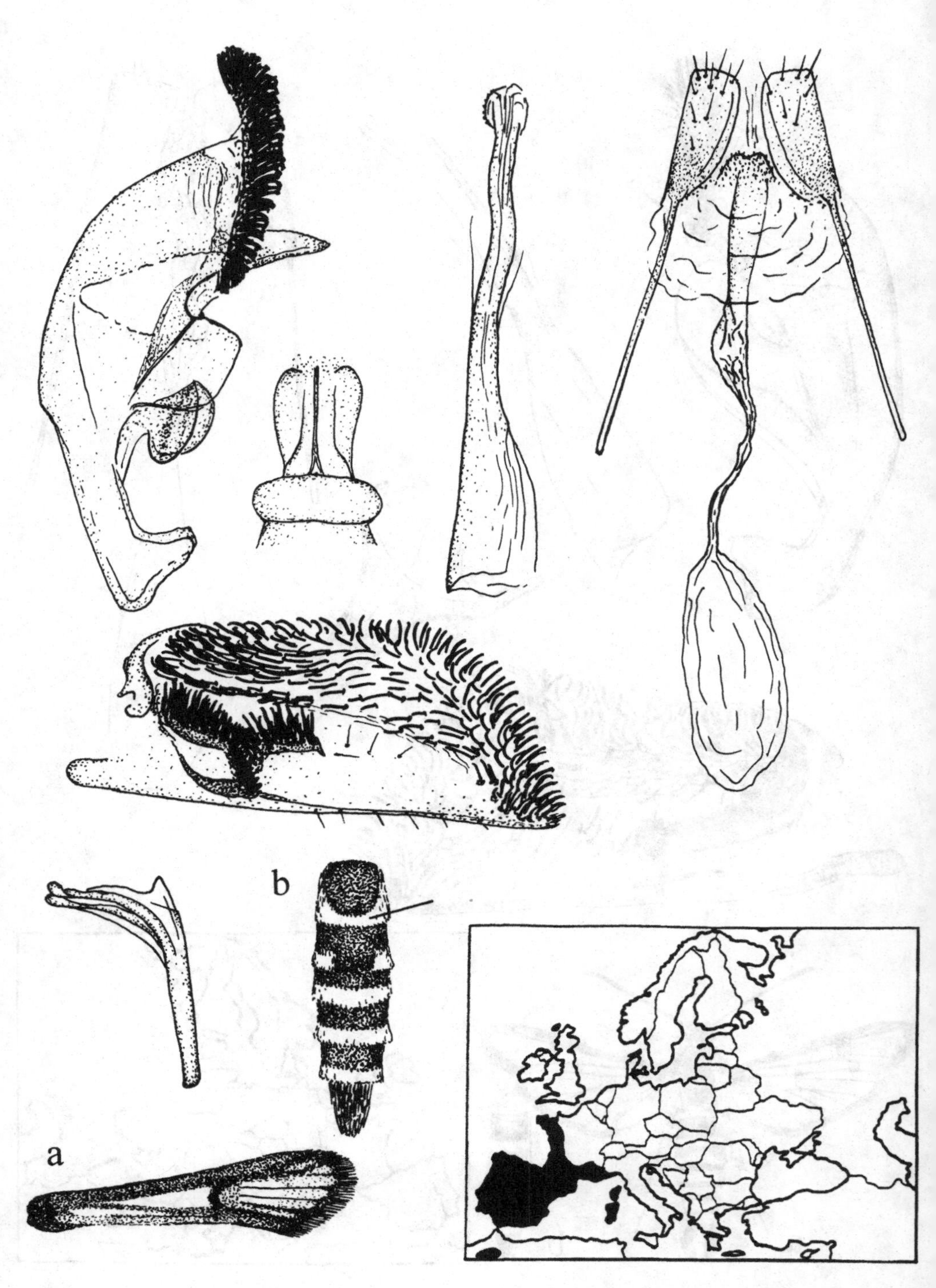

49. *Bembecia iberica* Špatenka – Spain, a: male forewing, b: female abdomen in ventral view (ZL).

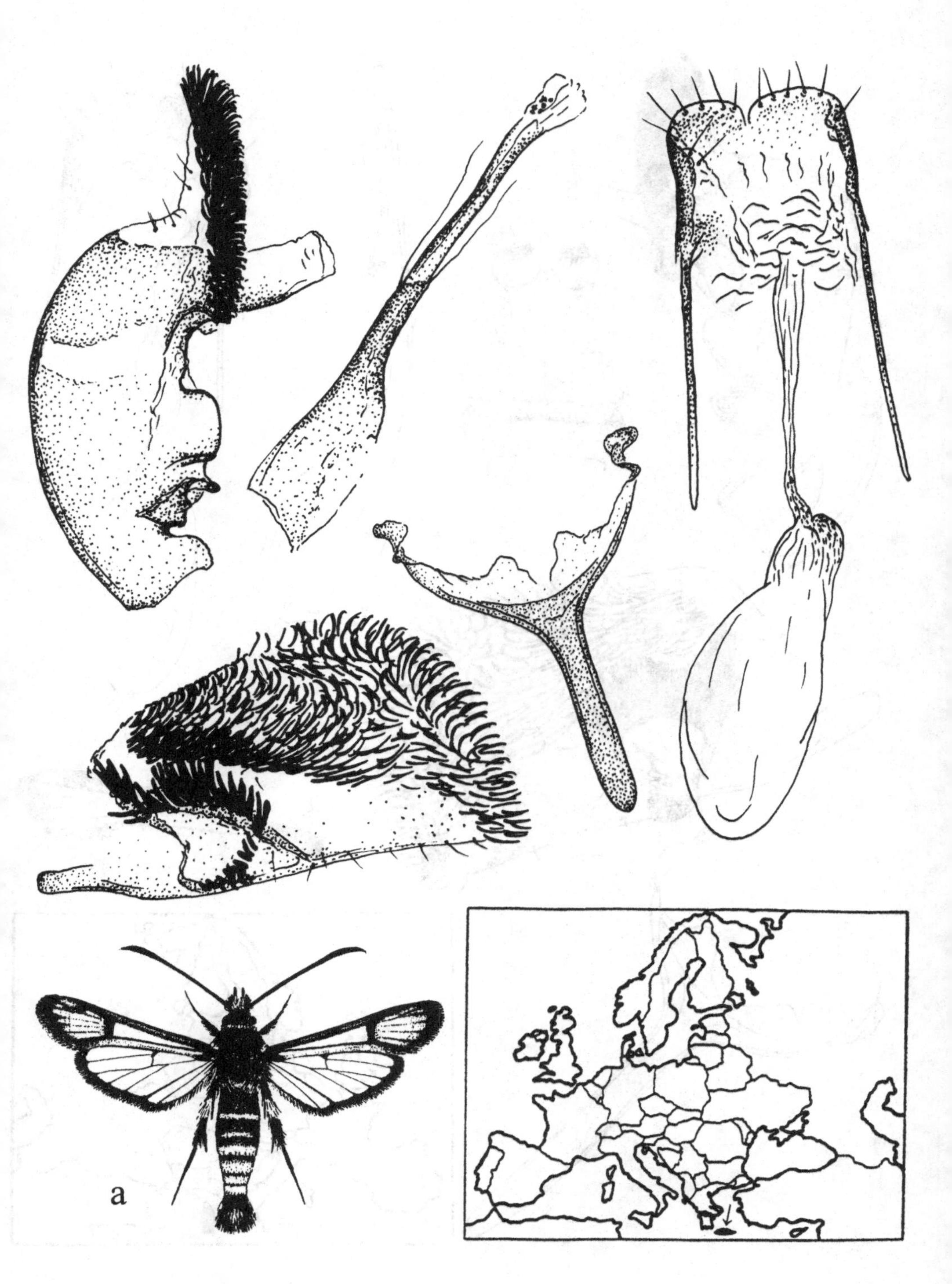

50. *Bembecia blanka* Špatenka – Crete, a: ♂ (KS).

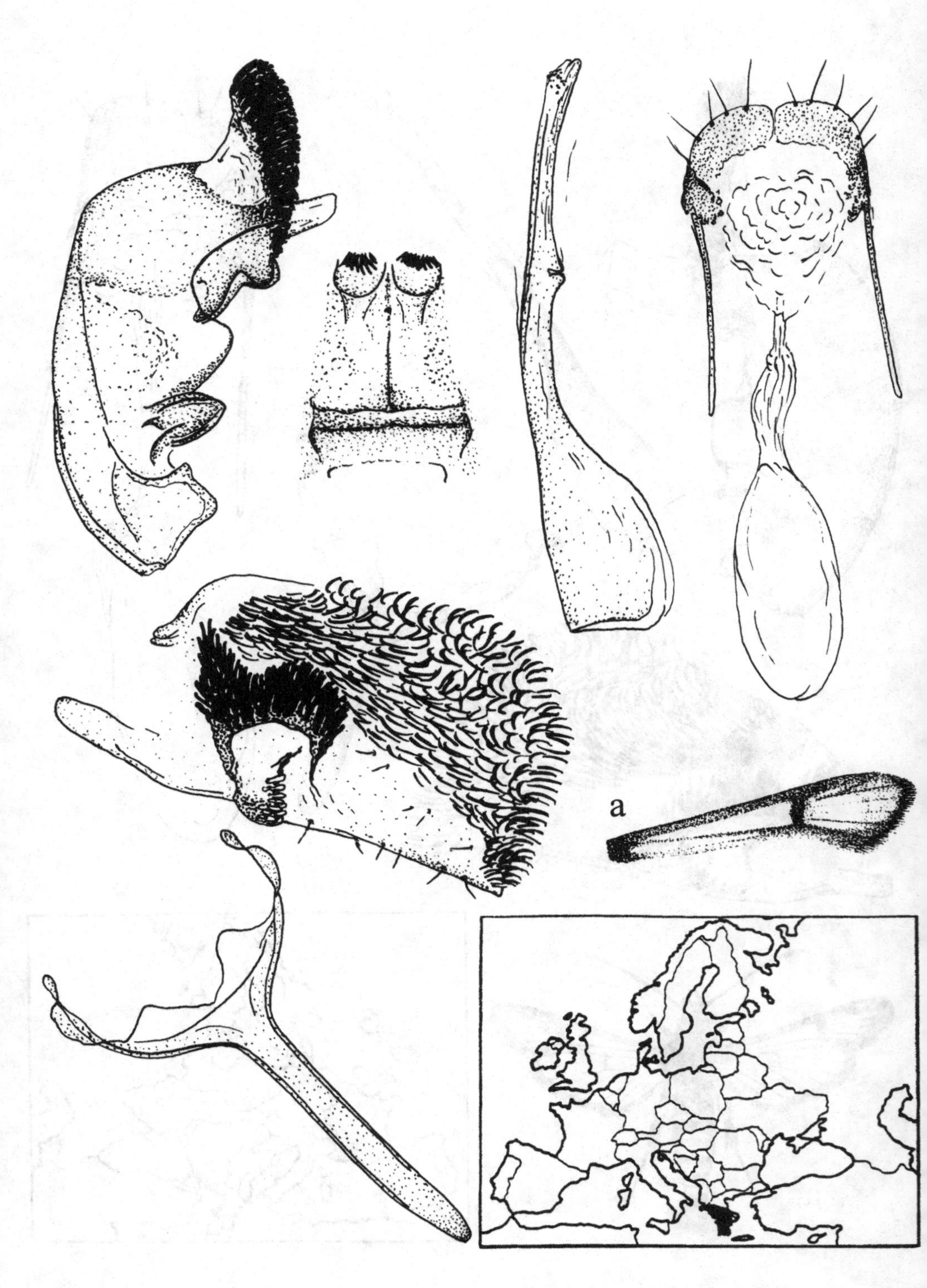

51. *Bembecia fokidensis* Toševski – Greece (♂ ZL, ♀ KS), a: male forewing (ZL).

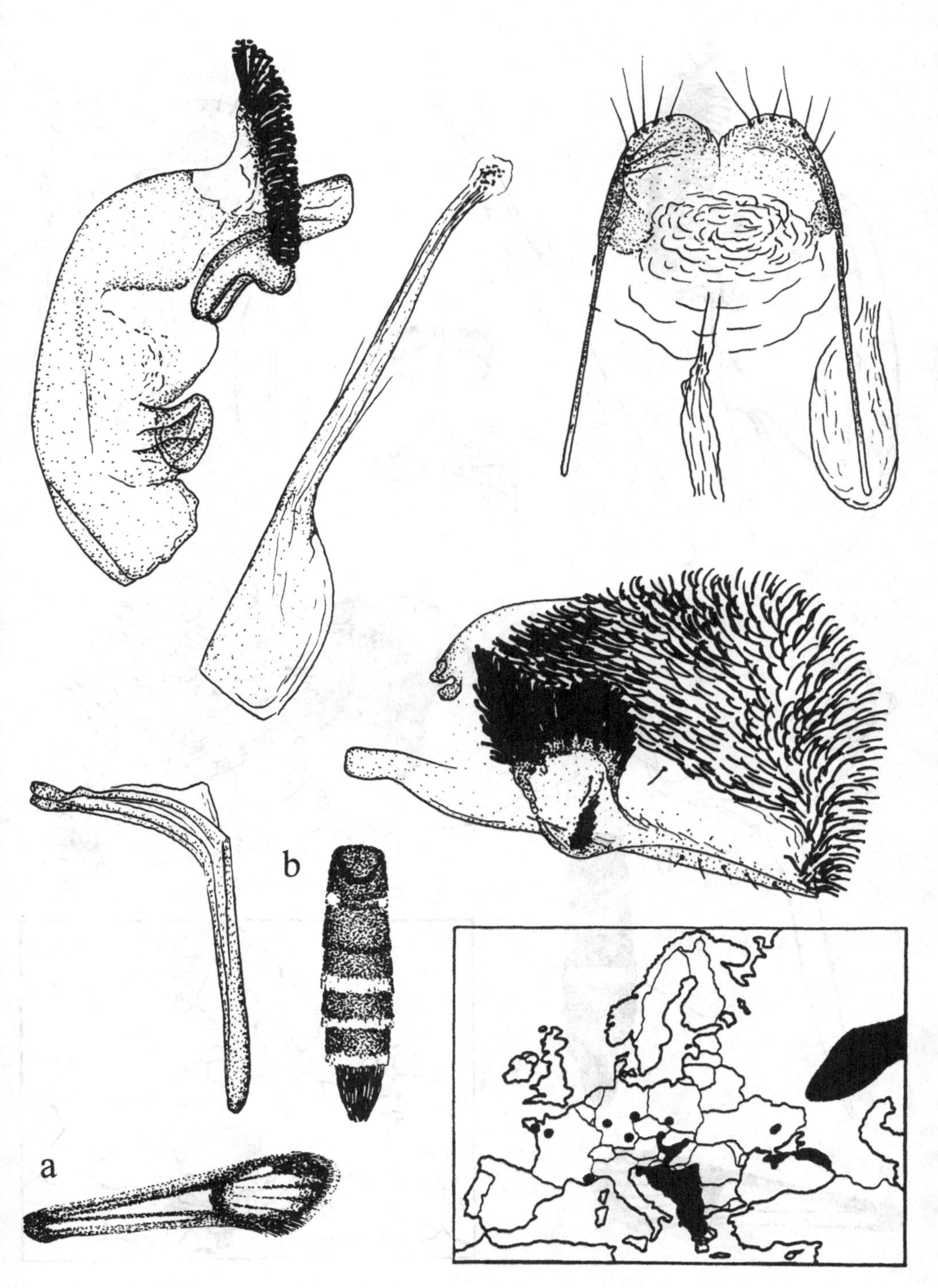

52. *Bembecia megillaeformis* (Hübner) – Slovakia, a: male forewing, b: female abdomen in ventral view (ZL).

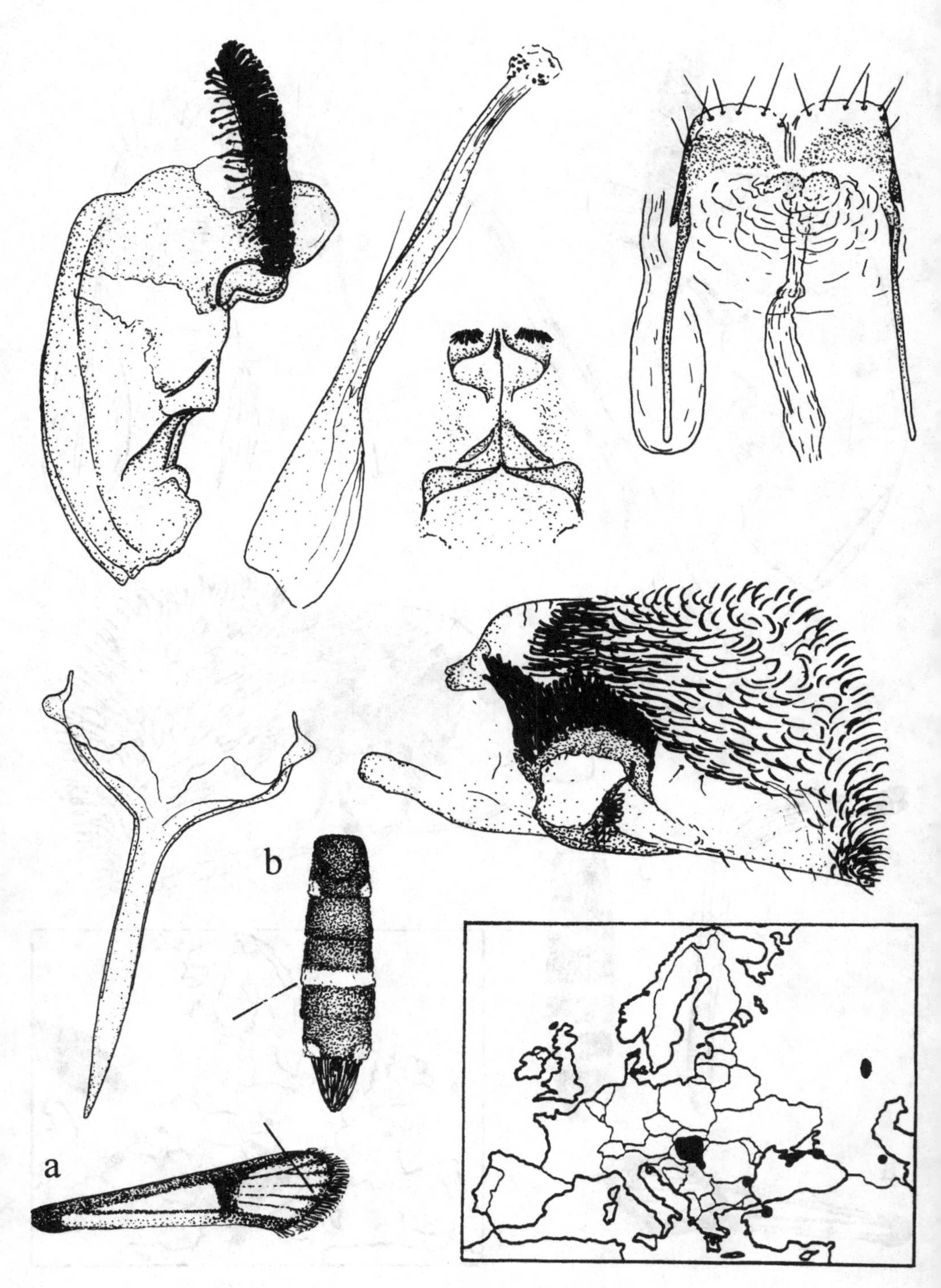

53. *Bembecia puella* Laštůvka – Slovakia, a: male forewing, b: female abdomen in ventral view (ZL).

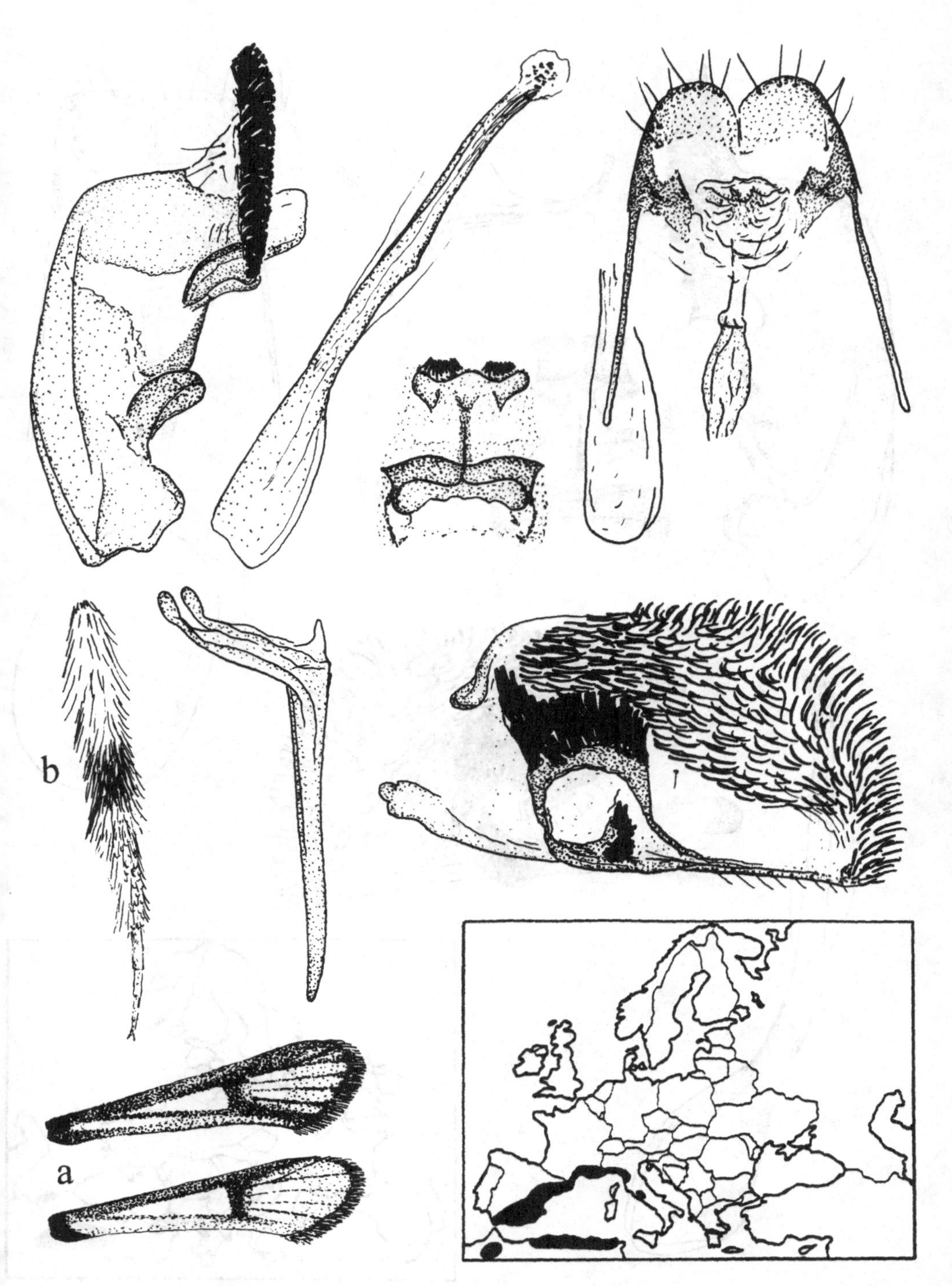

54. *Bembecia sirphiformis* (Lucas) – Spain, a: variability of forewing, b: hind tibia (ZL).

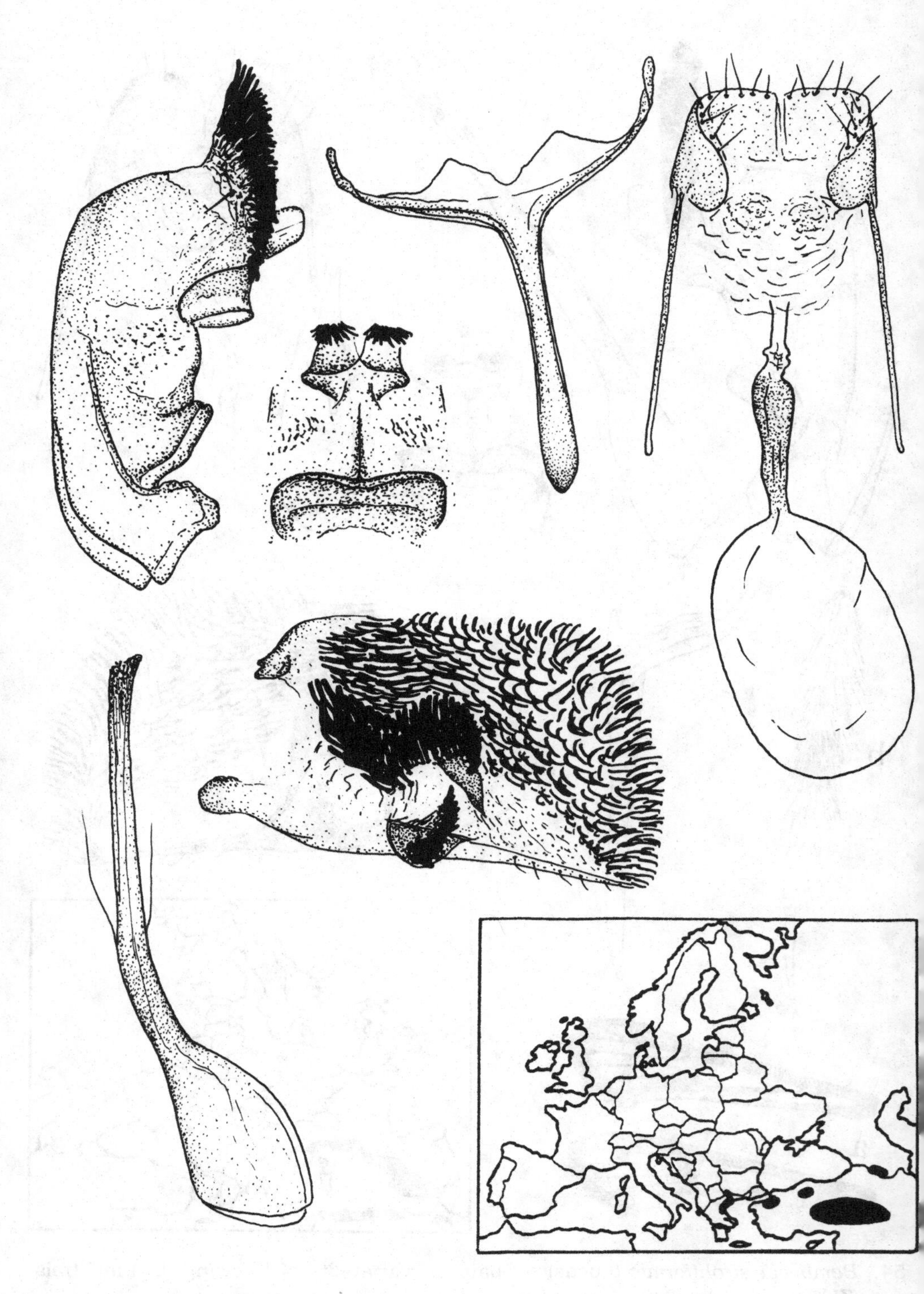

55. *Bembecia sanguinolenta* (Lederer) – Turkey (ZL, ex KS and M. Petersen).

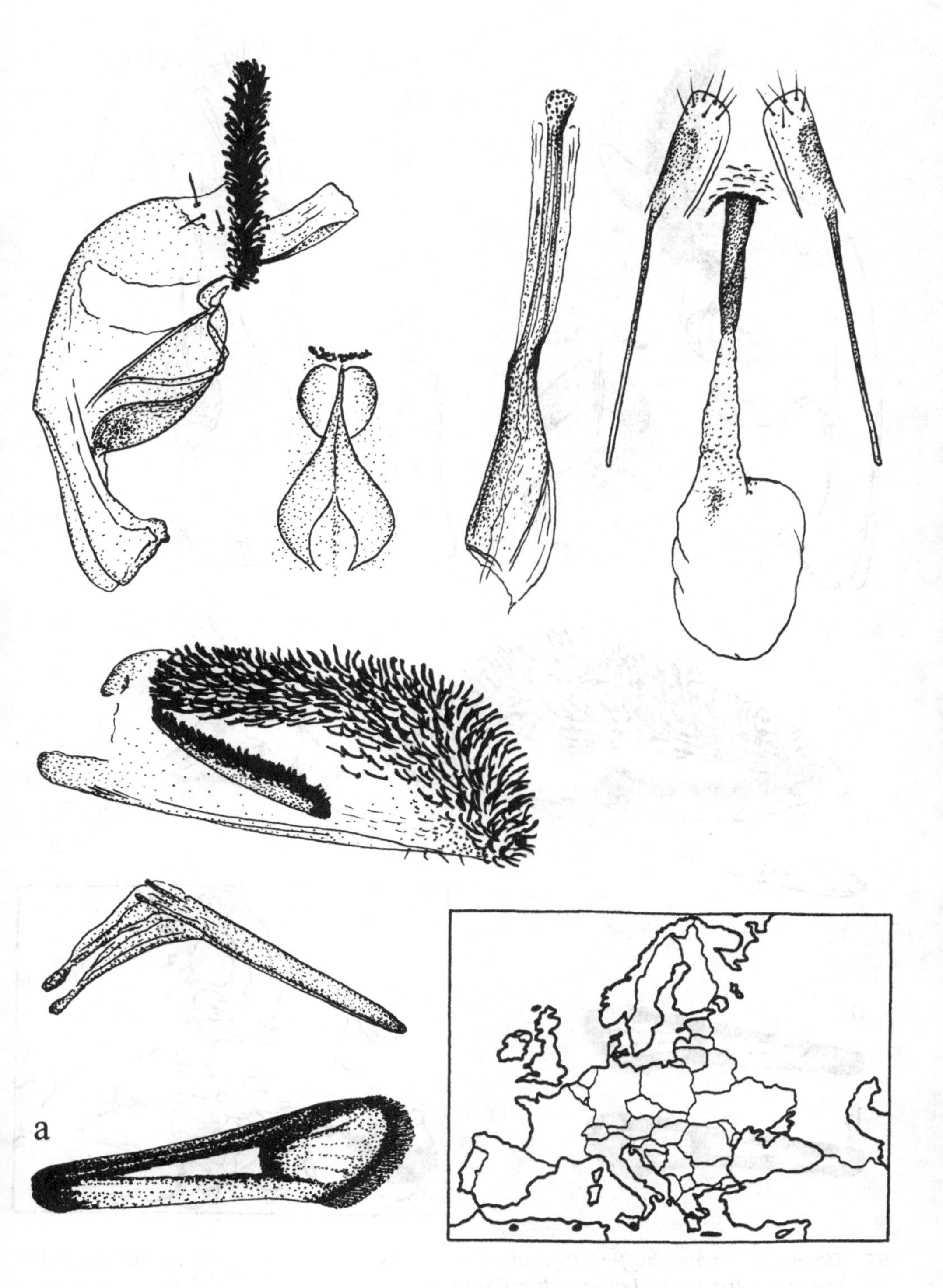

56. *Bembecia flavida* (Oberthür) – Sicily, a: male forewing (ZL).

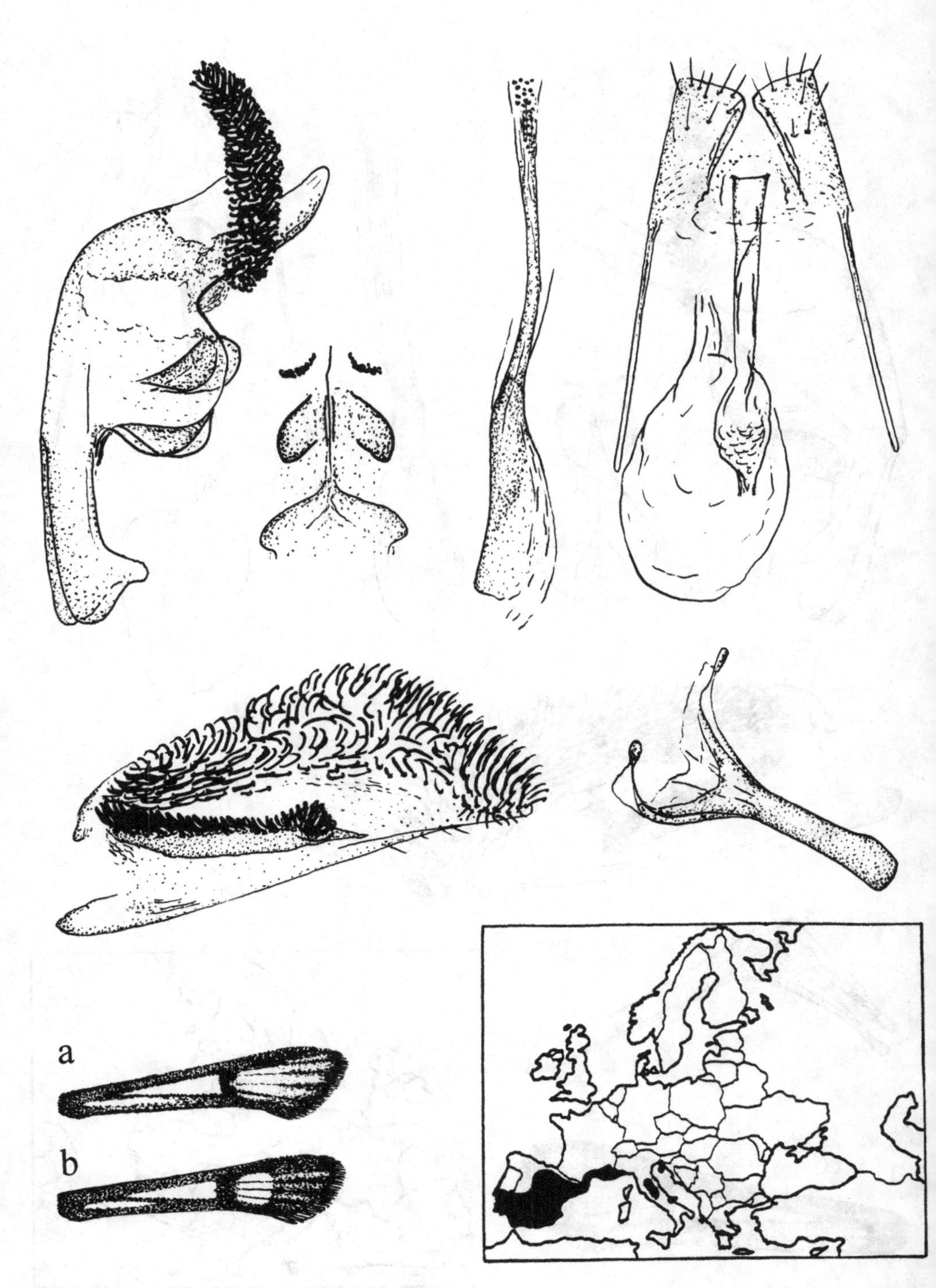

57. *Bembecia himmighoffeni* (Staudinger) – Spain, a: male forewing, b: male forewing of the form "*baumgartneri*" (ZL).

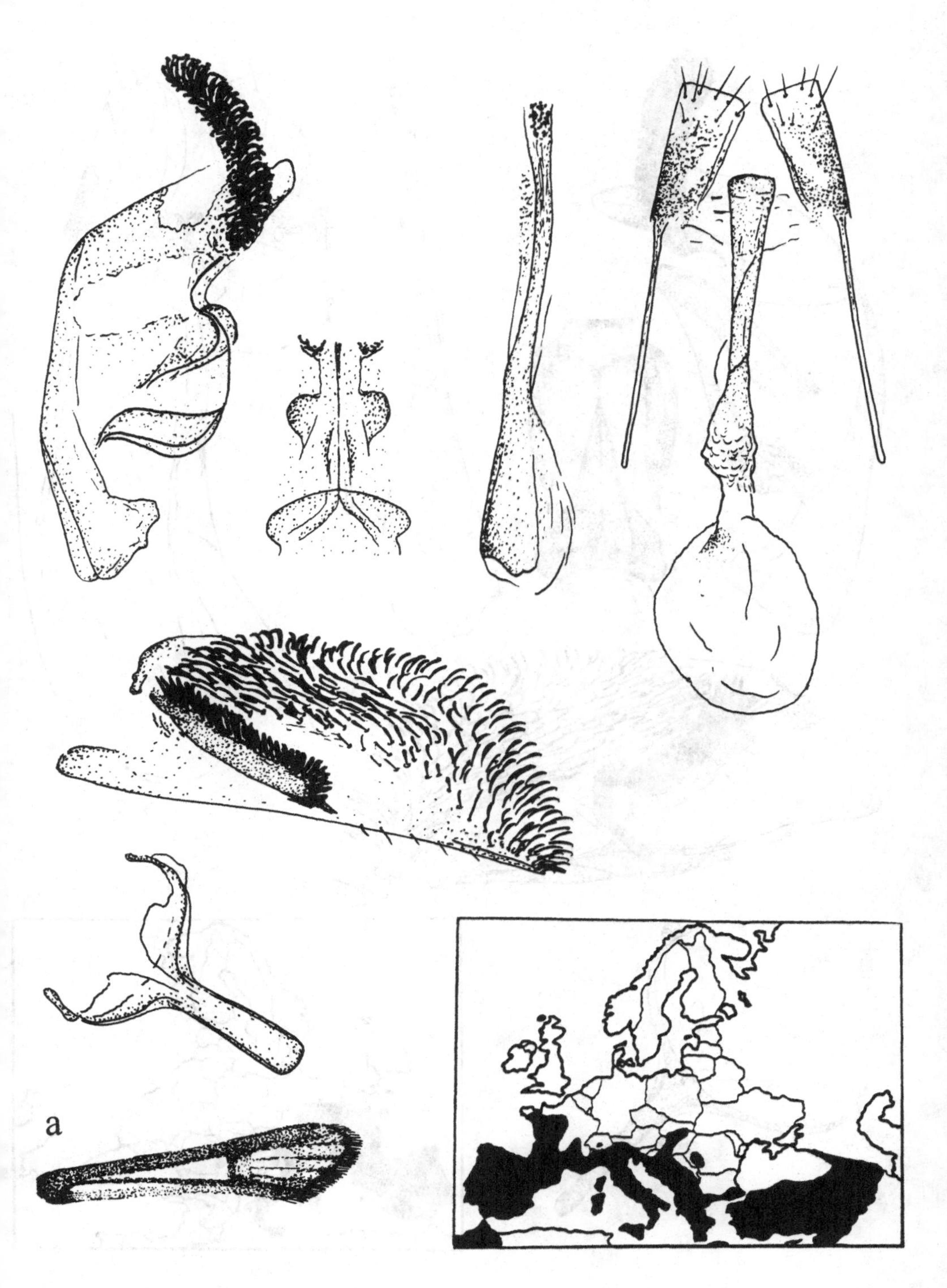

58. *Bembecia uroceriformis* (Treitschke) – Slovakia, a: male forewing (ZL).

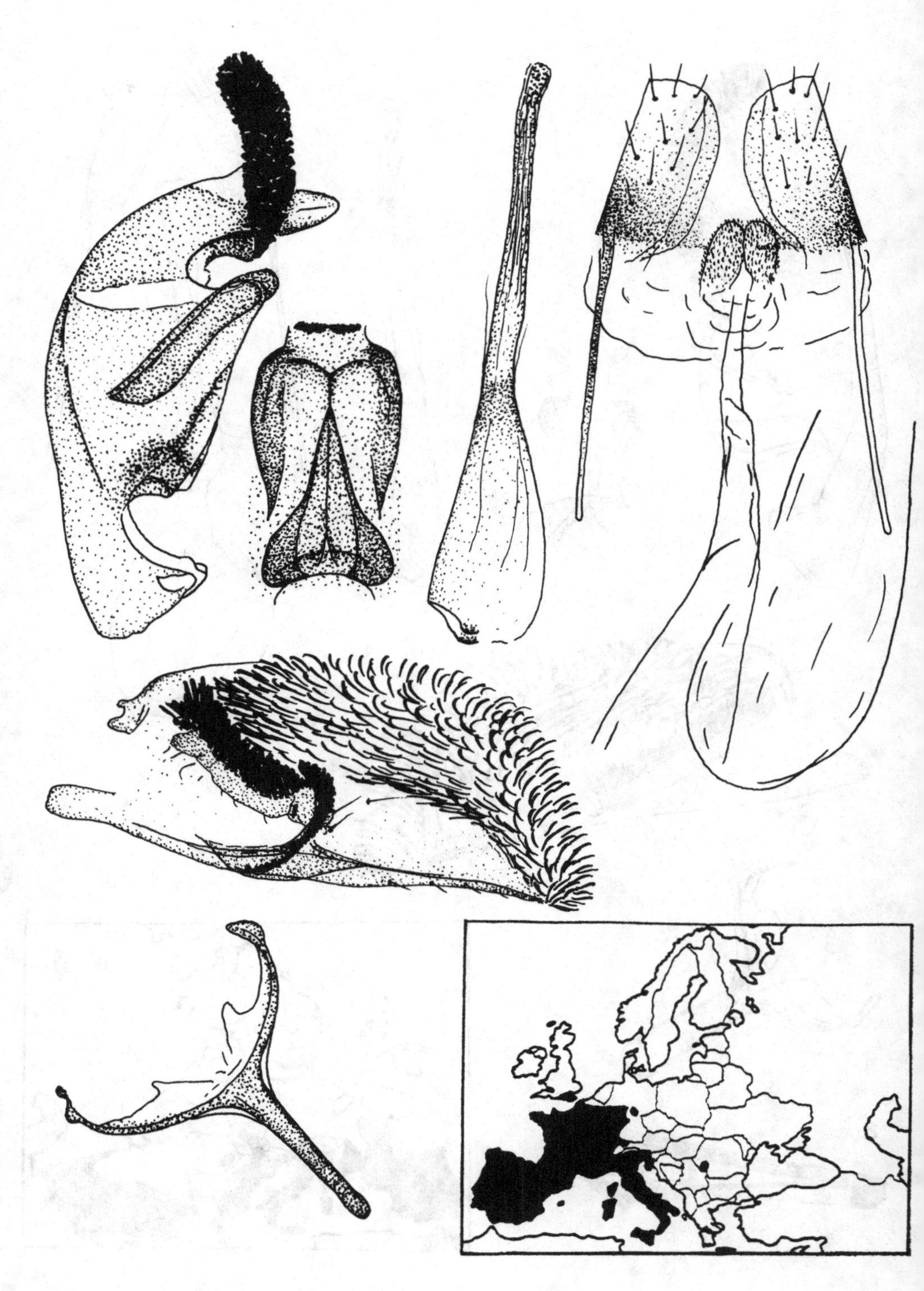

59. *Pyropteron chrysidiformis* (Esper) – Spain (ZL).

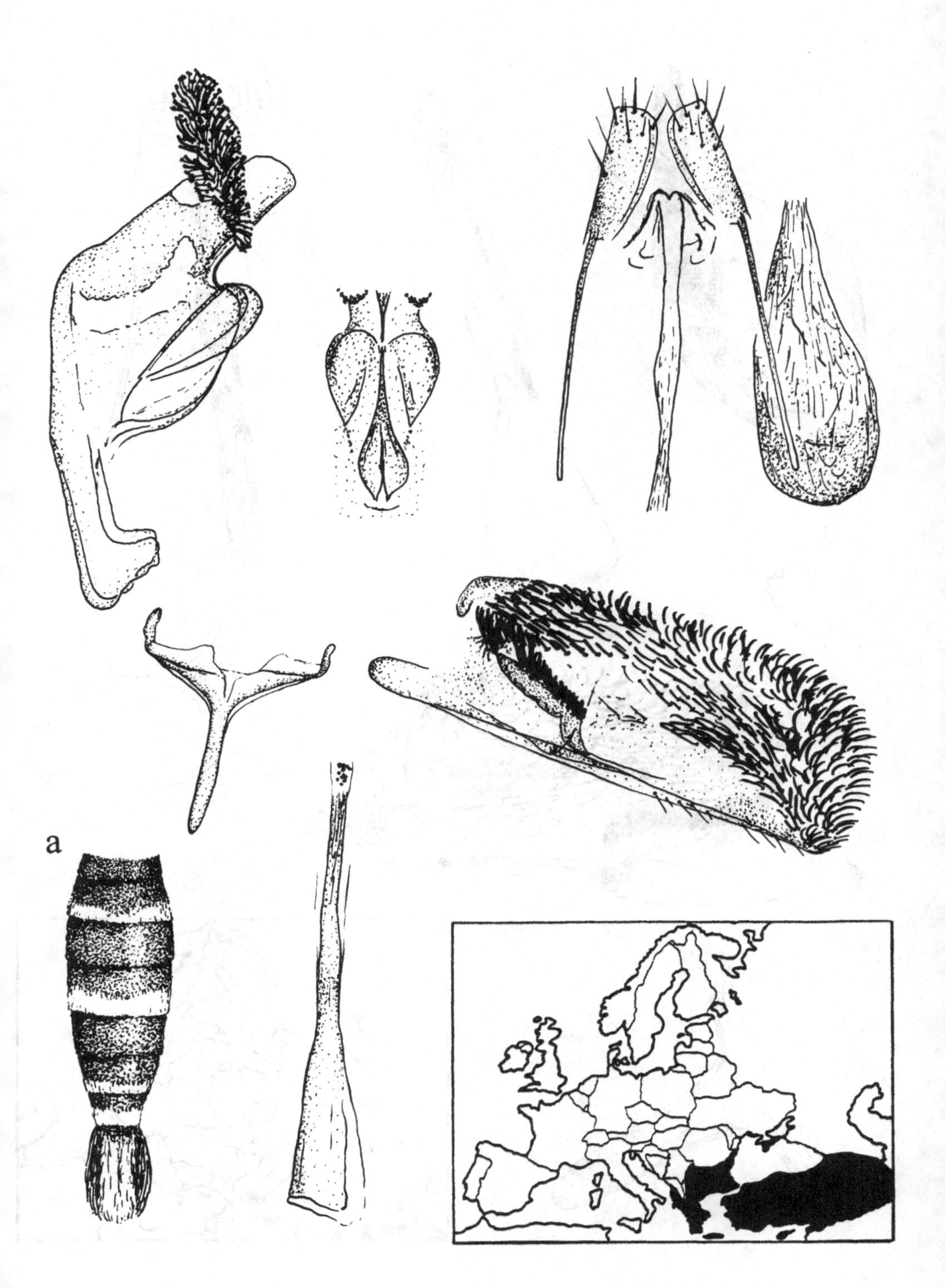

60. *Pyropteron minianiformis* (Freyer) – Bulgaria, a: male abdomen (ZL).

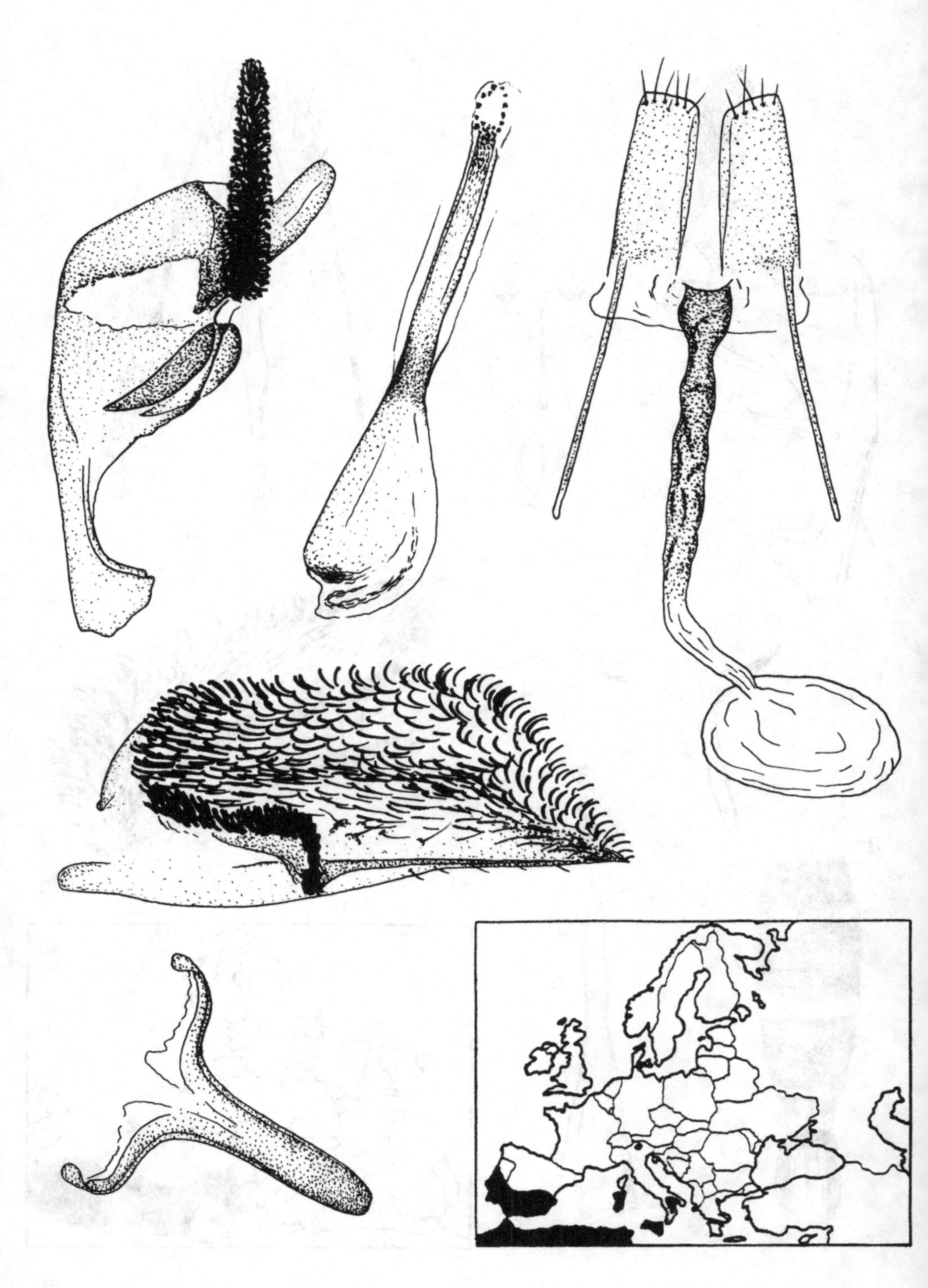

61. *Pyropteron doryliformis* (Ochsenheimer) – Spain (ZL).

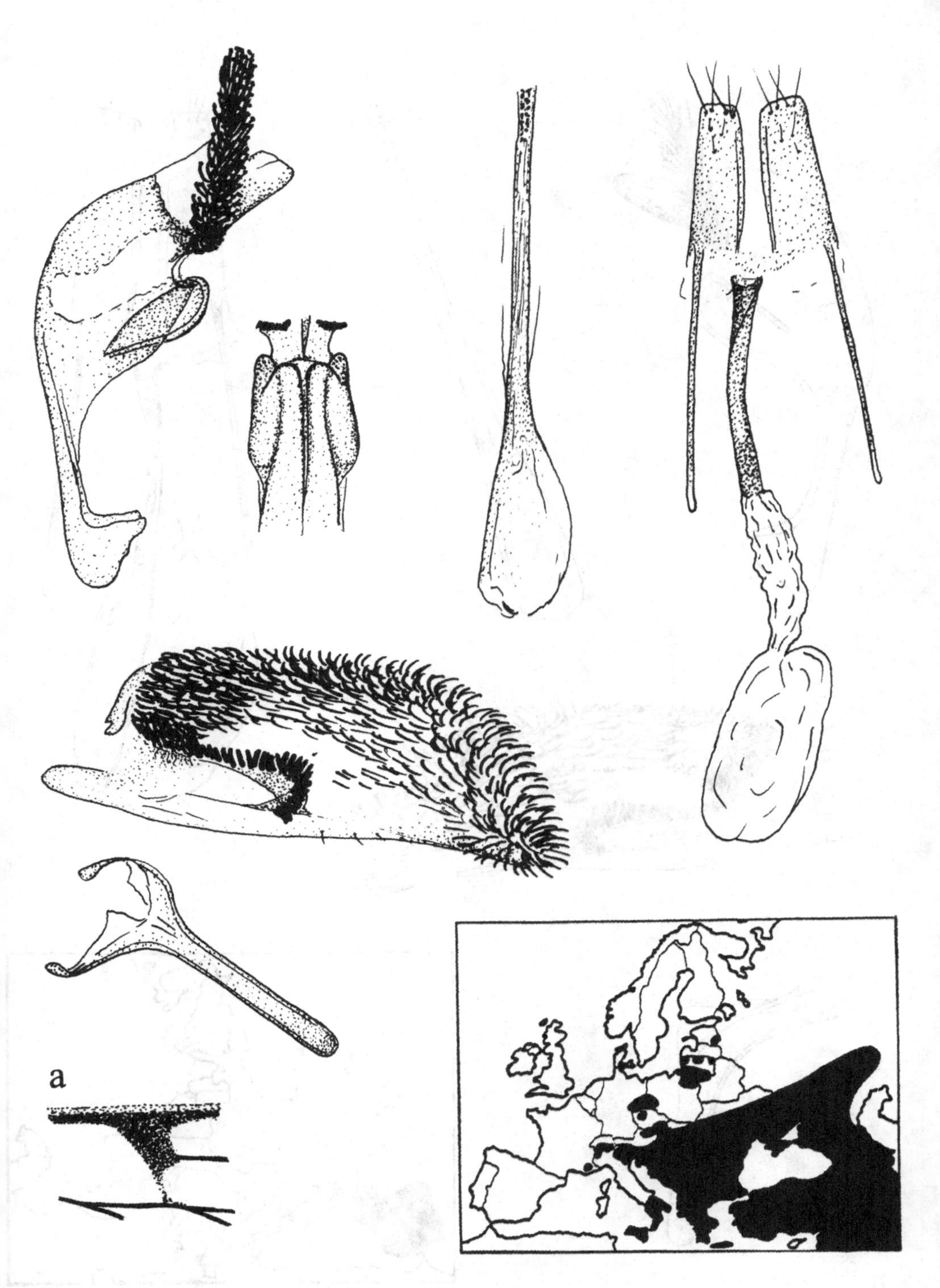

62. *Pyropteron triannuliformis* (Freyer) – Czech Republic, a: discal spot of hindwing (ZL).

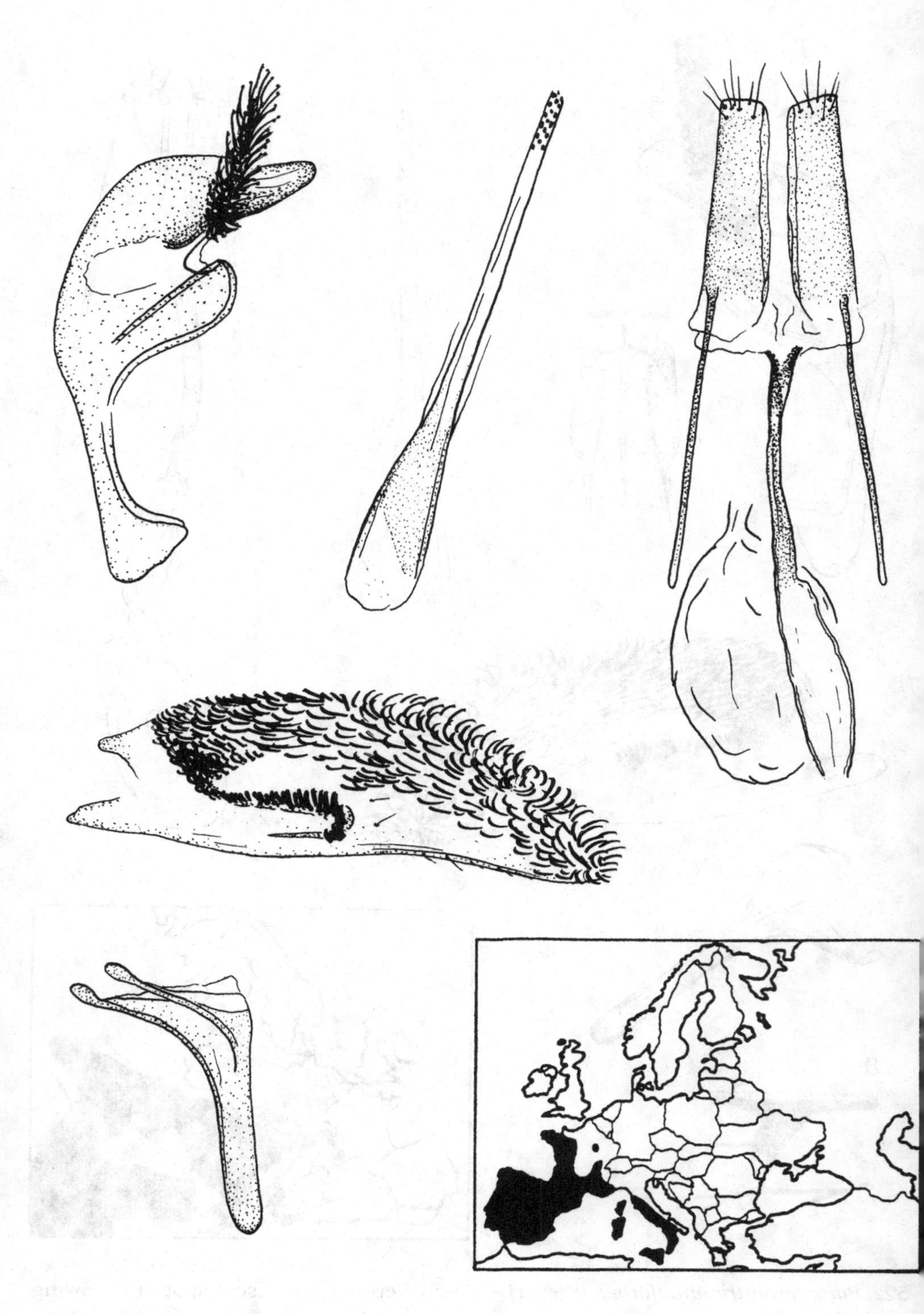

63. *Pyropteron meriaeformis* (Boisduval) – Spain (ZL).

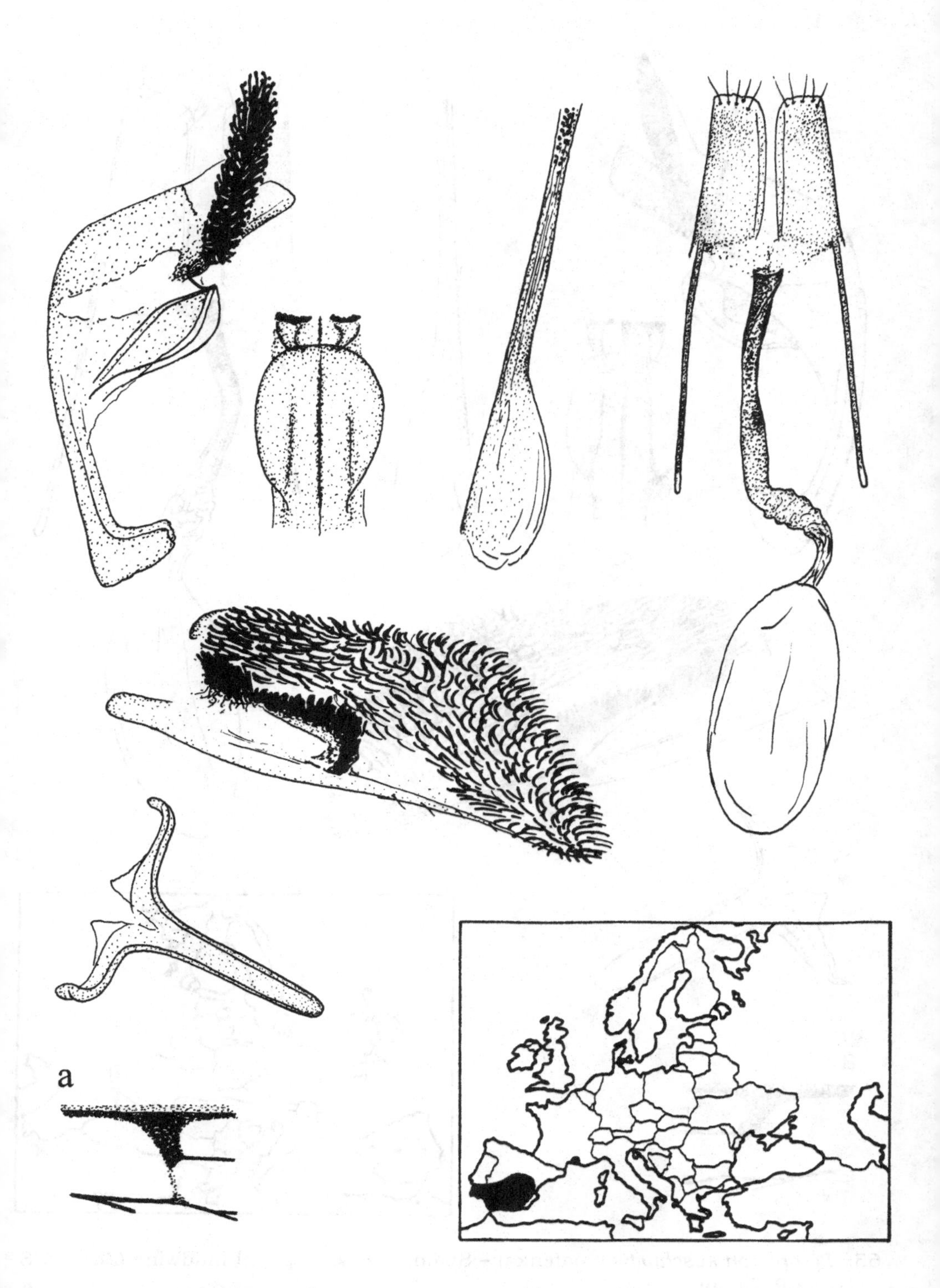

64. *Pyropteron hispanica* (Kallies) – Spain, a: discal spot of hindwing (ZL).

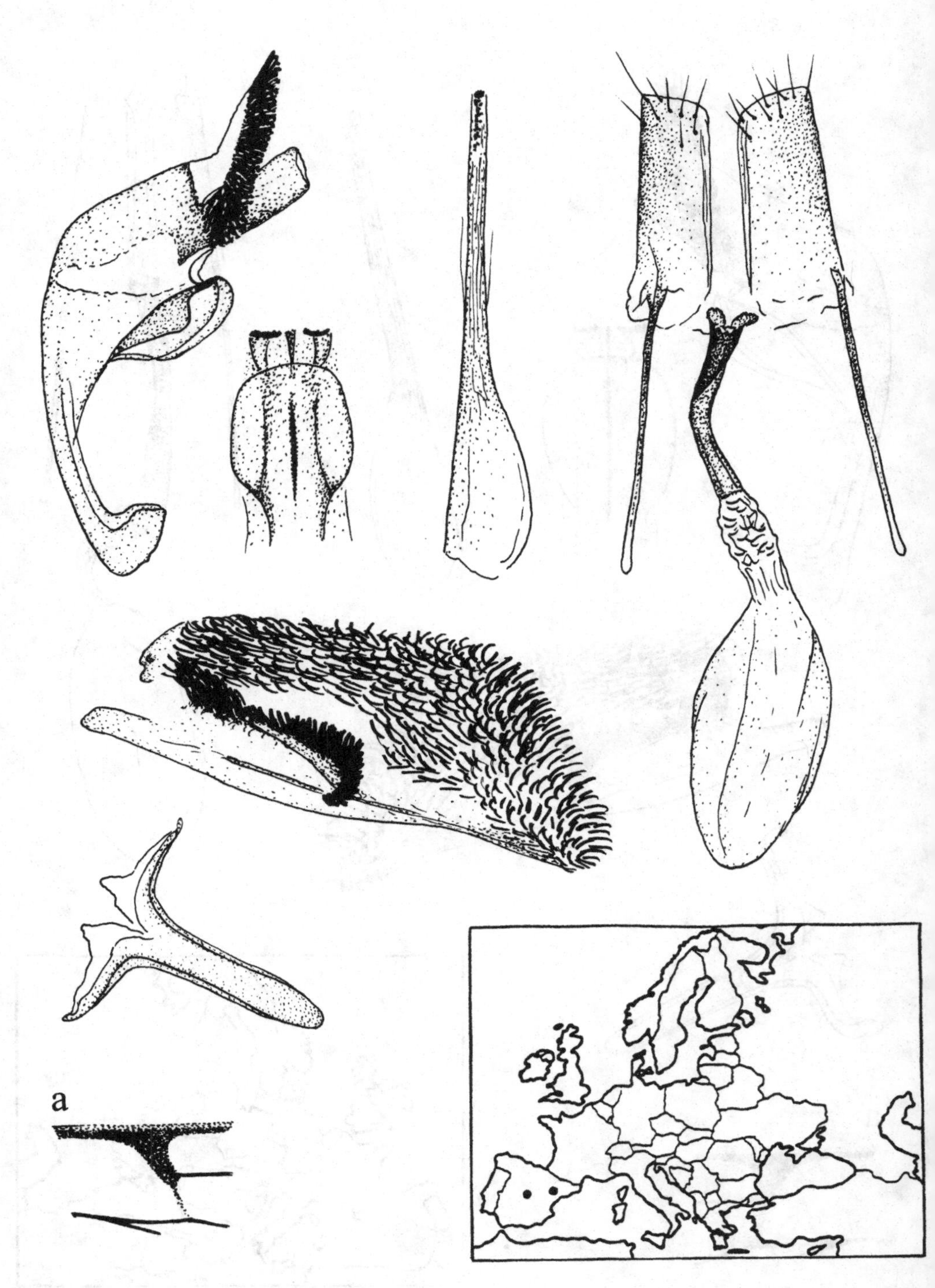

65. *Pyropteron koschwitzi* (Špatenka) – Spain, a: discal spot of hindwing (ZL, ex KS and R. Bläsius).

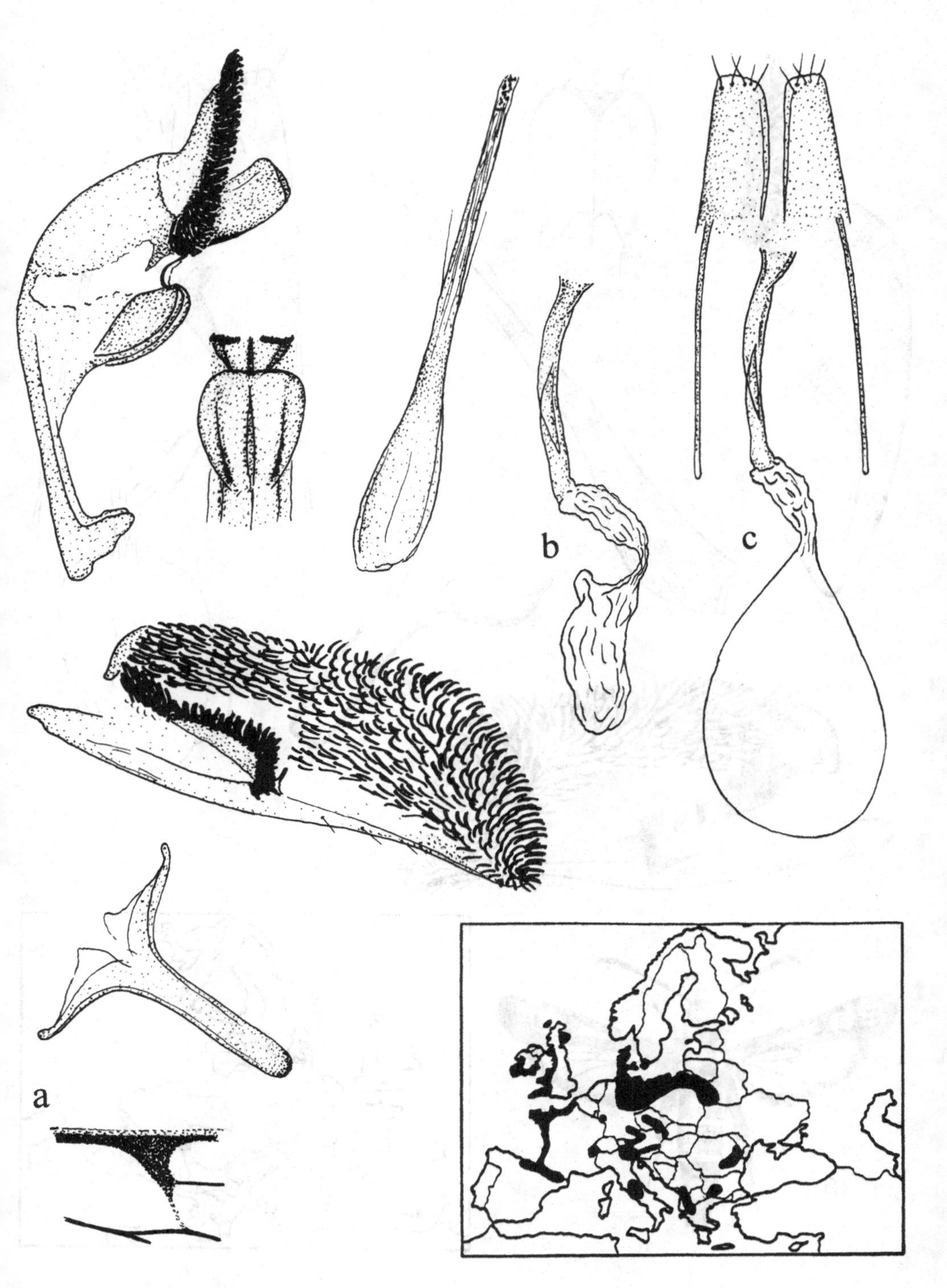

66. *Pyropteron muscaeformis* (Esper) – Slovakia, a: discal spot of hindwing, b: unfertilised ♀, c: fertilised ♀ (ZL).

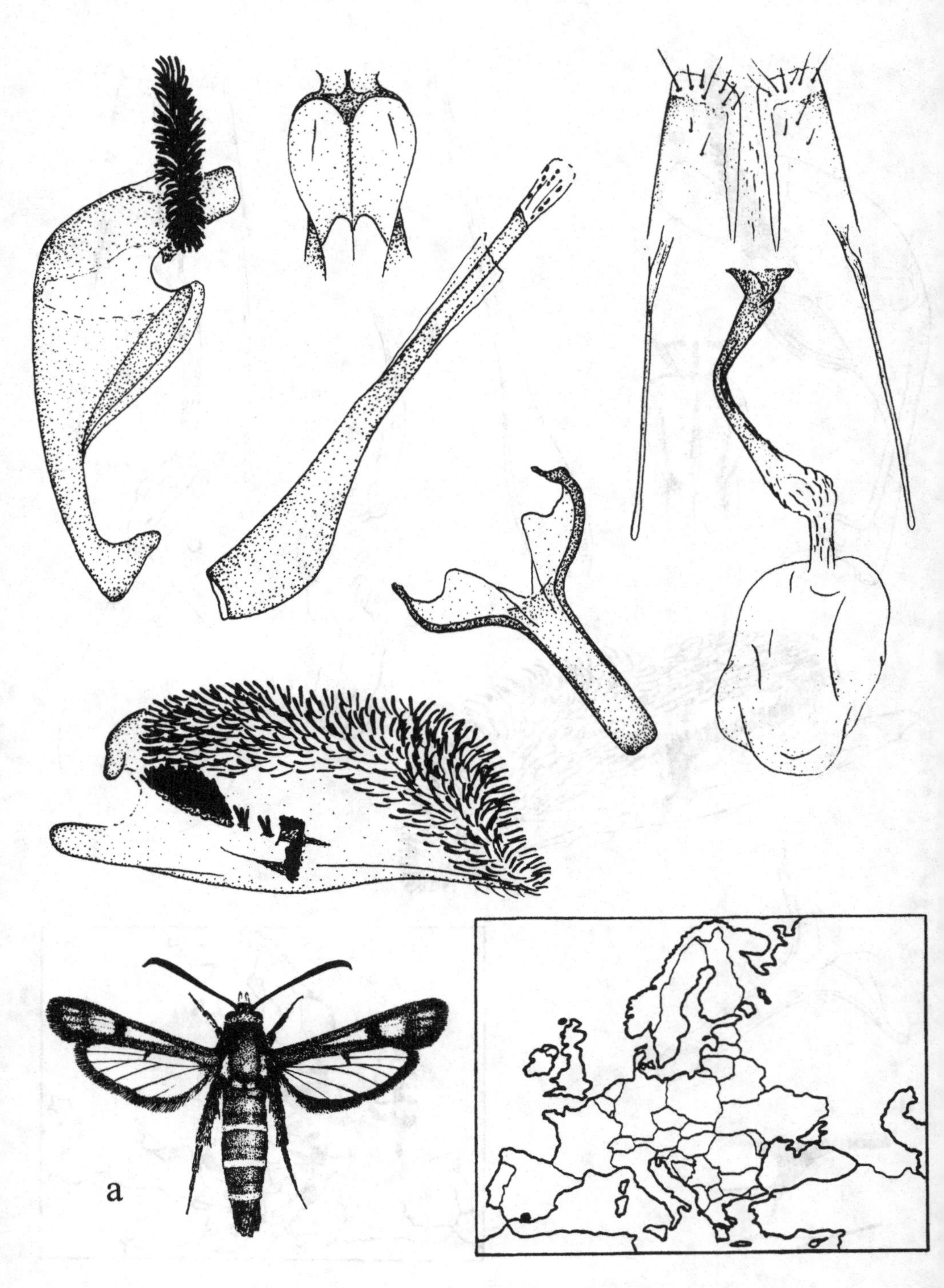

67. *Pyropteron kautzi* (Reisser) – Spain (♂ after Sobczyk in Pühringer, 1999, ♀, paralectotype, NHMV), a: ♀ (ditto).

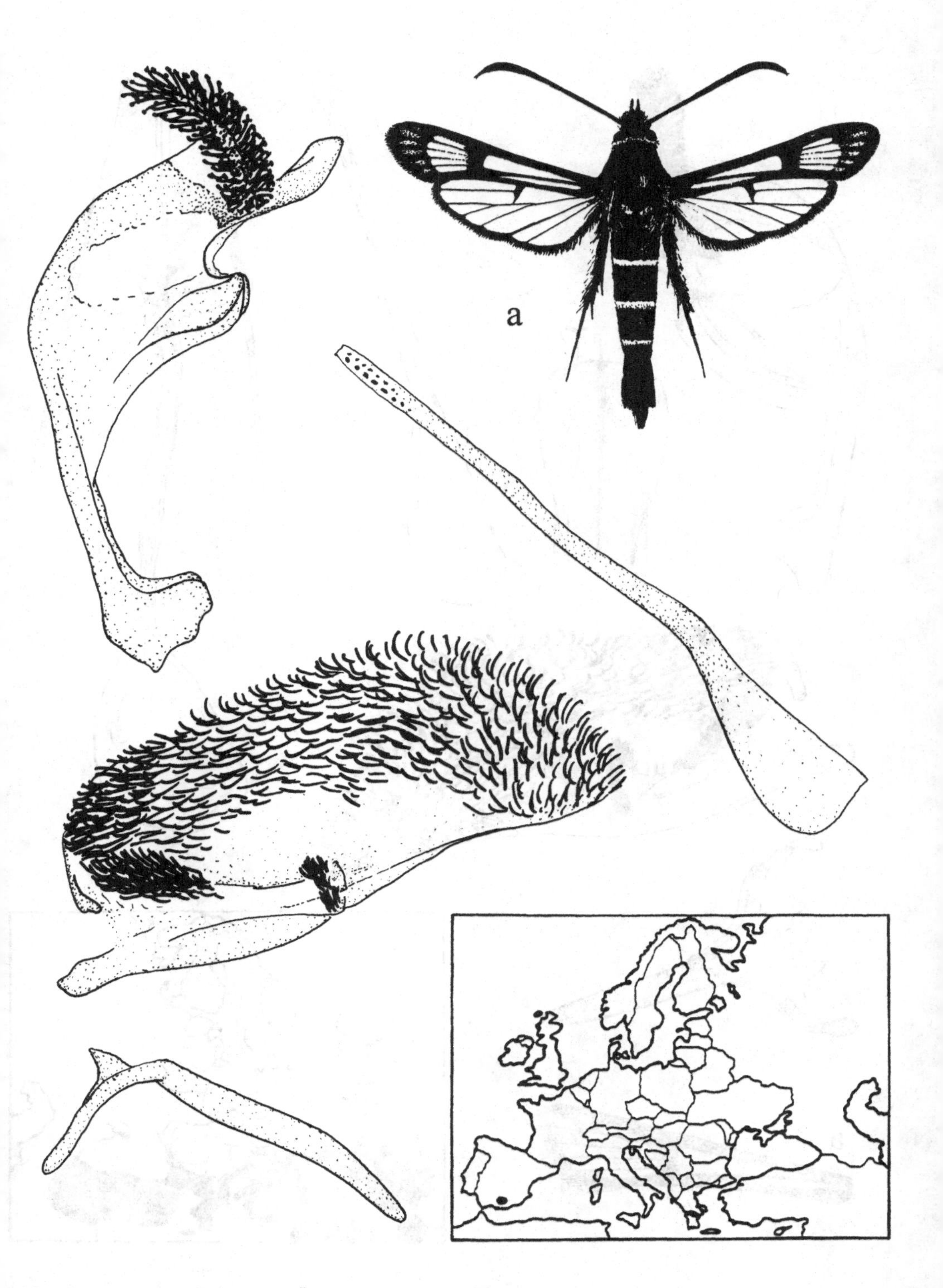

68. *Pyropteron aistleitneri* (Špatenka) – Spain (paratype, TW), a: ♂ (F. Pühringer).

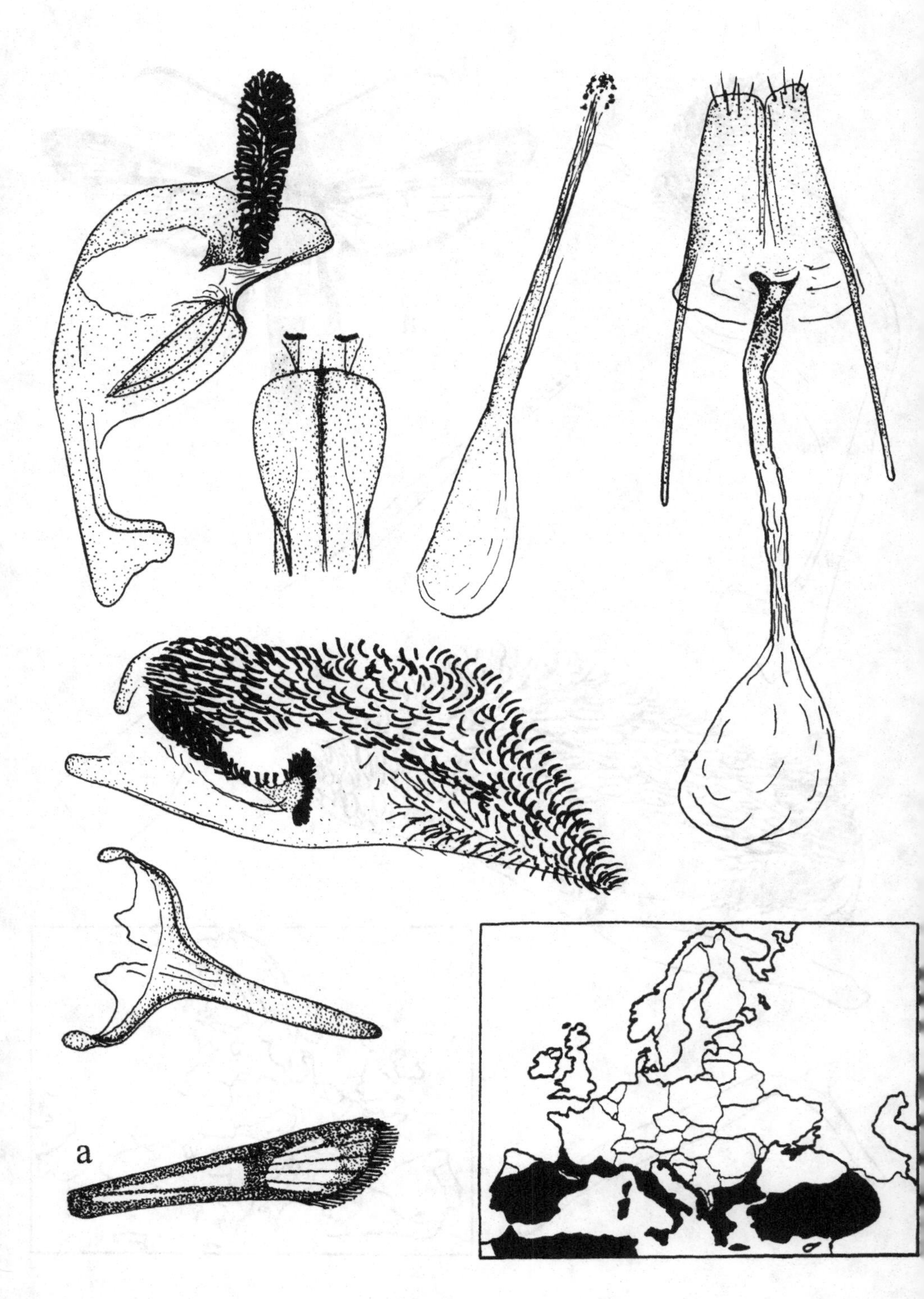

69. *Pyropteron leucomelaena* (Zeller) – Spain, a: male forewing (ZL).

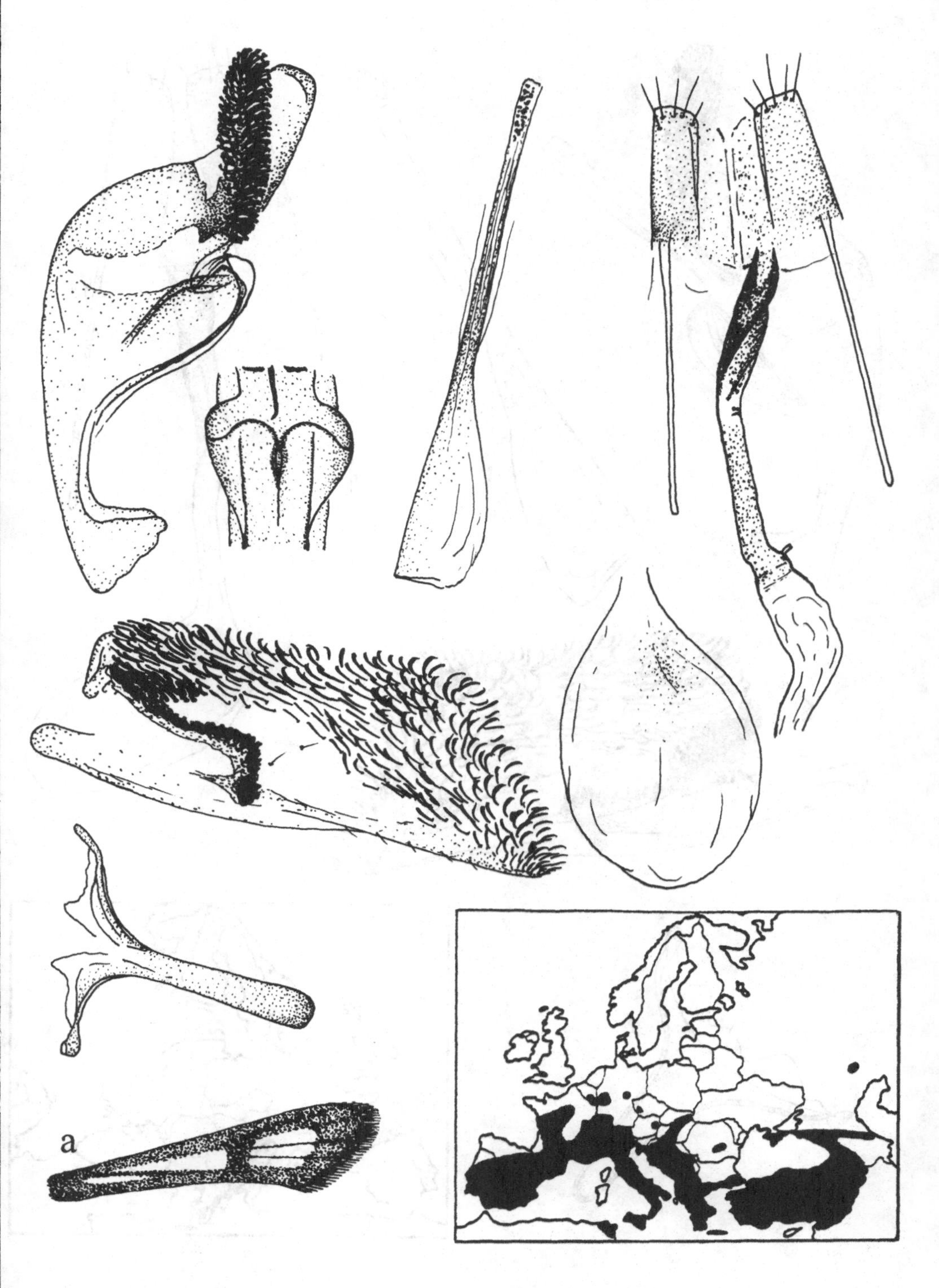

70. *Pyropteron affinis* (Staudinger) – Hungary, a: male forewing (ZL).

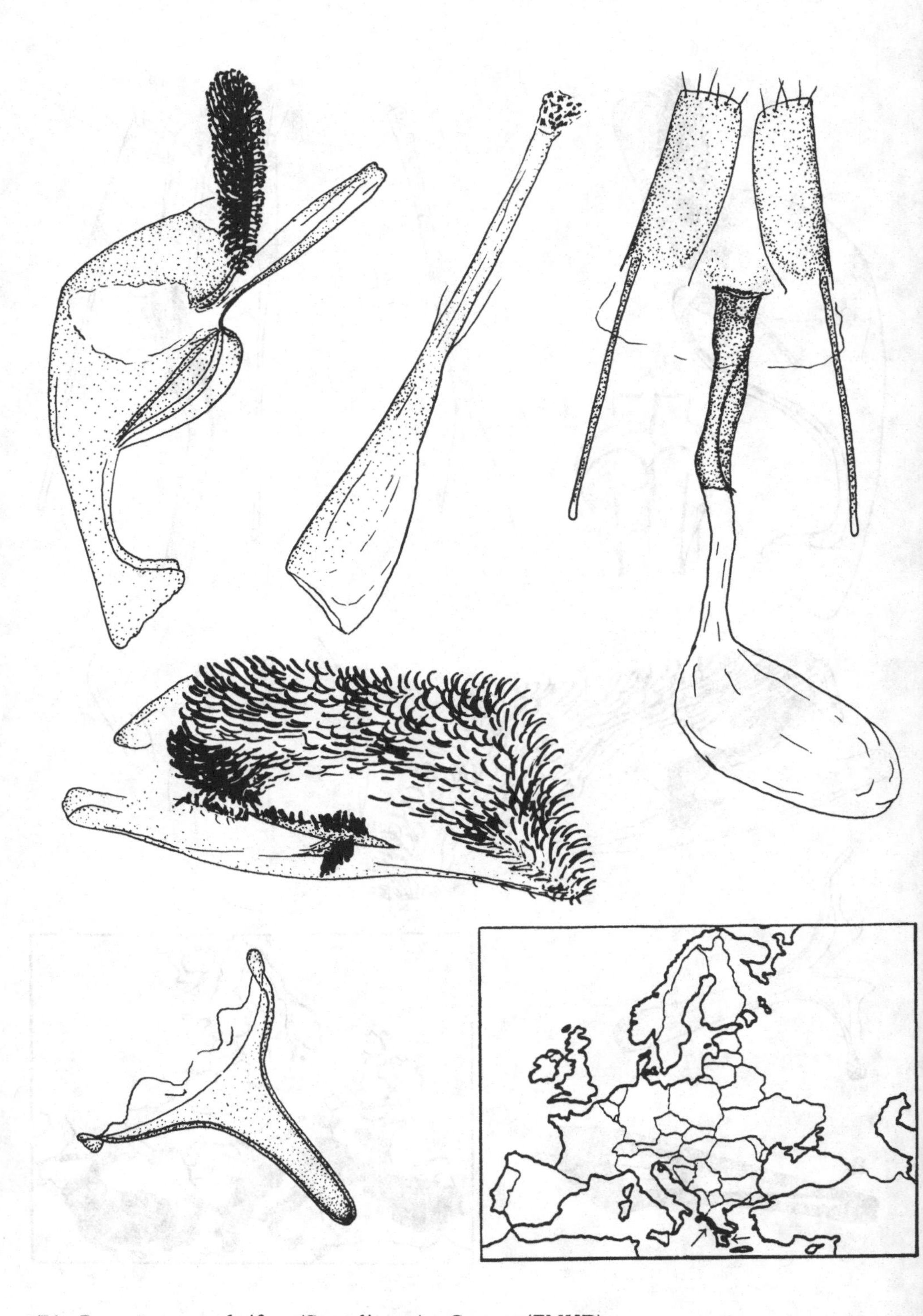

71. *Pyropteron umbrifera* (Staudinger) – Greece (ZMHB).

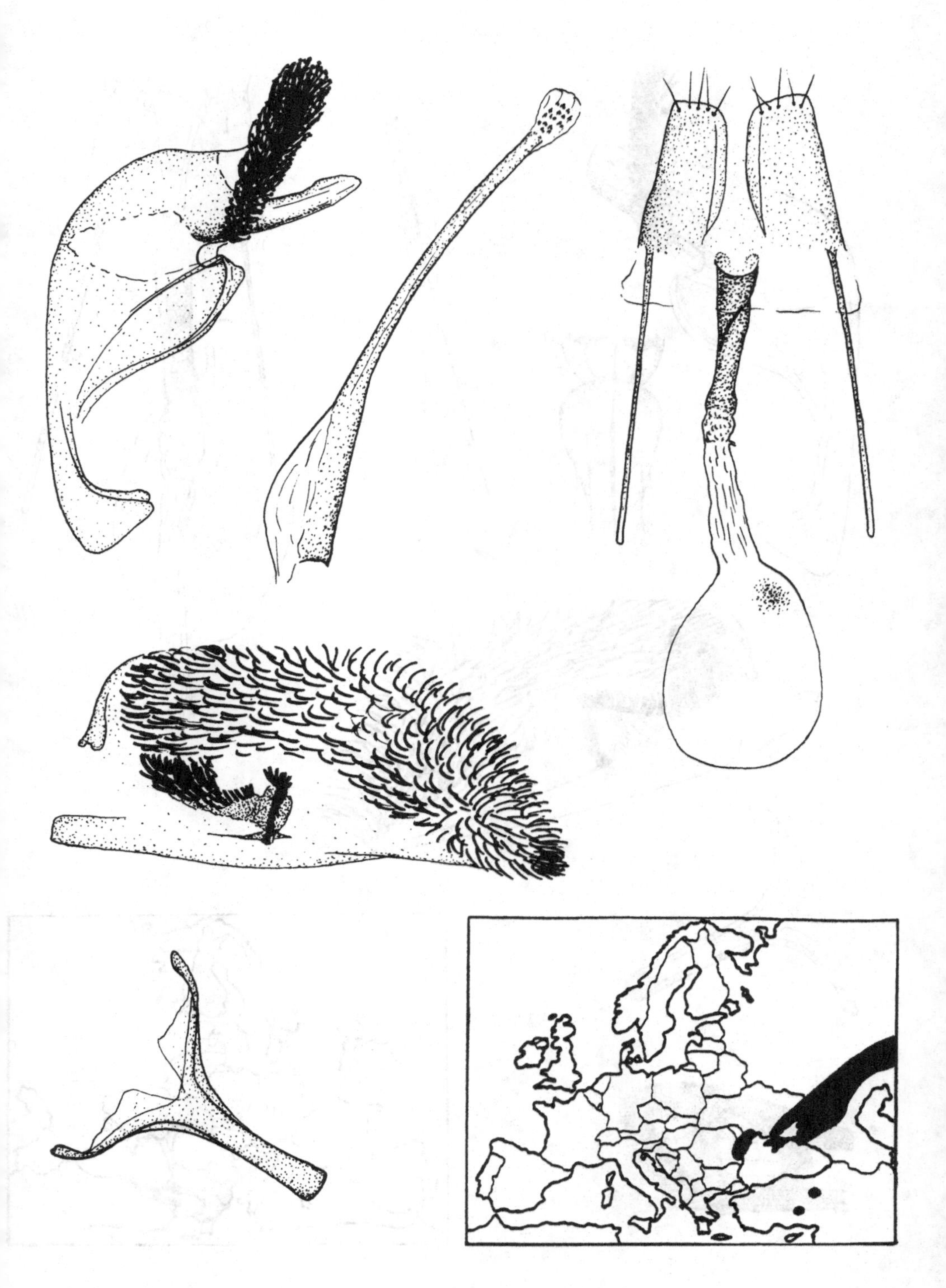

72. *Pyropteron cirgisa* (Bartel) – Russia (ZMHB).

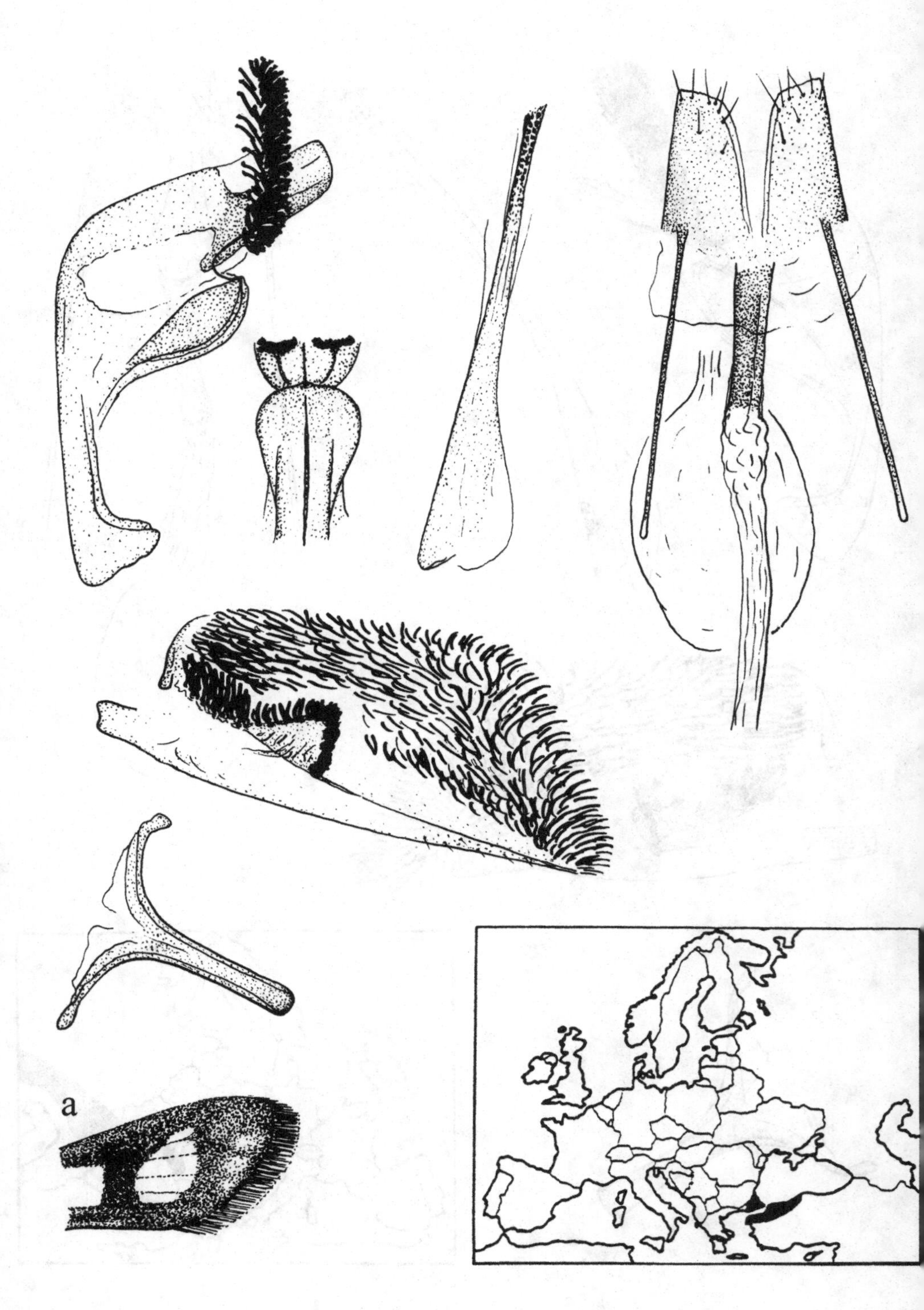

73. *Pyropteron mannii* (Lederer) – Bulgaria, a: apex of forewing (ZL).

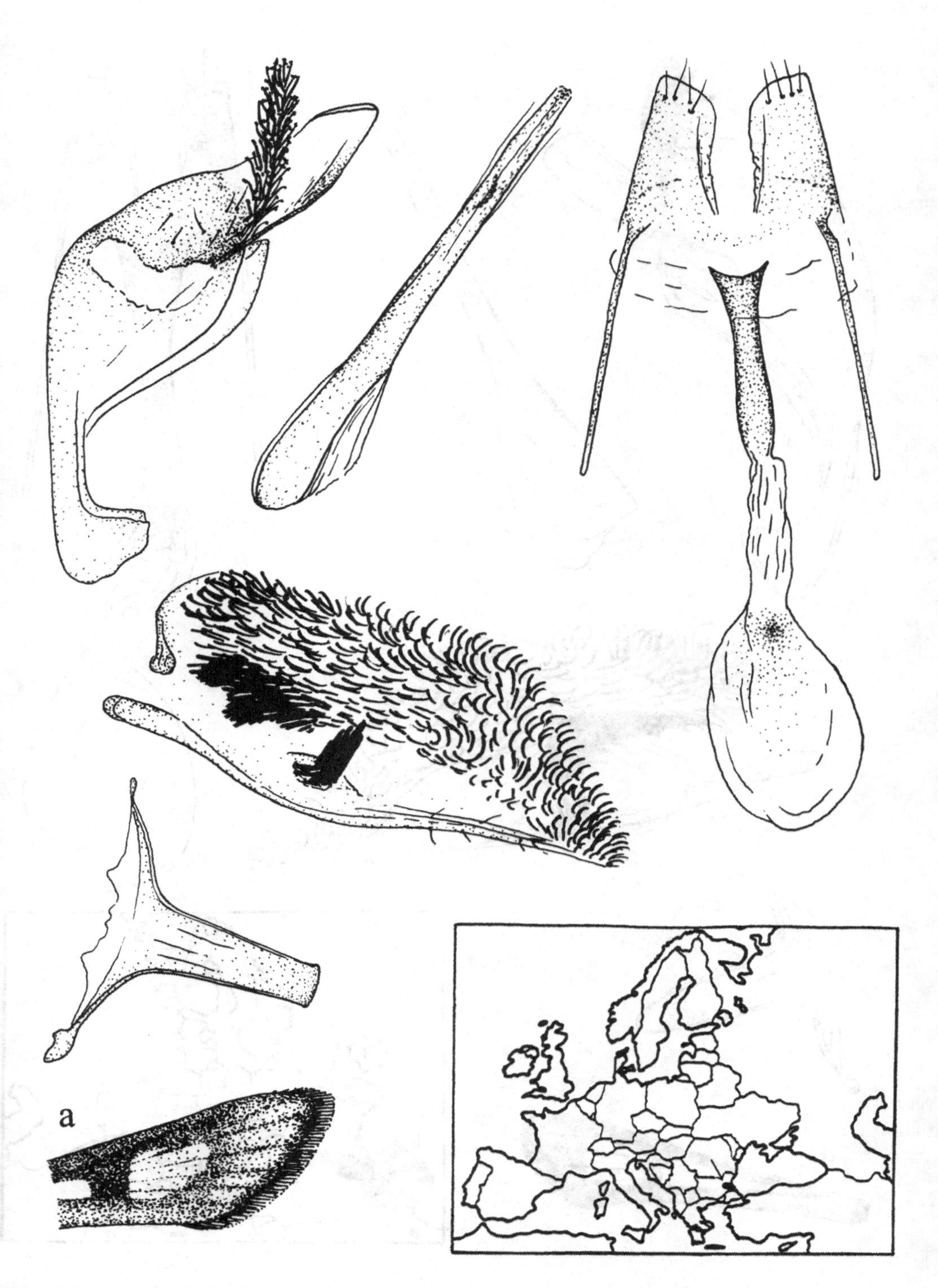

74. *Dipchasphecia lanipes* (Lederer) – Bulgaria, a: apex of forewing (ZMHB).

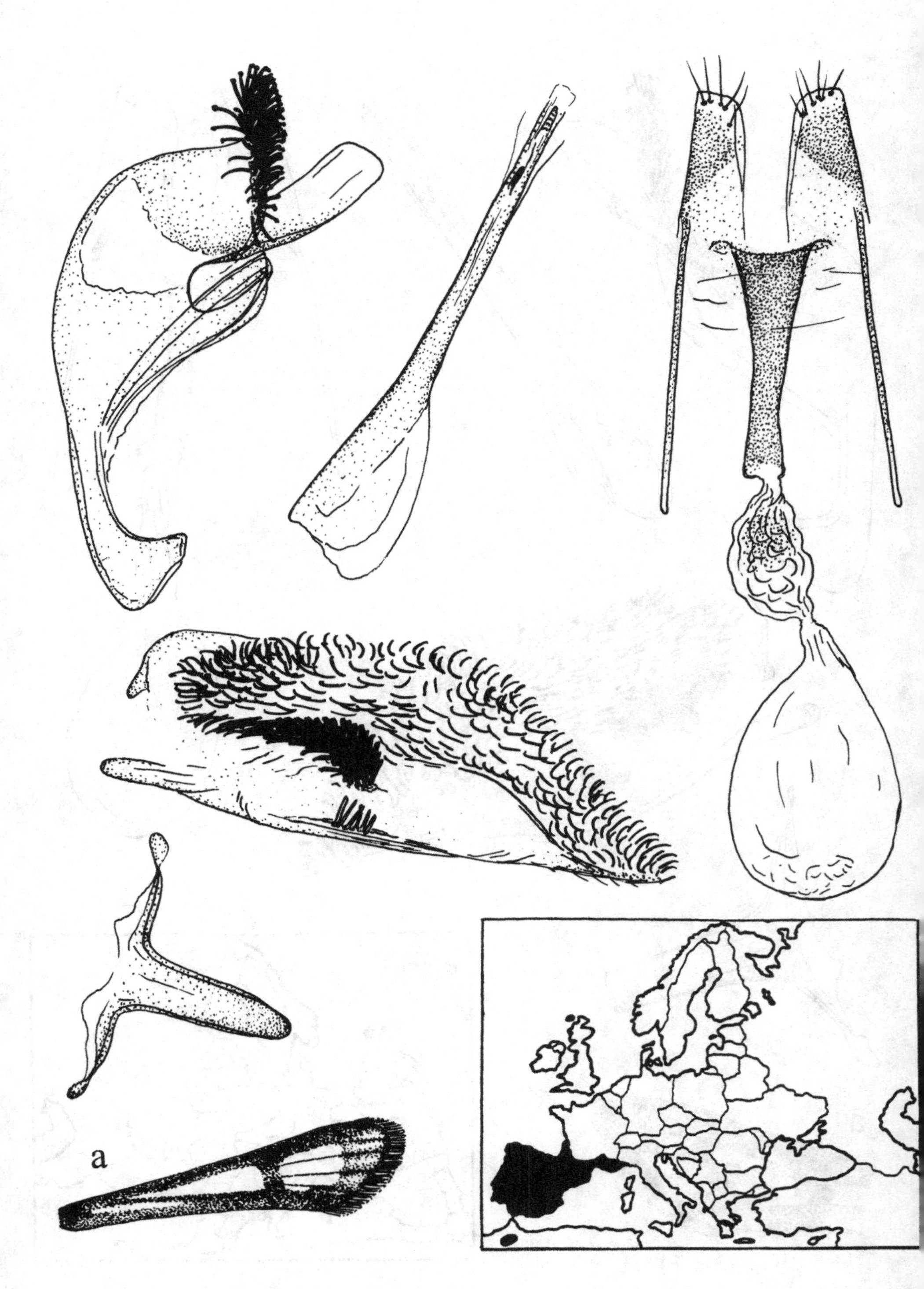

75. *Chamaesphecia mysiniformis* (Boisduval) – Spain, a: forewing (ZL).

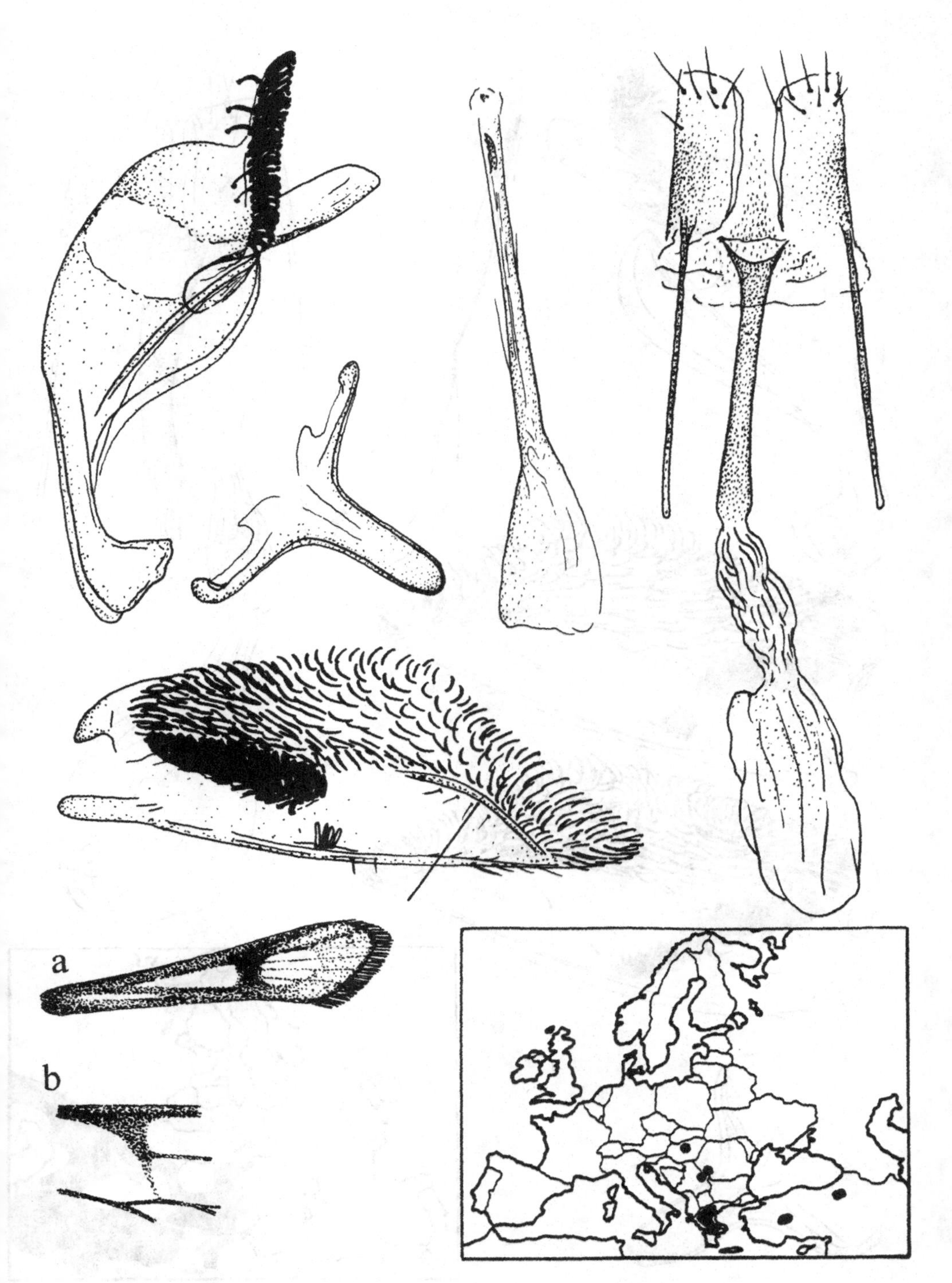

76. *Chamaesphecia anatolica* Schwingenschuss – Turkey, a: forewing, b: discal spot of hindwing (♂ NHMV, ♀ ZL).

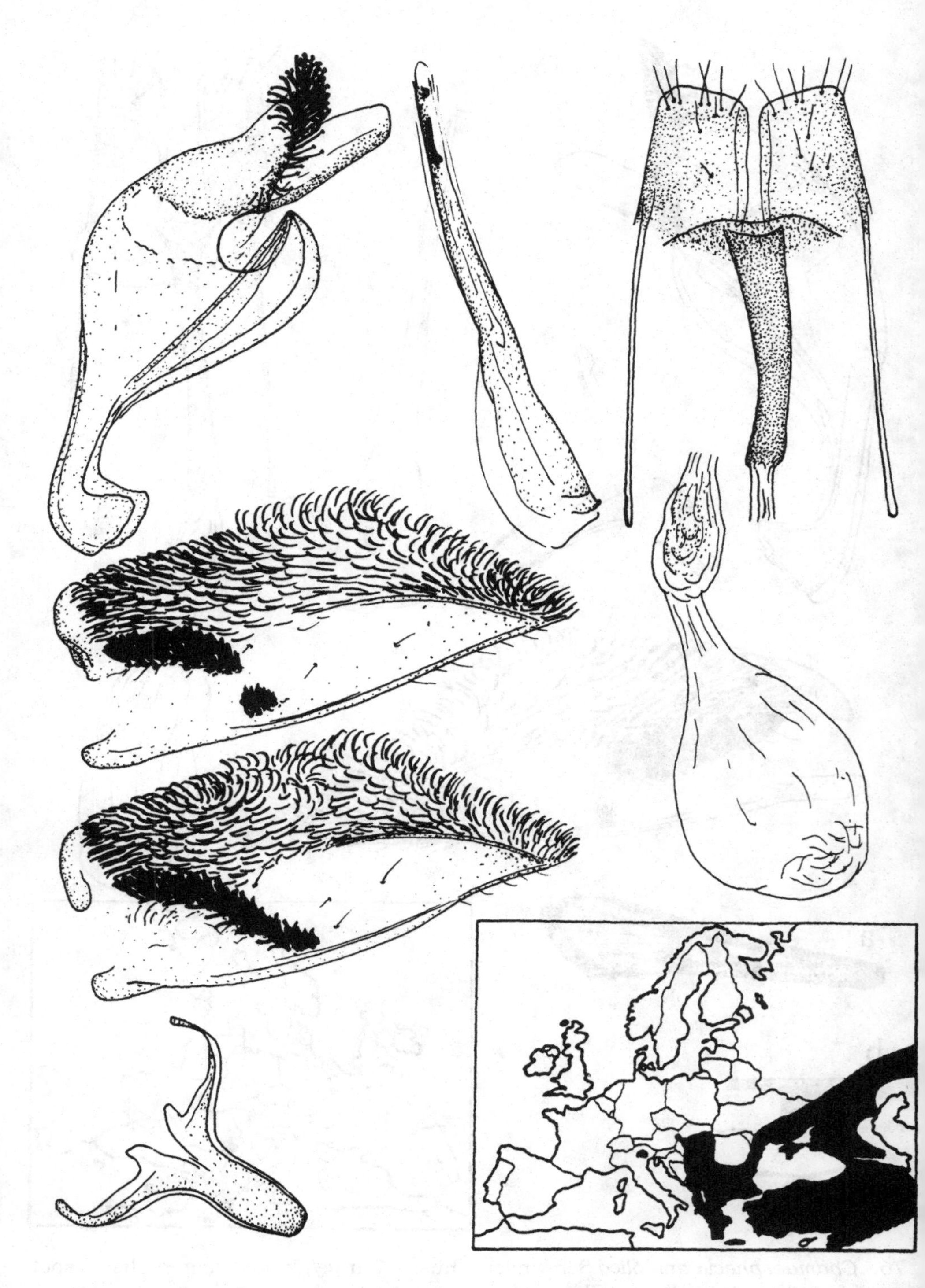

77. *Chamaesphecia chalciformis* (Esper) – Bulgaria, variability of crista sacculi (ZL).

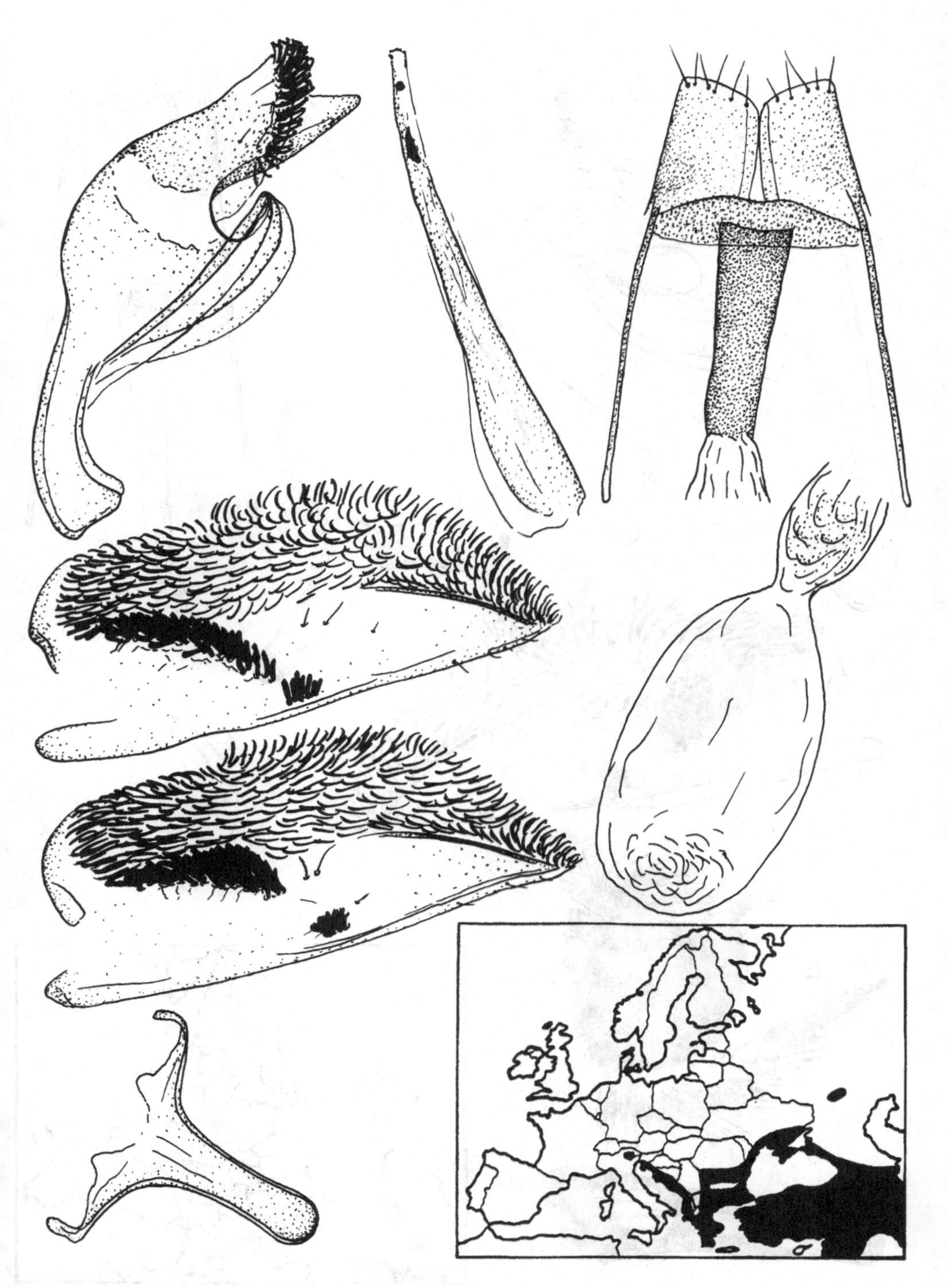

78. *Chamaesphecia schmidtiiformis* (Freyer) – Turkey, variability of crista sacculi (ZMHB).

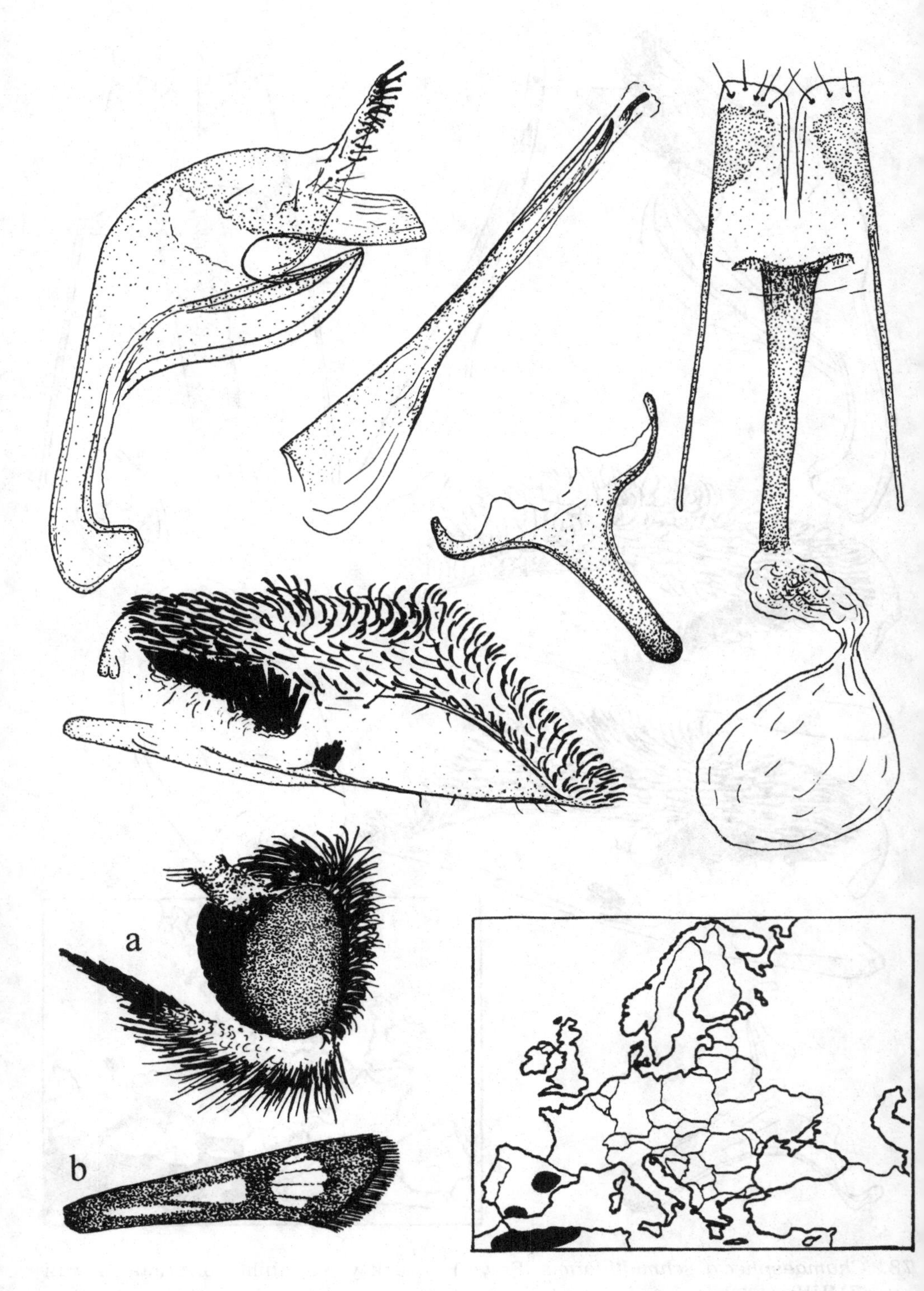

79. *Chamaesphecia anthrax* Le Cerf – Spain, a: head, b: male forewing (ZL).

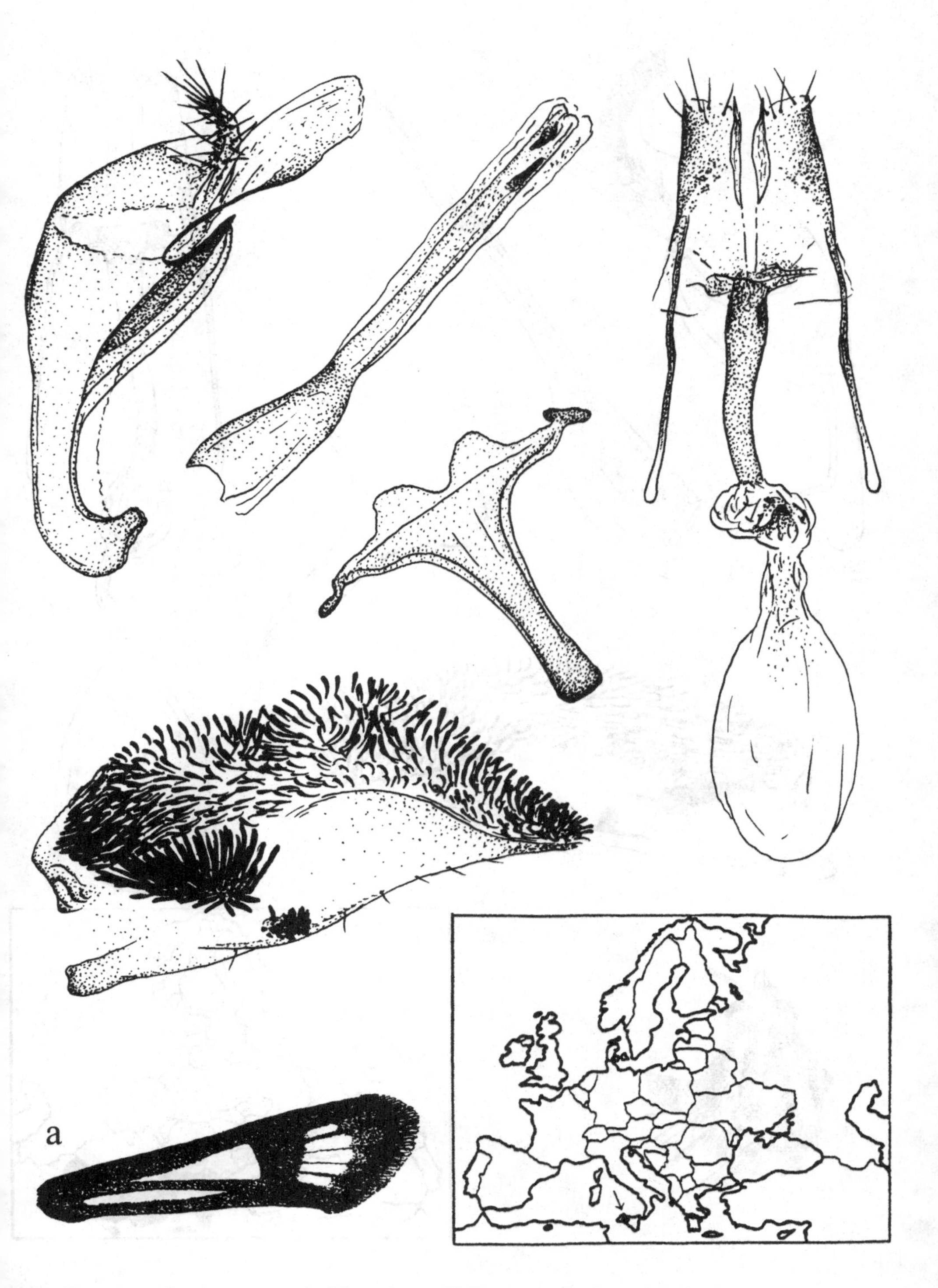

80. *Chamaesphecia maurusia* Püngeler – Sicily, a: male forewing (ZL).

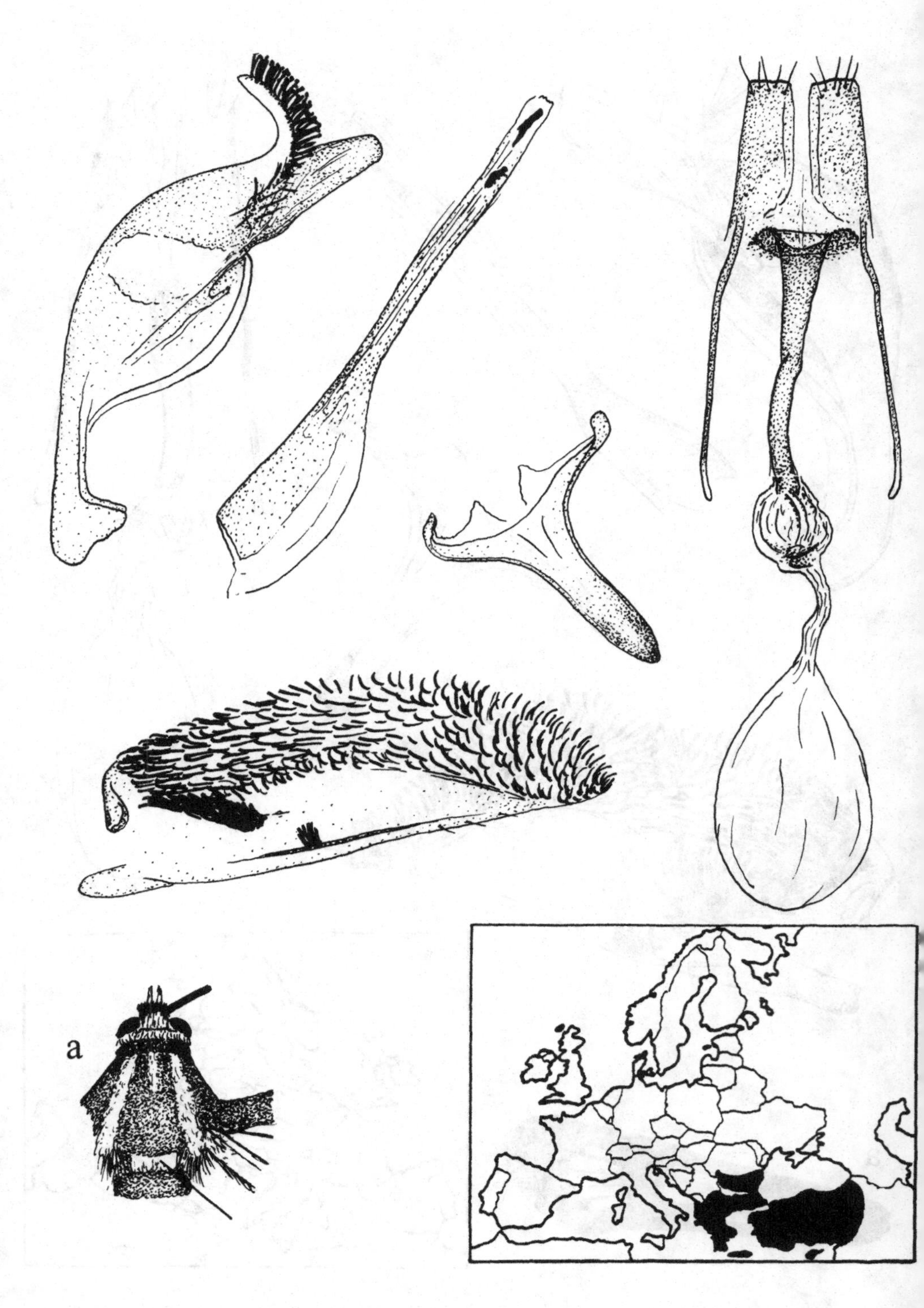

81. *Chamaesphecia alysoniformis* (Herrich-Schäffer) – Bulgaria, a: thorax (ZL).

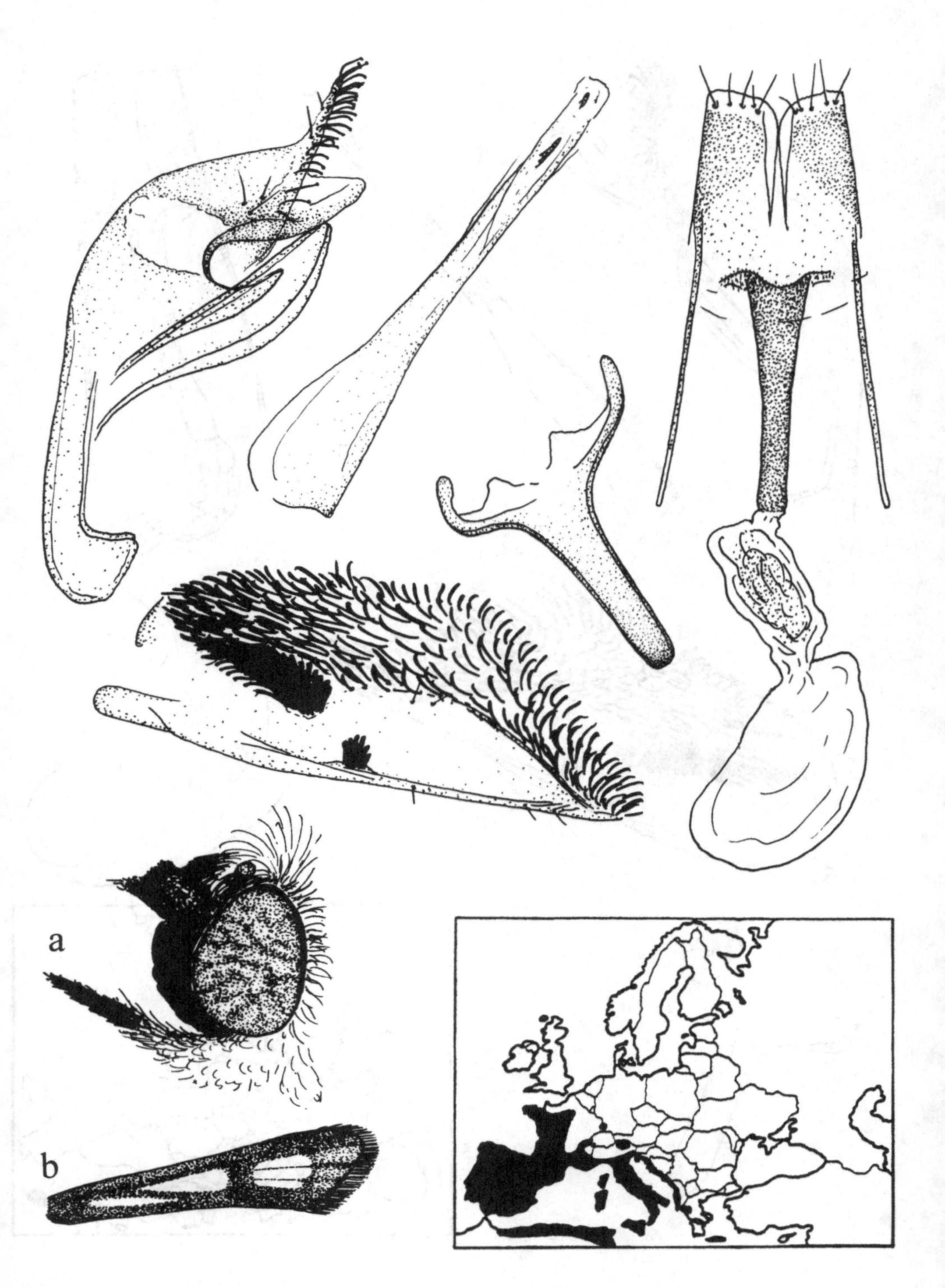

82. *Chamaesphecia aerifrons* (Zeller) – Spain, a: head, b: forewing (ZL).

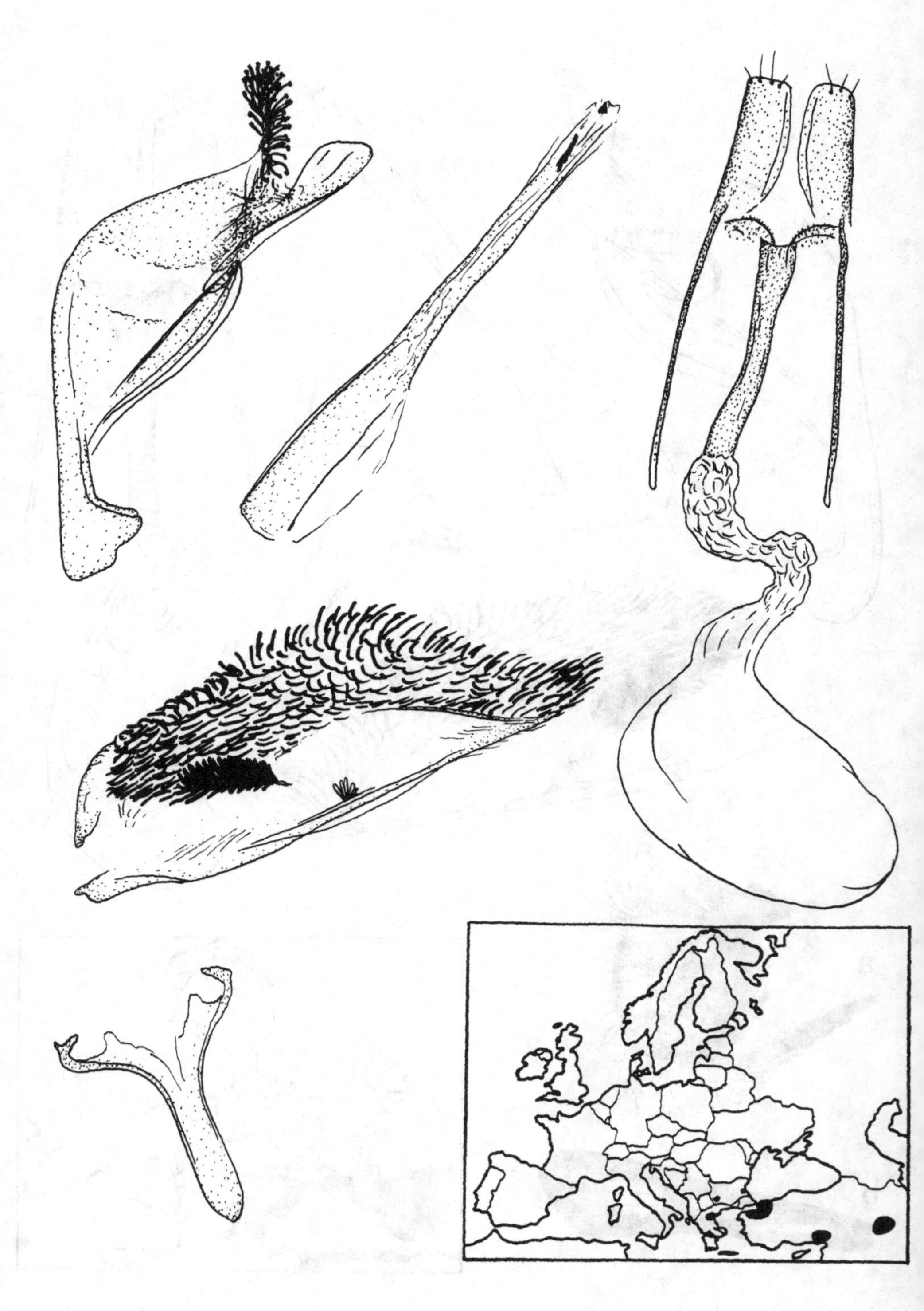

83. *Chamaesphecia albiventris* (Lederer) – Turkey (ZMHB).

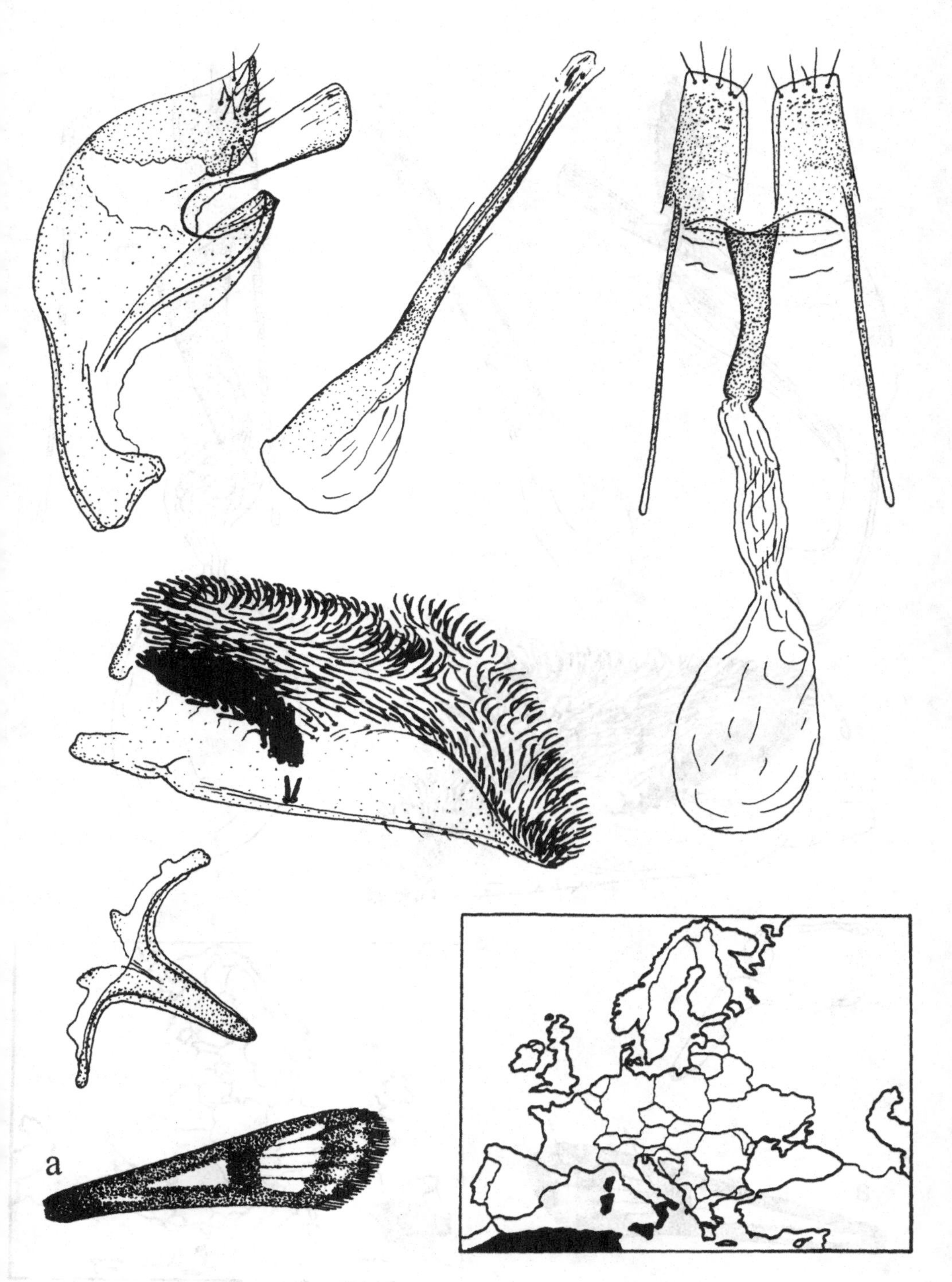

84. *Chamaesphecia osmiaeformis* (Herrich-Schäffer) – Morocco, a: male forewing (ZL, ex KS and H. Malicky).

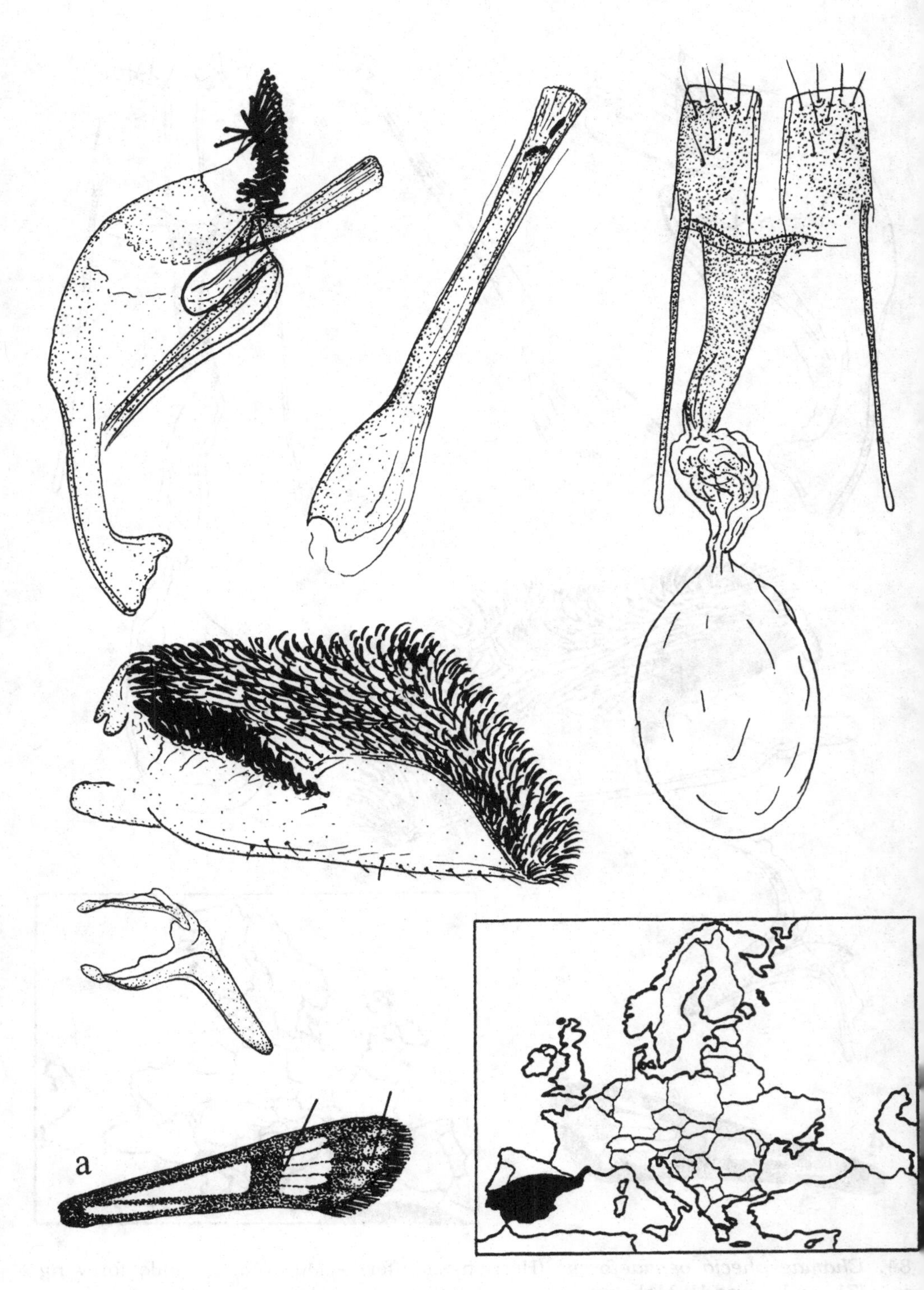

85. *Chamaesphecia ramburi* (Staudinger) – Spain, a: forewing (ZL).

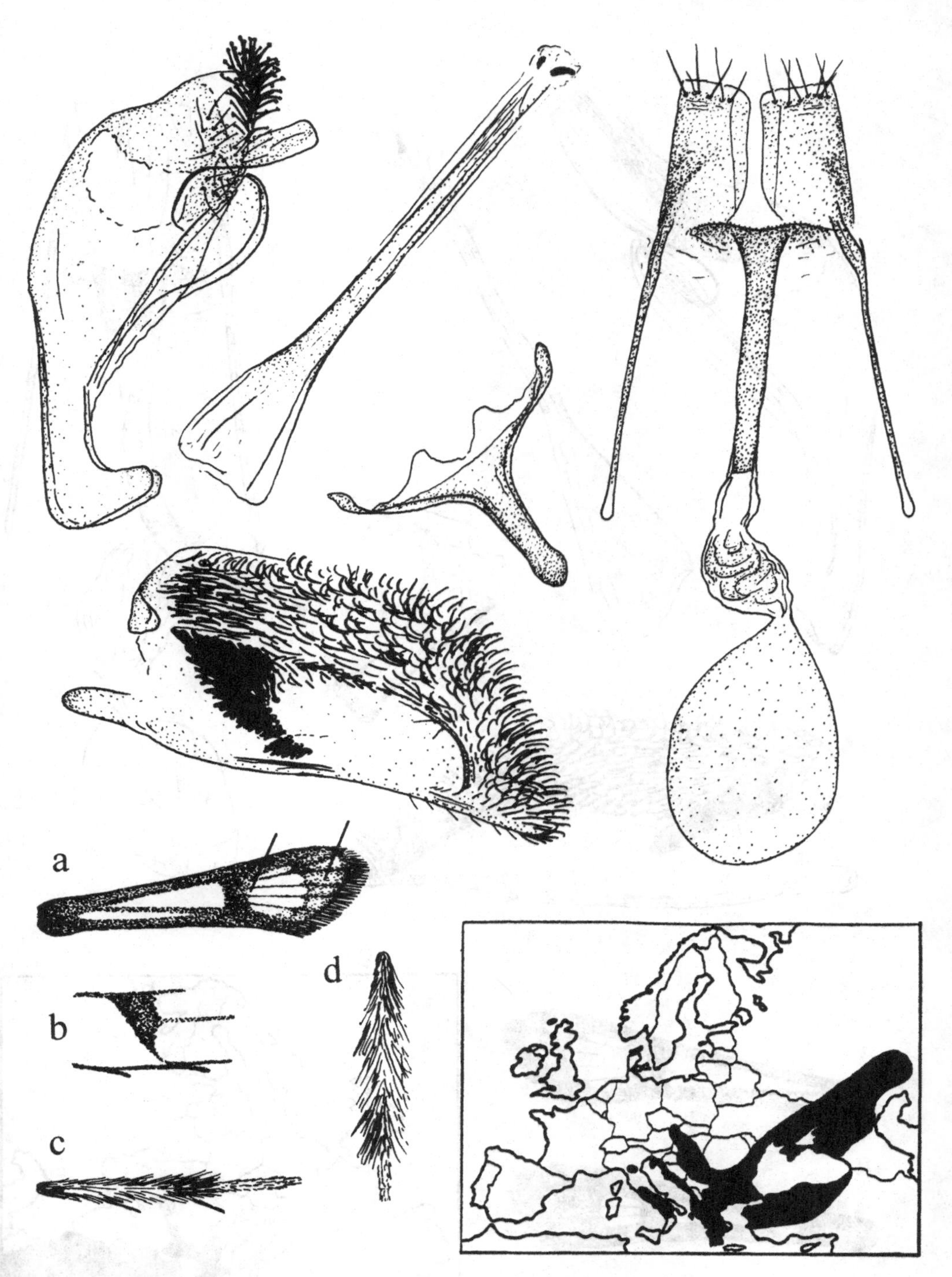

86. *Chamaesphecia doleriformis* (Herrich-Schäffer) – Slovakia (genitalia and a, b, c – ZL), Croatia (d – ZL, ex IT), a: forewing, b: discal spot of hindwing, c: hind tibia of ssp. *colpiformis* (Staudinger), d: hind tibia of ssp. *doleriformis*.

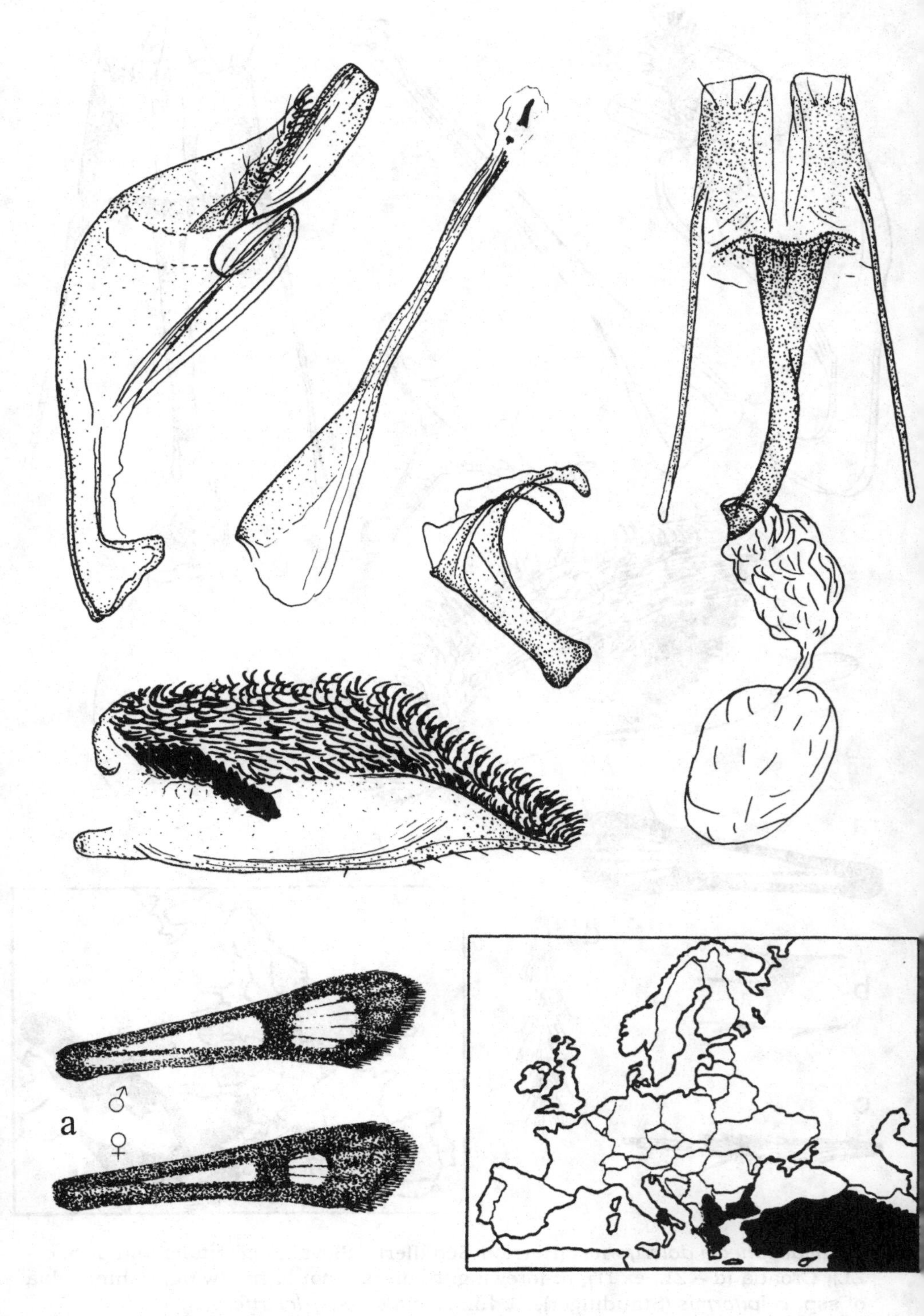

87. *Chamaesphecia thracica* Laštůvka – Bulgaria, a: forewings (ZL).

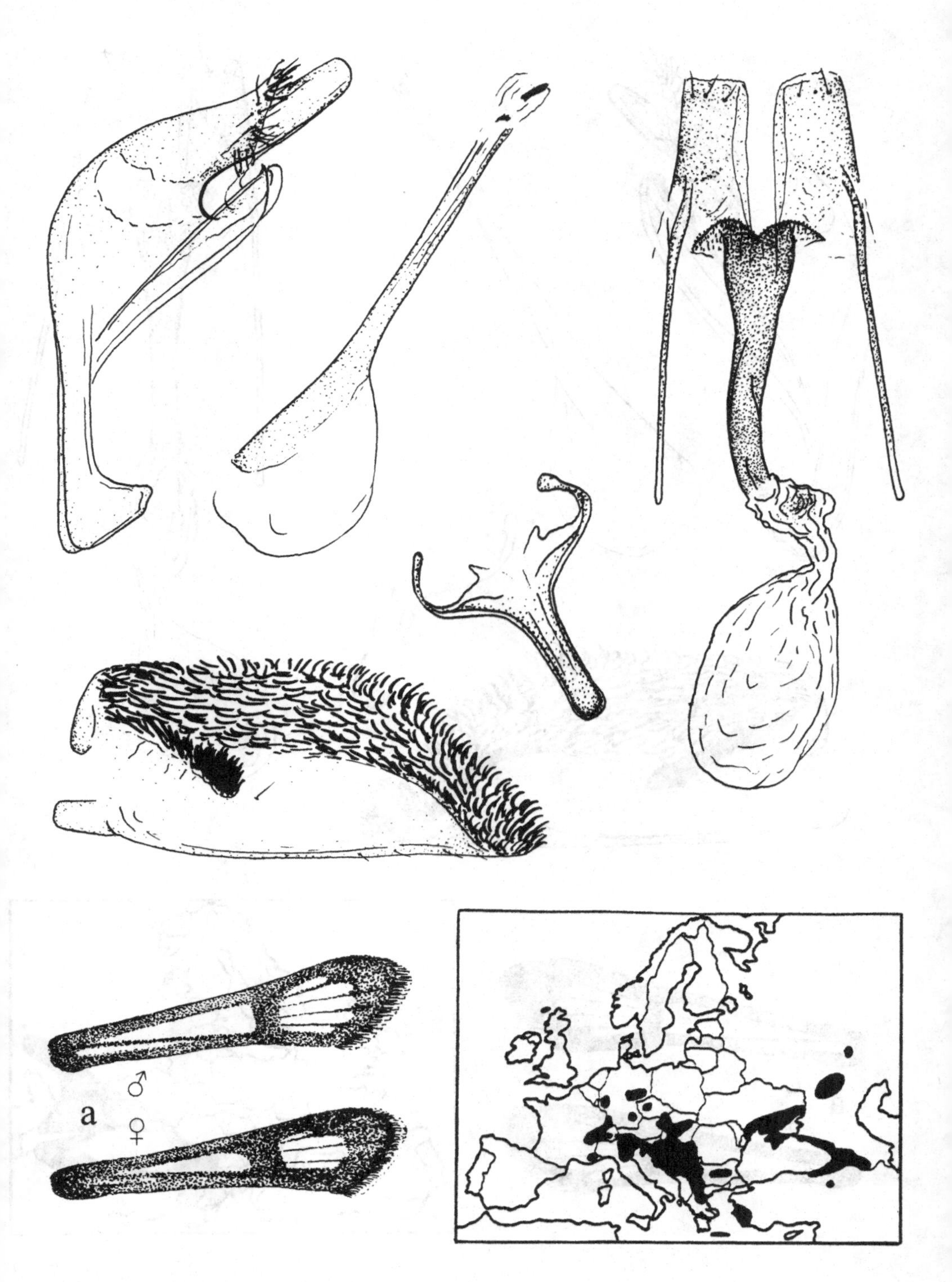

88. *Chamaesphecia dumonti* Le Cerf – Czech Republic, a: forewings (ZL).

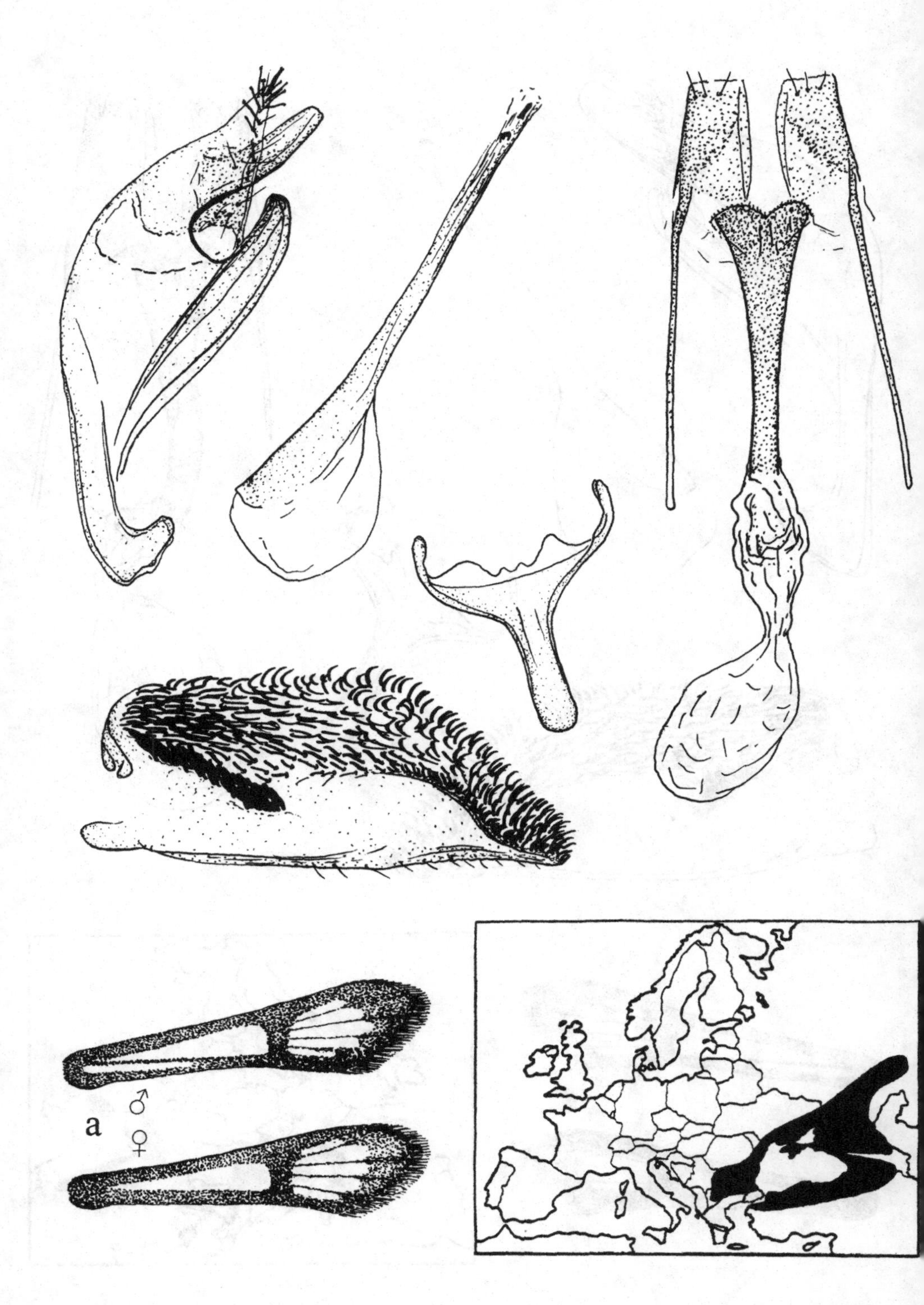

89. *Chamaesphecia oxybeliformis* (Herrich-Schäffer) – Bulgaria, a: forewings (ZL).

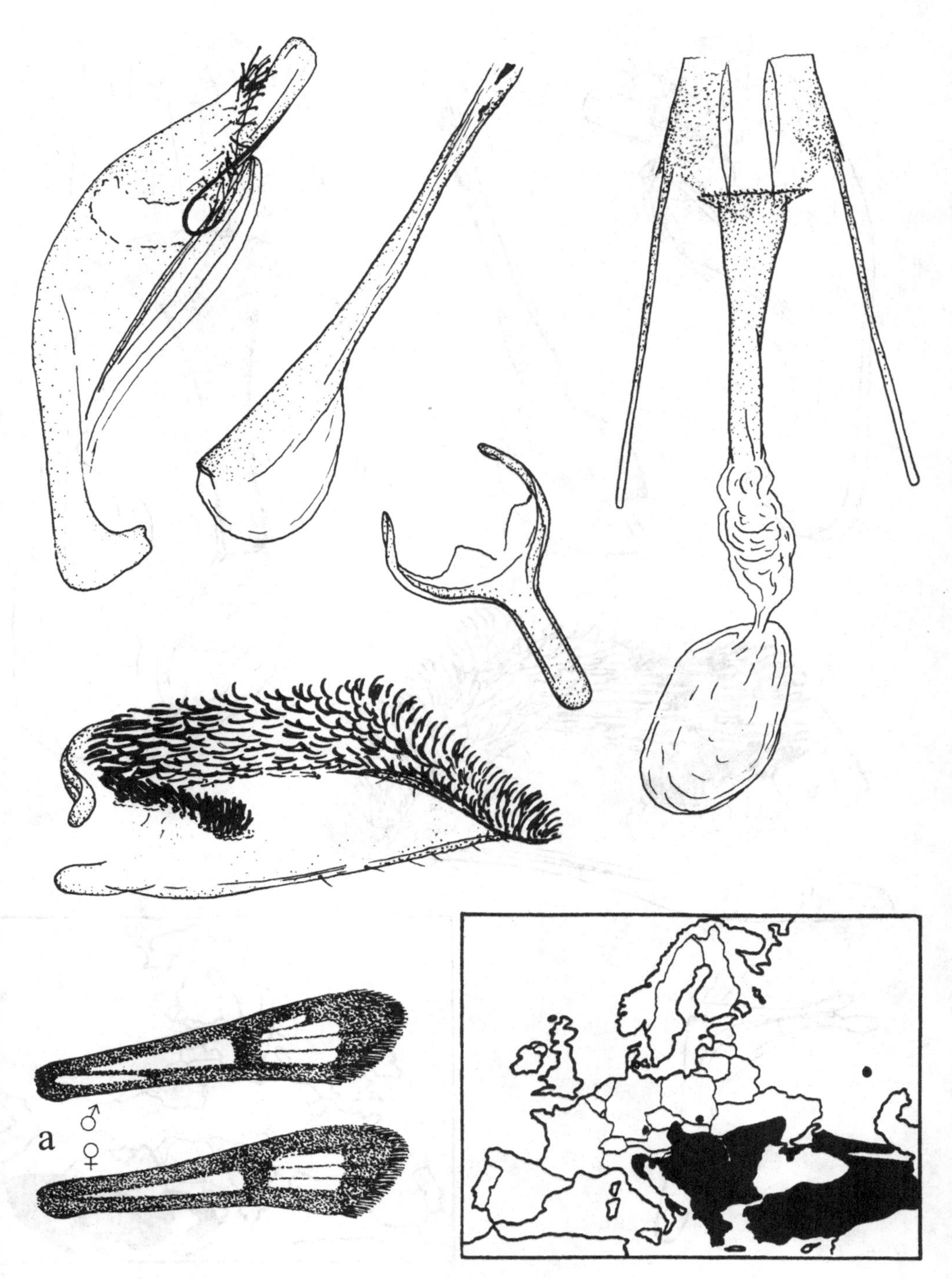

90. *Chamaesphecia annellata* (Zeller) – Slovakia, a: forewings (ZL).
(the genitalia of 91. *C. staudingeri* are not figured)

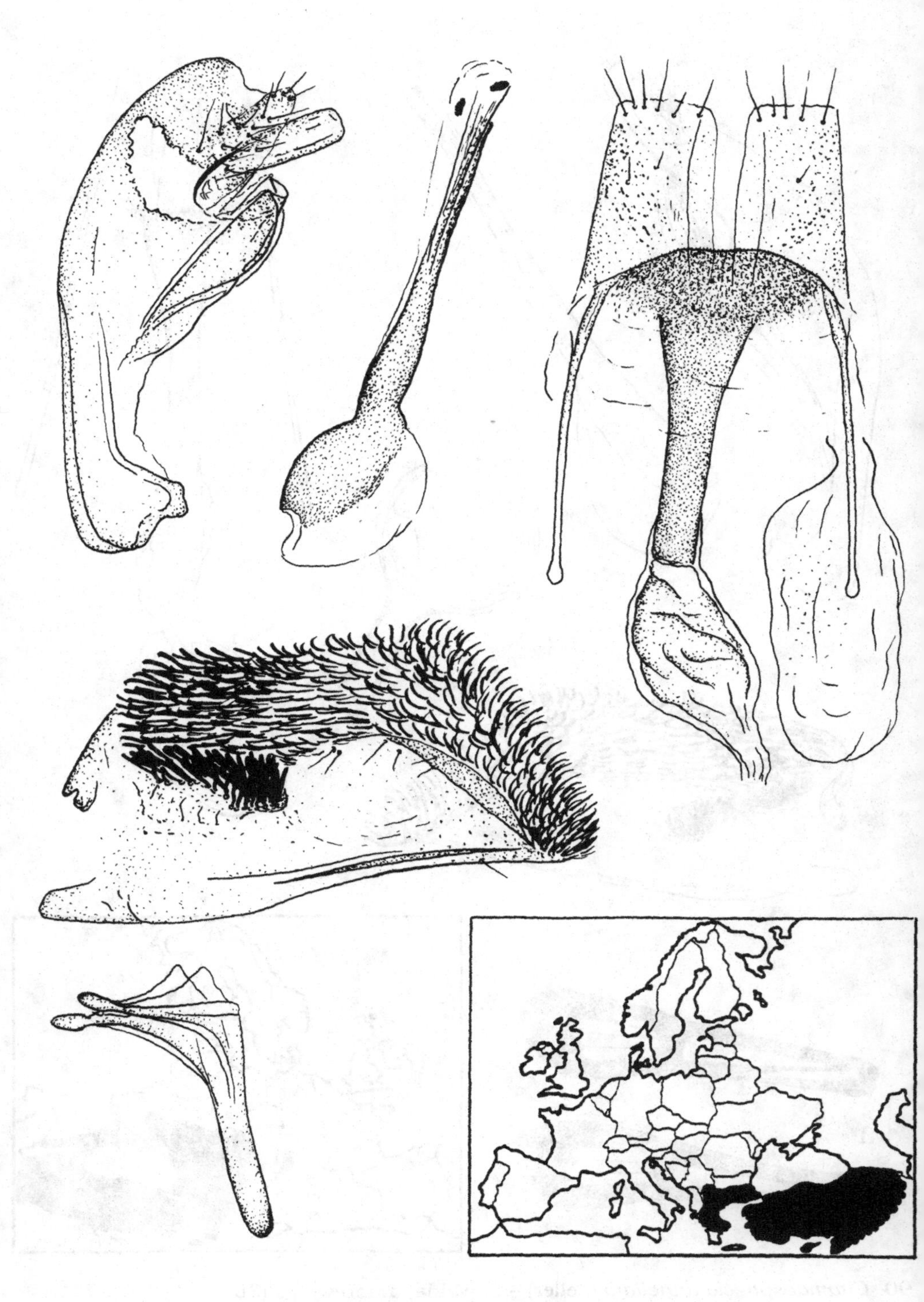

92. *Chamaesphecia proximata* (Staudinger) – Bulgaria (ZL).

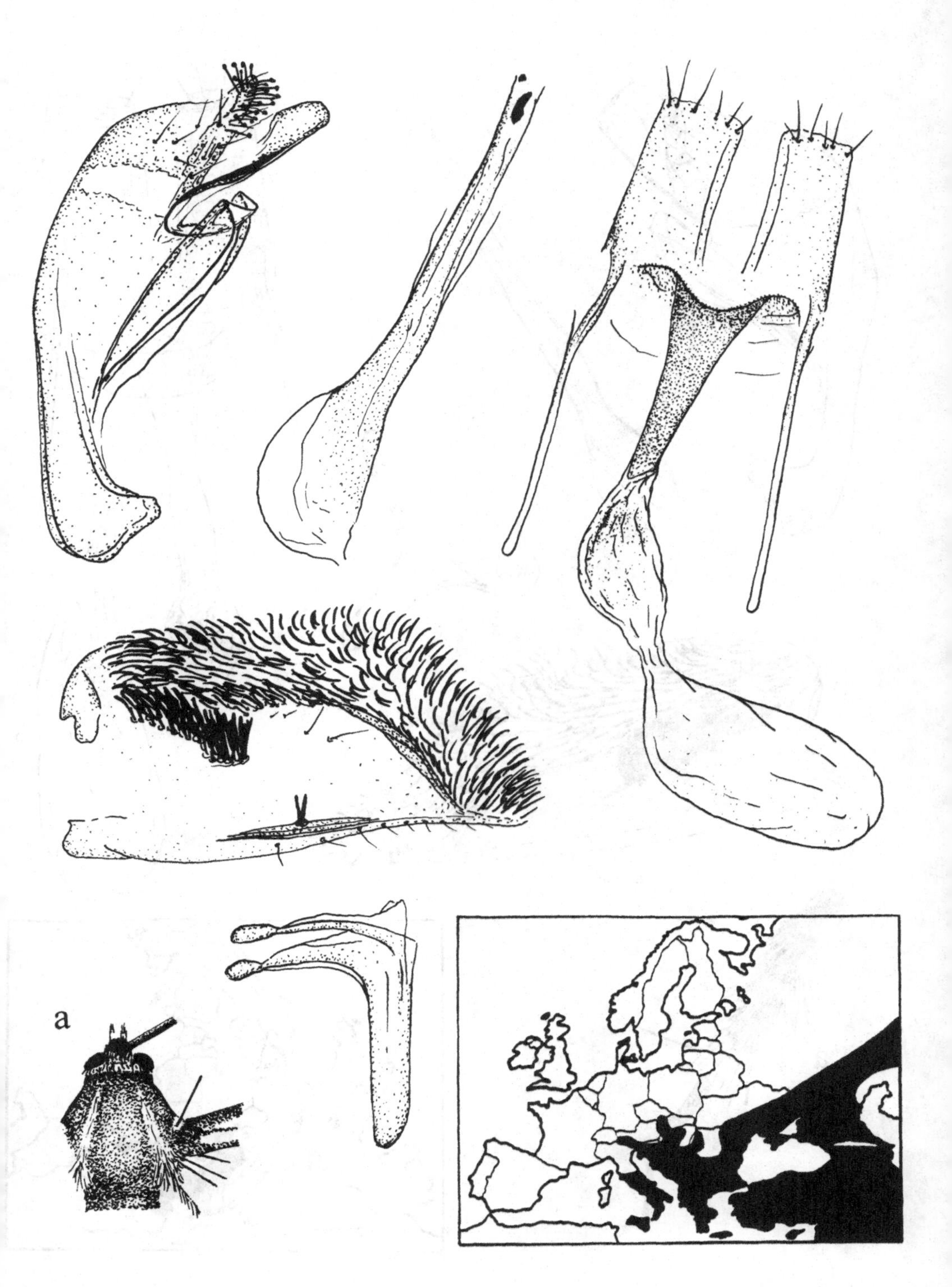

93. *Chamaesphecia masariformis* (Ochsenheimer) – Slovakia, a: thorax (ZL).

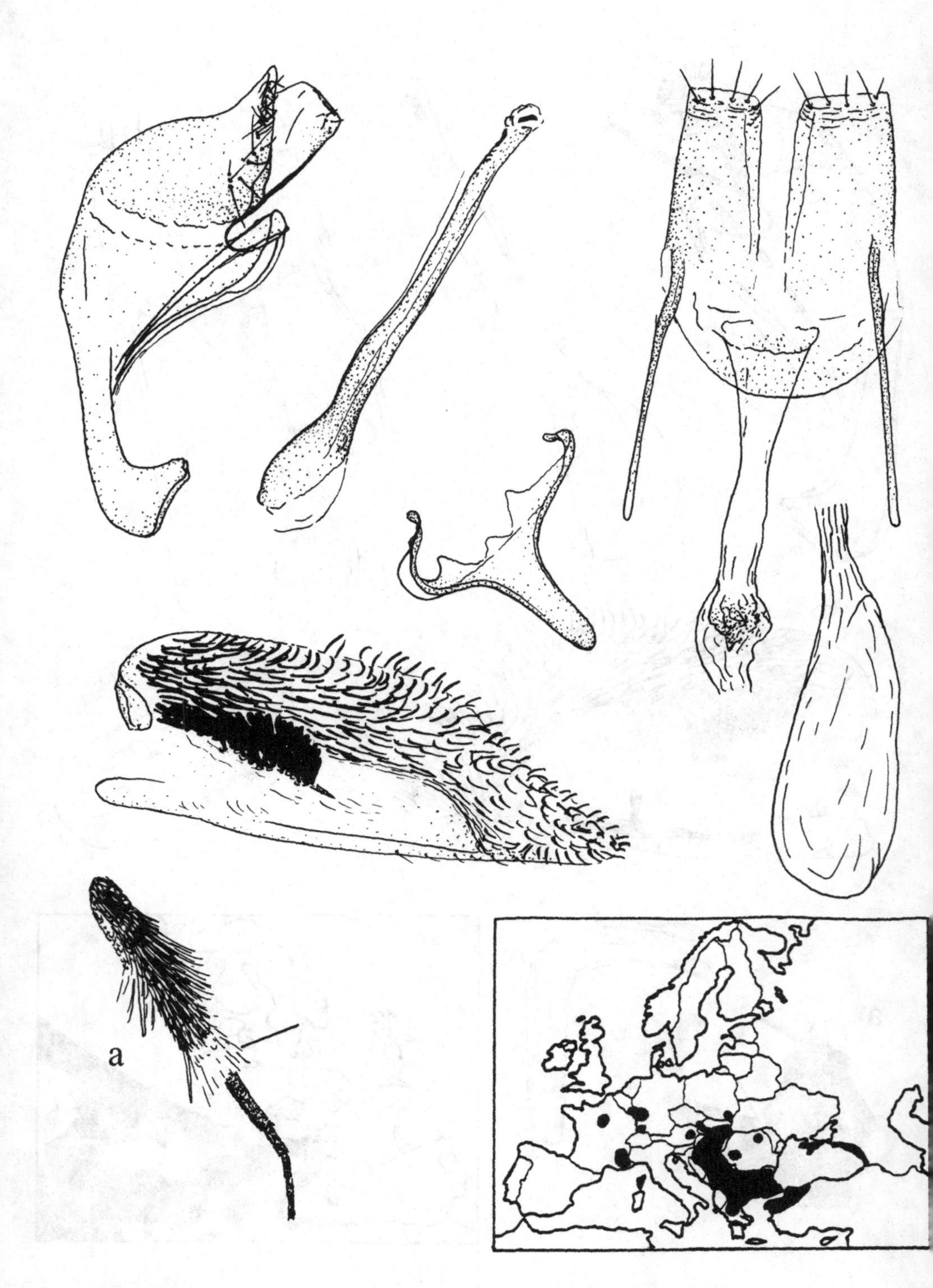

94. *Chamaesphecia nigrifrons* (Le Cerf) – Slovakia, a: hind tibia (ZL).

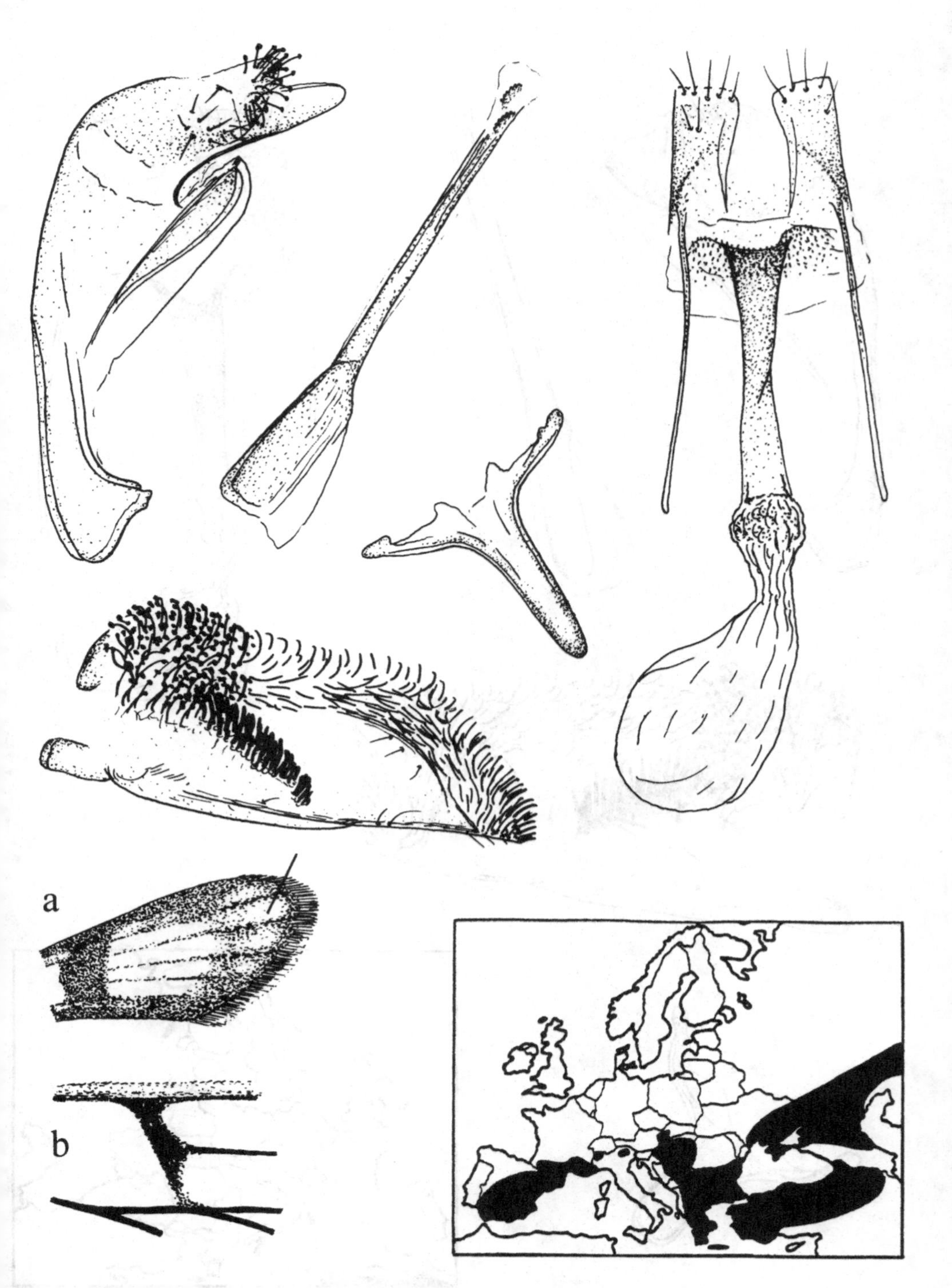

95. *Chamaesphecia bibioniformis* (Esper) – Slovakia, a: external transparent area and apex of forewing, b: discal spot of hindwing (ZL).

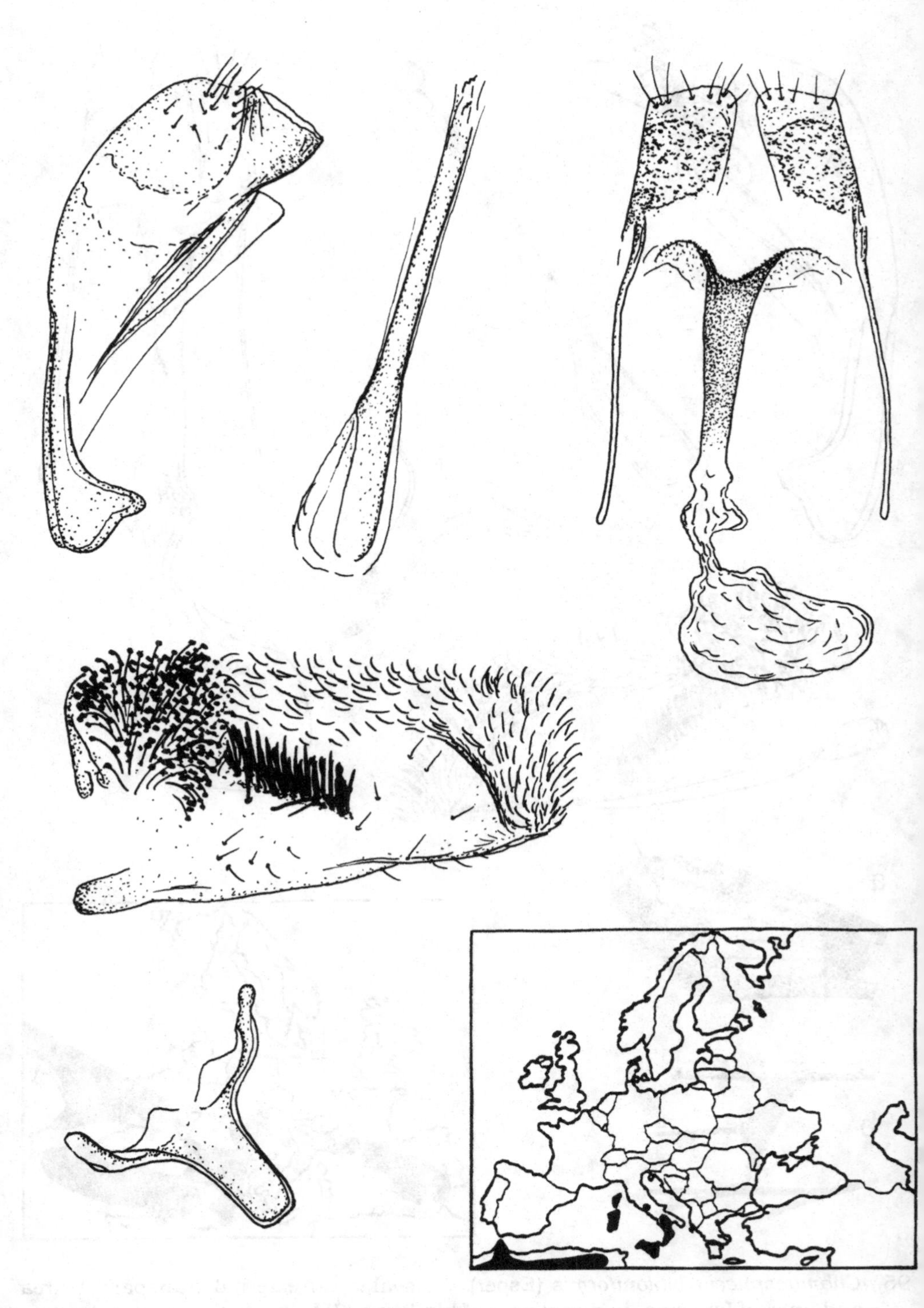

96. *Chamaesphecia anthraciformis* (Rambur) – Morocco (ZL, ex KS).

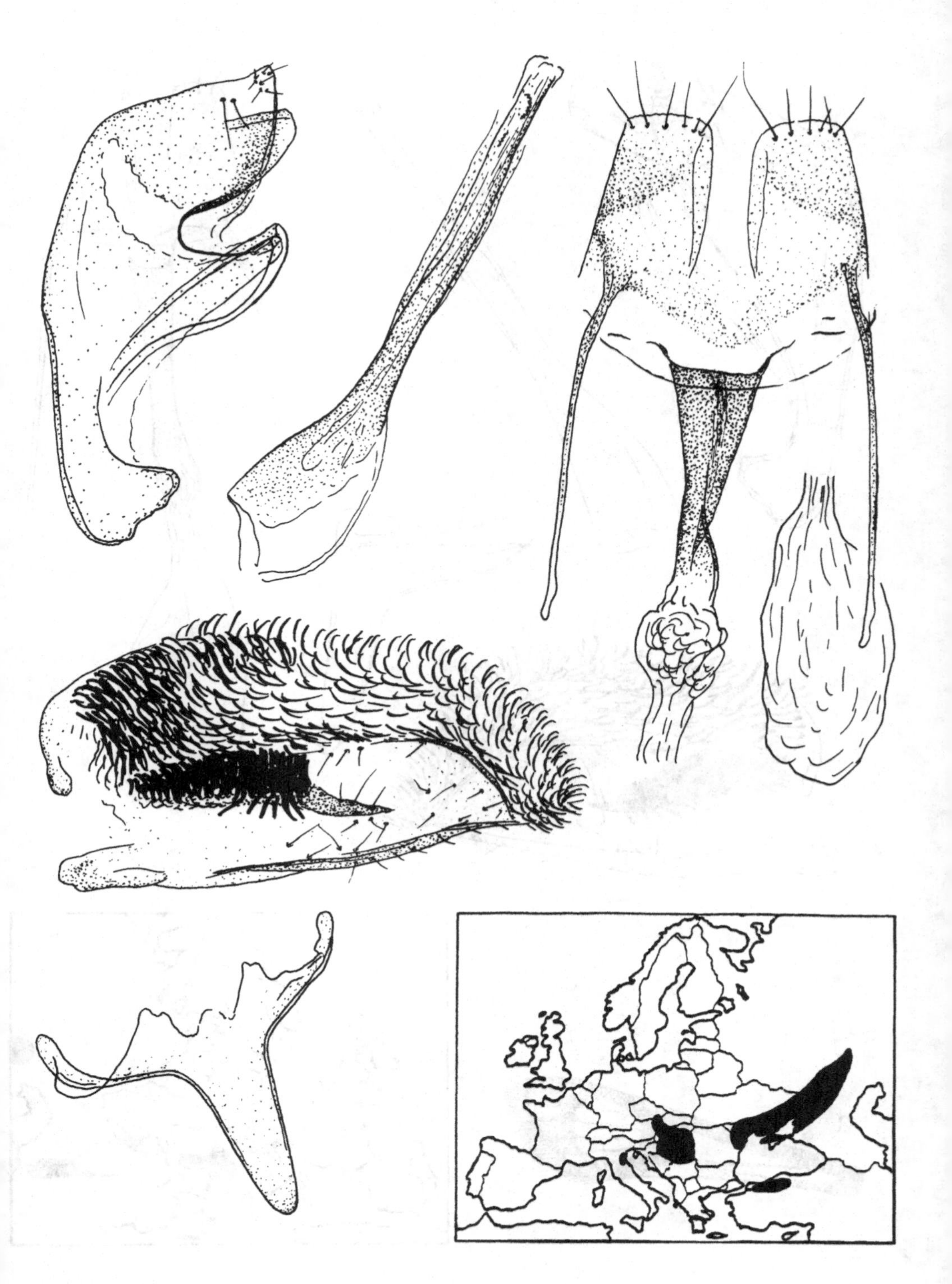

97. *Chamaesphecia palustris* Kautz – Slovakia (ZL).

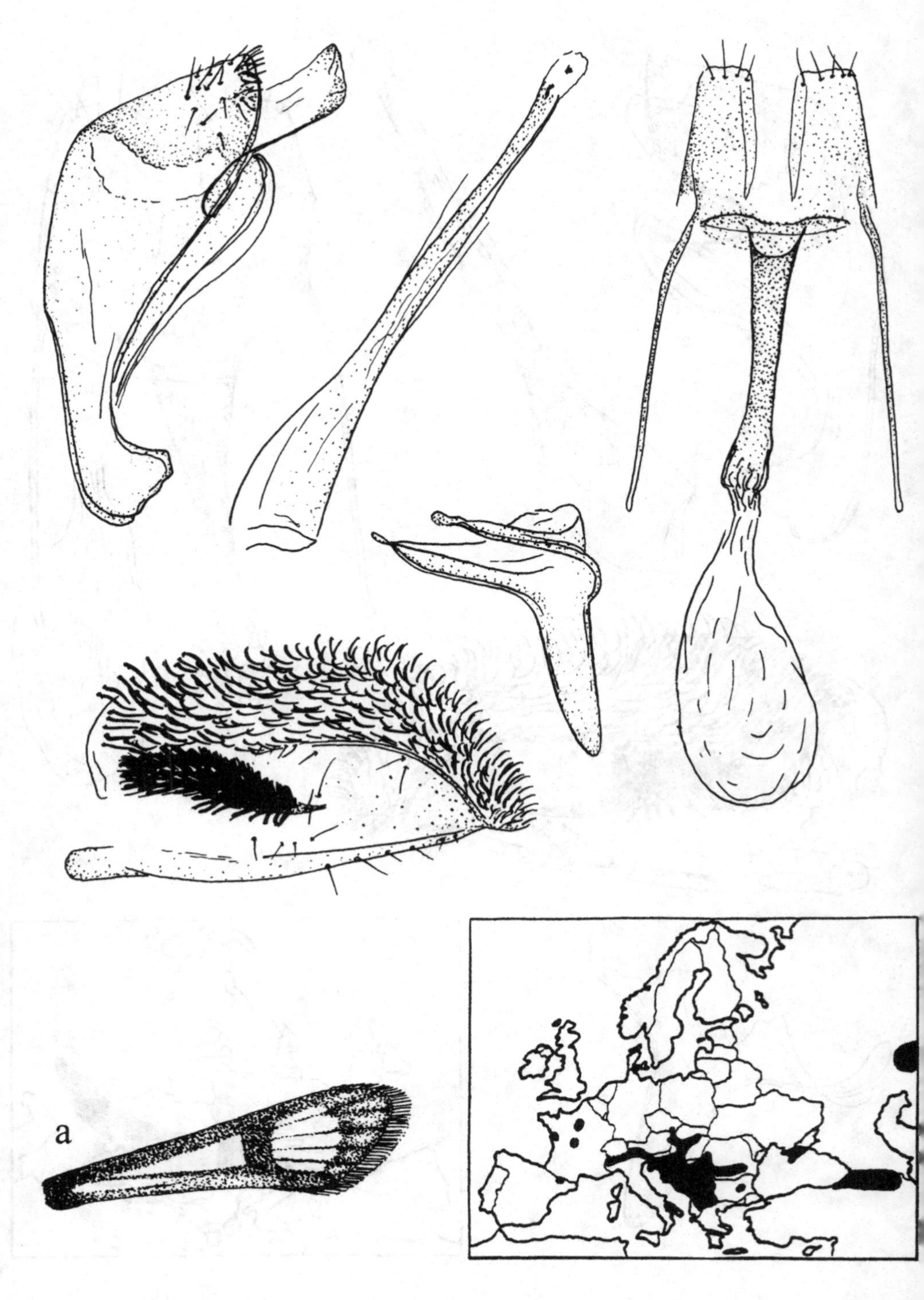

98. *Chamaesphecia euceraeformis* (Ochsenheimer) – Slovakia, a: male forewing (ZL).

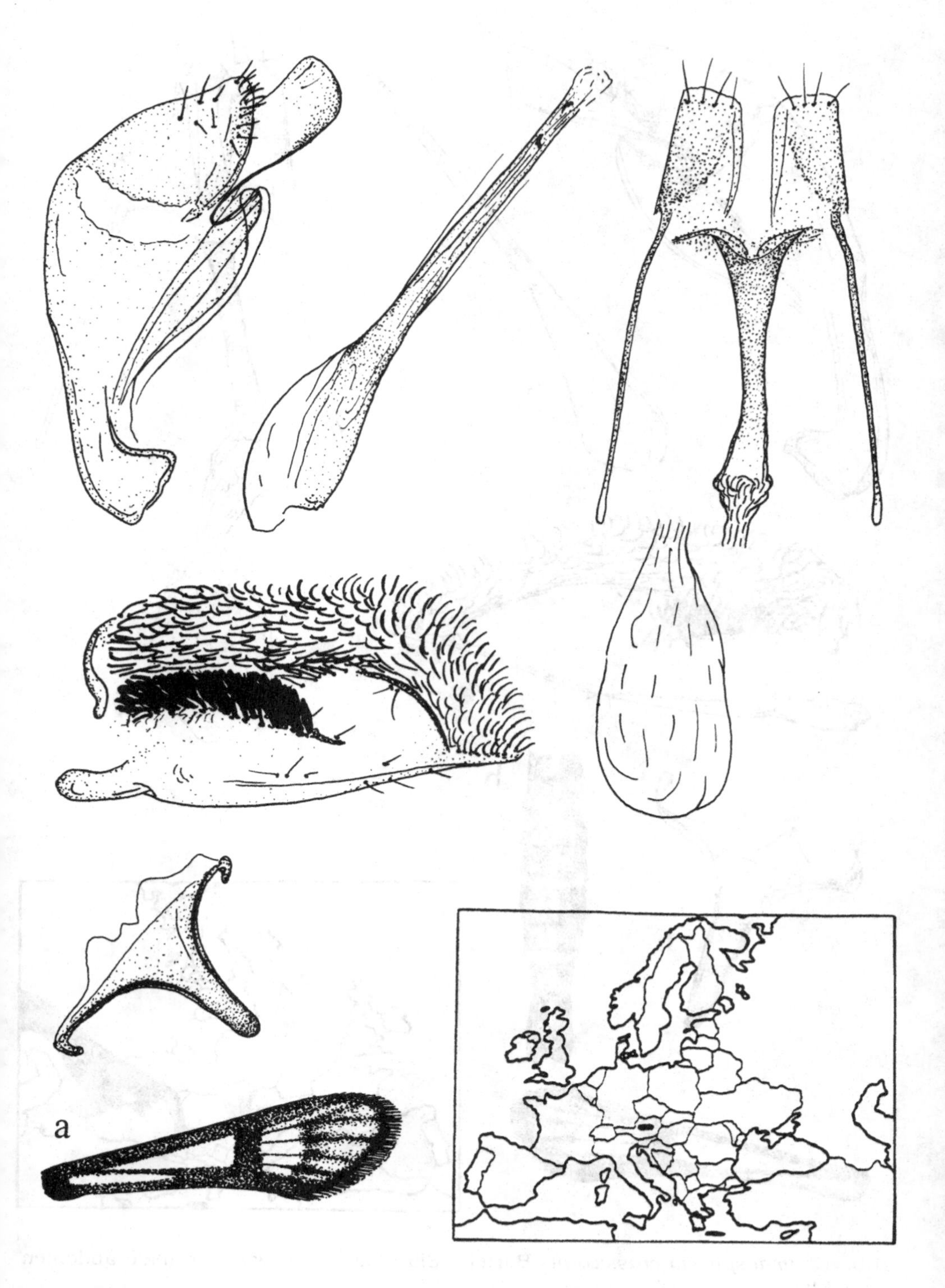

99. *Chamaesphecia amygdaloidis* Schleppnik – Austria, a: male forewing (ZL).

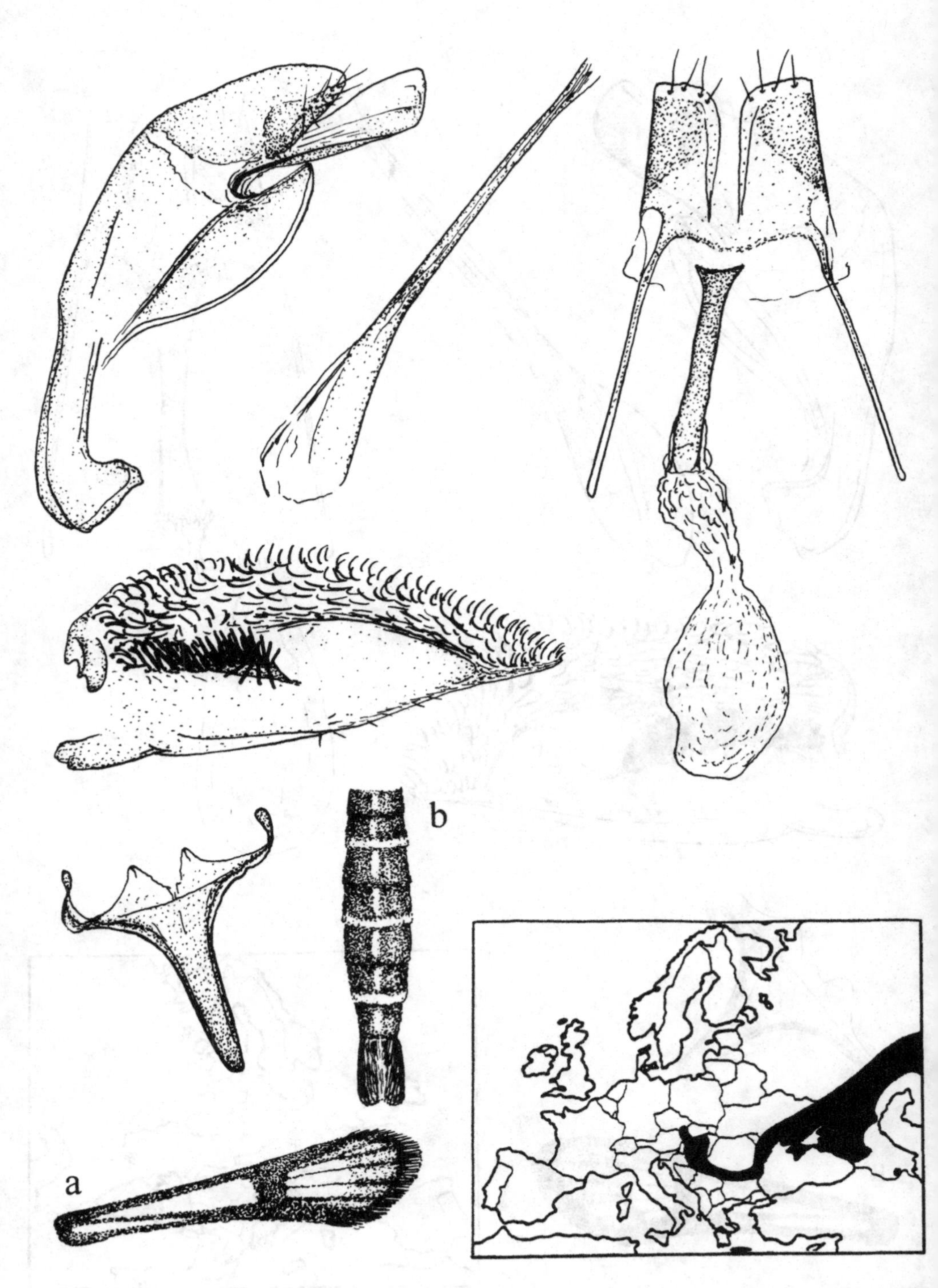

100. *Chamaesphecia crassicornis* Bartel – Slovakia, a: forewing, b: male abdomen (ZL).

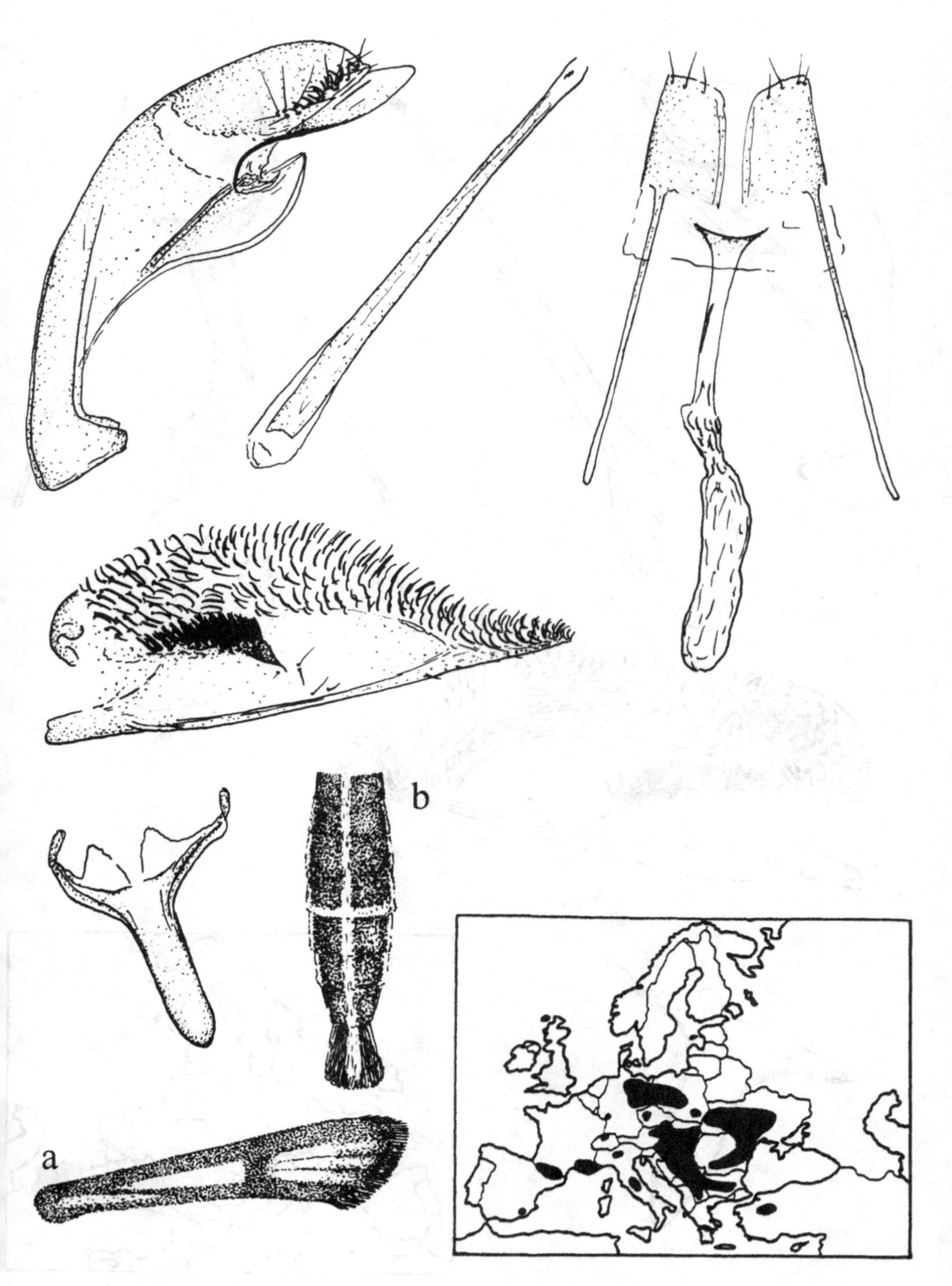

101. *Chamaesphecia leucopsiformis* (Esper) – Slovakia, a: forewing, b: male abdomen (ZL).

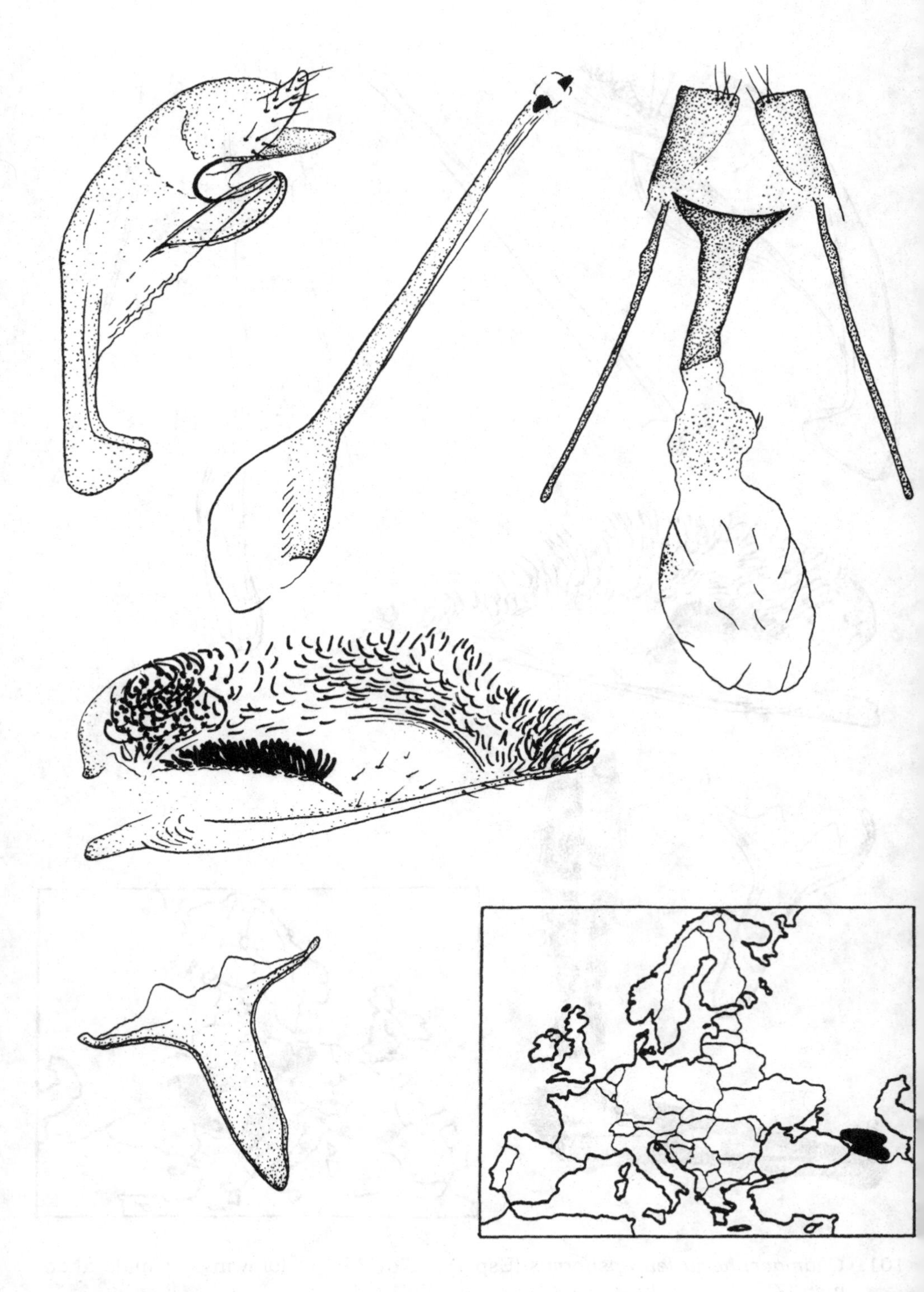

102. *Chamaesphecia guriensis* (Emich) – Georgia (ZMHB).

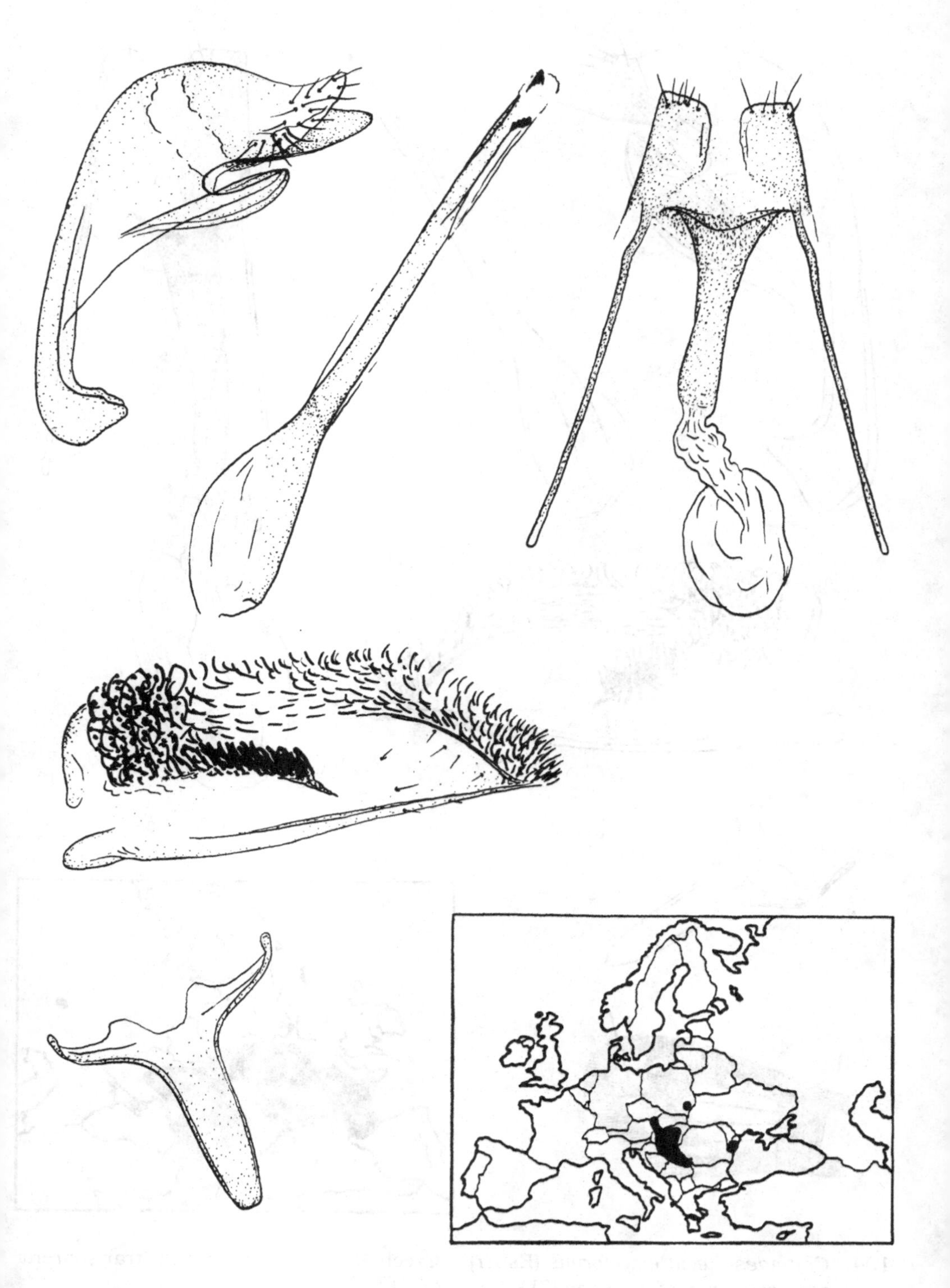

103. *Chamaesphecia hungarica* (Tomala) – Slovakia (ZL).

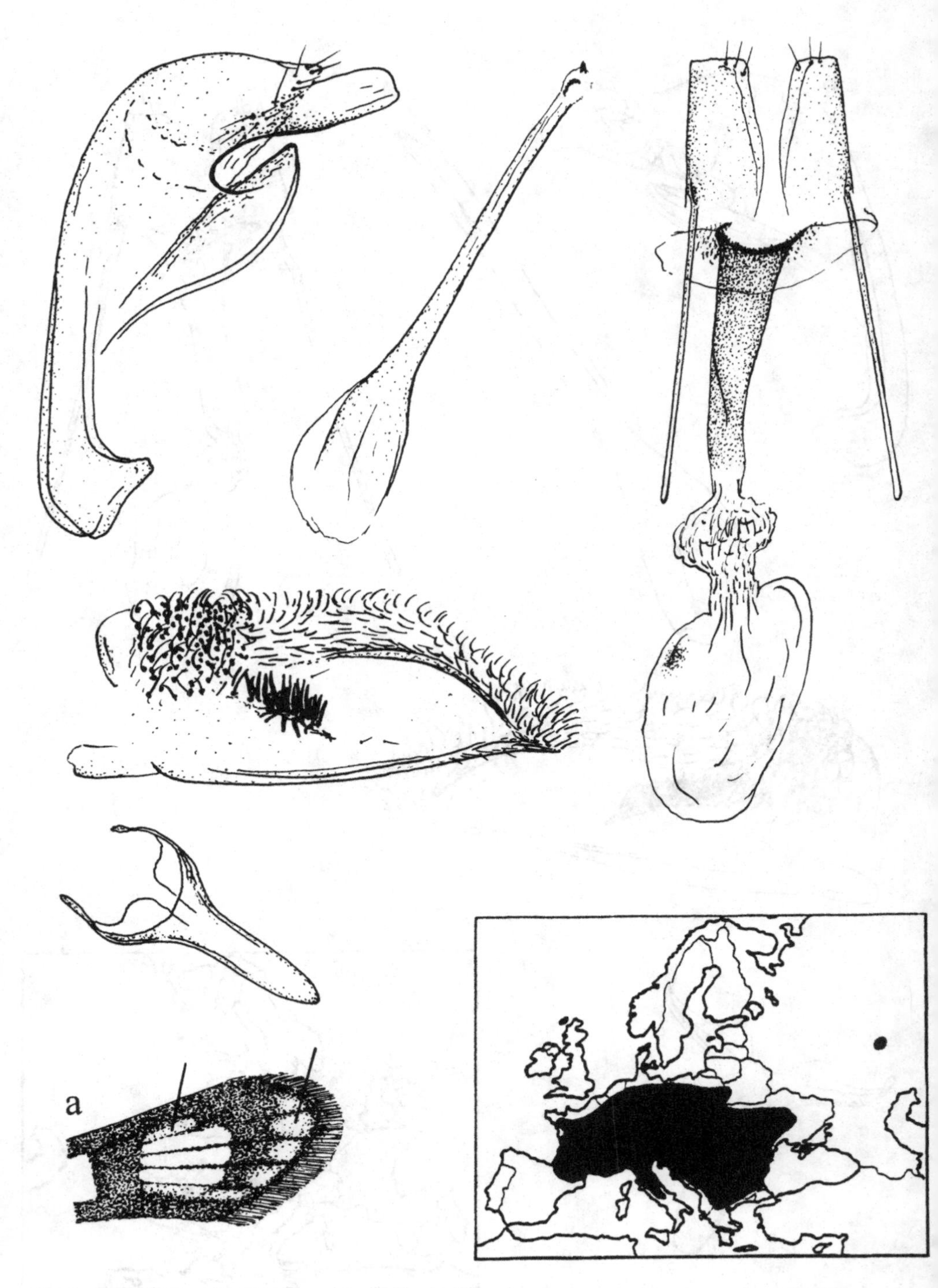

104. *Chamaesphecia empiformis* (Esper) – Czech Republic, a: external transparent area and apex of forewing (ZL).

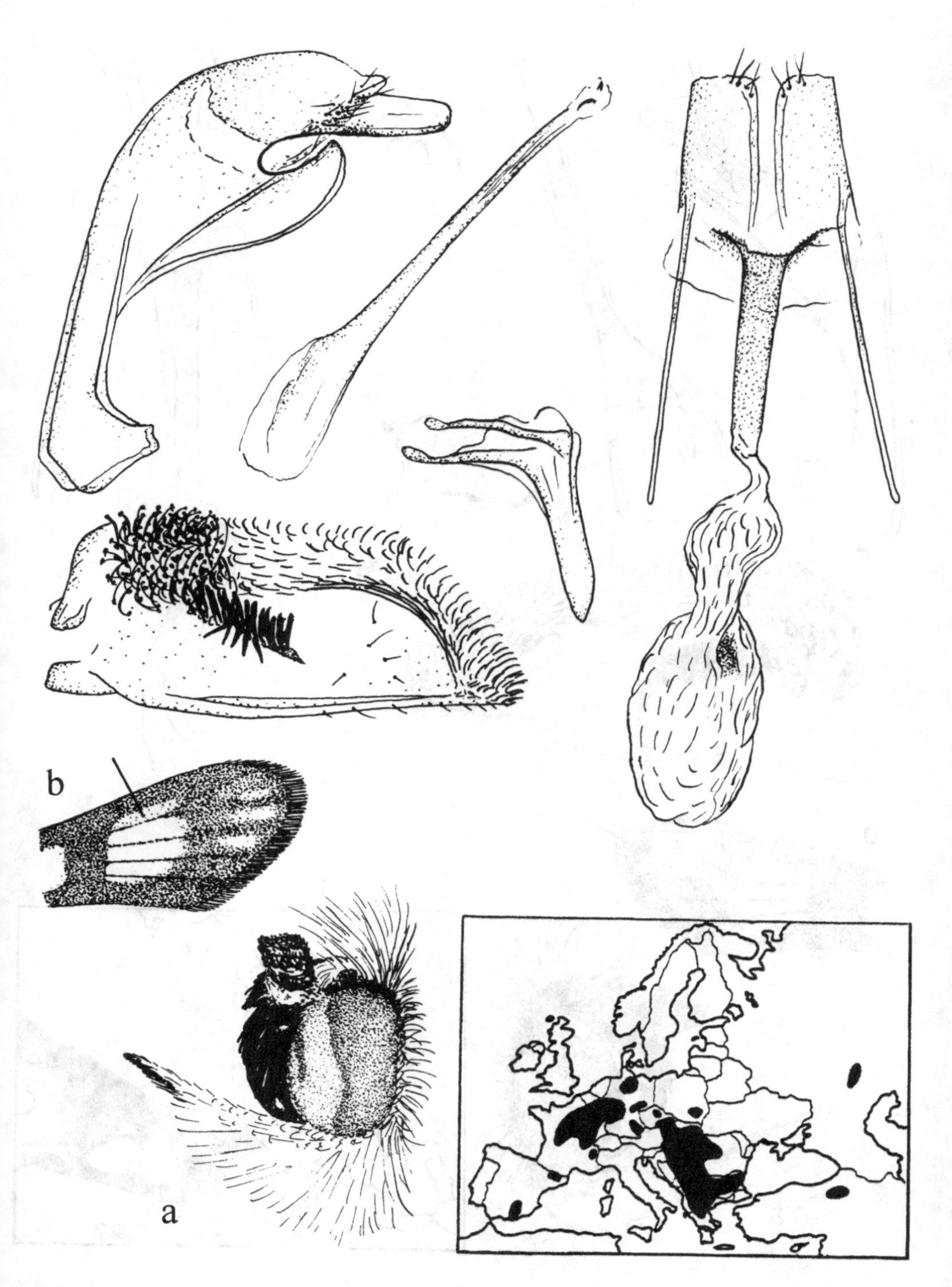

105. *Chamaesphecia tenthrediniformis* ([Denis & Schiffermüller]) – Czech Republic, a: head, b: external transparent area and apex of forewing (ZL).

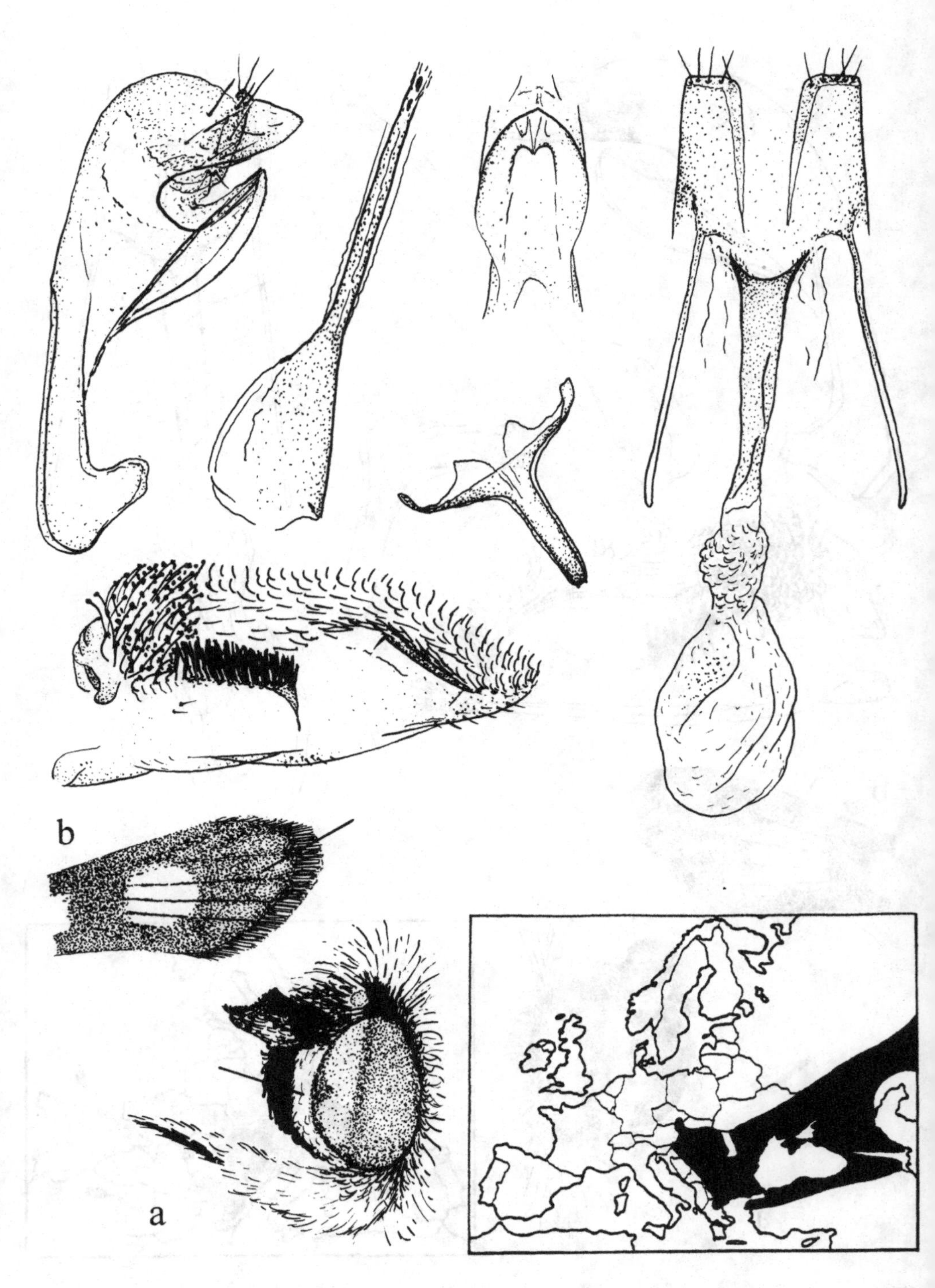

106. *Chamaesphecia astatiformis* (Herrich-Schäffer) – Czech Republic, a: head, b: external transparent area and apex of forewing (ZL).

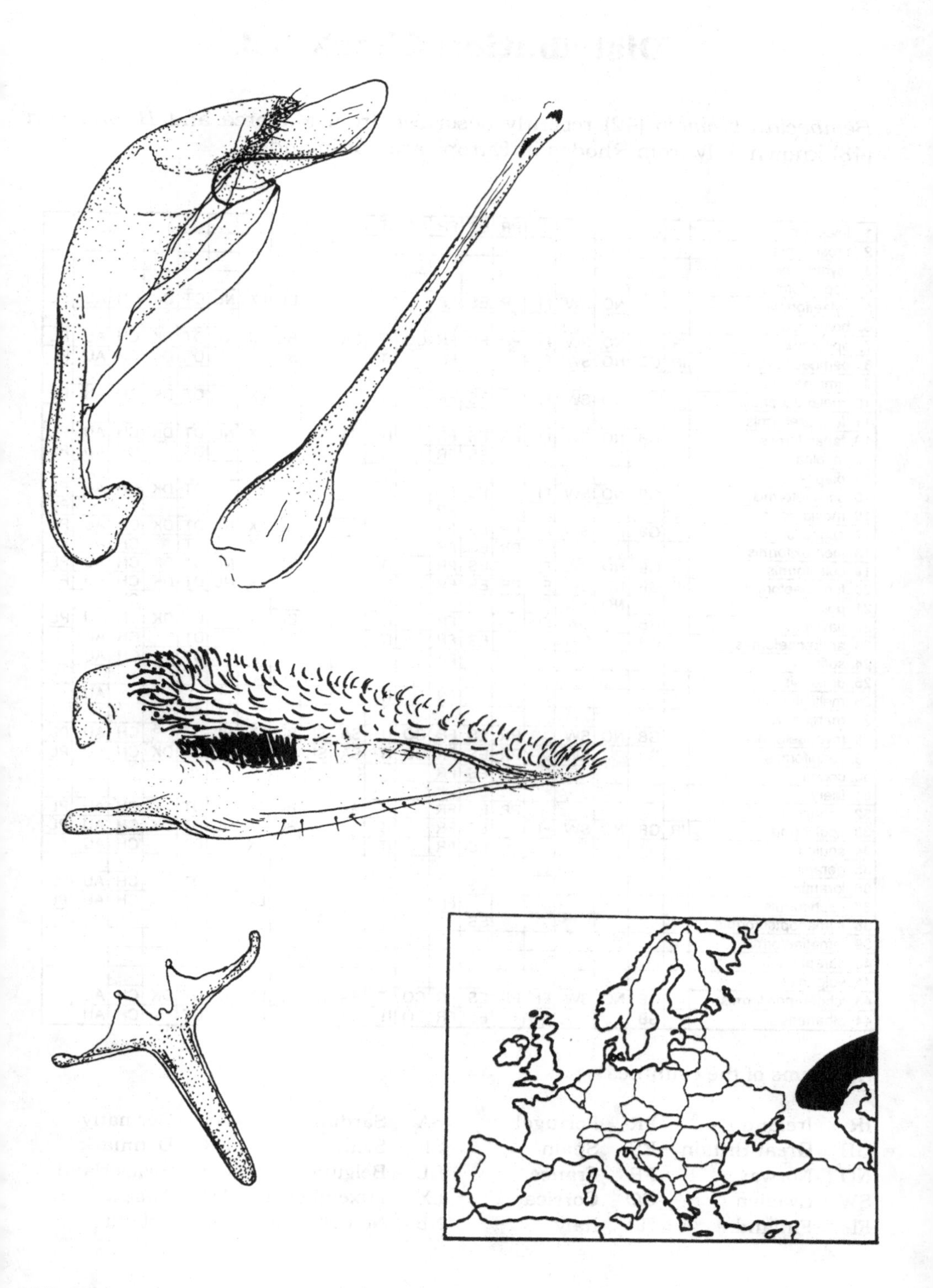

107. *Weismanniola agdistiformis* (Staudinger) – Russia (ZMHB).

Distribution Check-list

Bembecia abromeiti (42) recently described from Mallorca and *B. priesneri* (48) known only from Rhodes in Europe are not included.

1. tineiformis						PR	ES	FR	CO	IT	SA	SI								
2. brosiformis																				
3. myrmosaeformis																				
4. hoplisiformis																				
5. hylaeiformis			NO	SW	FI	PR	ES	FR		IT			BL	LX	NL	DT	DK	CH	AU	PL
6. bohemica																				
7. apiformis	IR	GB	NO	SW	FI	PR	ES	FR	CO	IT	SA	SI	BL	LX	NL	DT	DK	CH	AU	PL
8. bembeciformis	IR	GB	NO	SW	FI			FR		IT			BL	LX	NL	DT	DK	CH	AU	PL
9. pimplaeformis																				
10. melanocephala			NO	SW	FI		ES	FR		IT				LX		DT	DK	CH	AU	PL
11. fenusaeformis																				
12. tabaniformis		GB	NO	SW	FI	PR	ES	FR	CO	IT	SA	SI	BL	LX	NL	DT	DK	CH	AU	PL
13. insolita							ES	FR		IT		SI		LX		DT		CH	AU	PL
14. diaphana																				
15. scoliaeformis	IR	GB	NO	SW	FI		ES	FR		IT			BL	LX		DT	DK	CH	AU	PL
16. mesiaeformis								FR												PL
17. spheciformis		GB	NO	SW	FI	PR	ES	FR		IT			BL	LX	NL	DT	DK	CH	AU	PL
18. stomoxiformis						PR	ES	FR		IT				LX		DT		CH	AU	PL
19. culiciformis		GB	NO	SW	FI		ES	FR		IT			BL	LX	NL	DT	DK	CH	AU	PL
20. formicaeformis	IR	GB	NO	SW	FI	PR	ES	FR		IT			BL	LX	NL	DT	DK	CH	AU	PL
21. polaris			NO	SW	FI					IT								CH		
22. flaviventris		GB		SW	FI			FR		IT			BL	LX		DT	DK	CH	AU	PL
23. andrenaeformis		GB		SW			ES	FR		IT			BL	LX		DT		CH	AU	
24. soffneri								FR								DT		CH	AU	
25. uralensis																				
26. melliniformis								FR		IT									AU	
27. martjanovi																				
28. myopaeformis		GB	NO	SW		PR	ES	FR	CO	IT	SA	SI	BL	LX	NL	DT		CH	AU	PL
29. vespiformis		GB		SW		PR	ES	FR	CO	IT	SA	SI	BL	LX	NL	DT	DK	CH	AU	PL
30. codeti						PR	ES	FR			SA									
31. theryi							ES													
32. conopiformis						PR	ES	FR		IT	SA	SI	BL	LX		DT		CH	AU	PL
33. tipuliformis	IR	GB	NO	SW	FI		ES	FR		IT			BL	LX	NL	DT	DK	CH	AU	PL
34. spuleri							ES	FR		IT		SI		LX		DT		CH	AU	
35. geranii																				
36. loranthi							ES	FR		IT				LX		DT		CH	AU	PL
37. cephiformis								FR		IT				LX		DT		CH	AU	PL
38. hymenopteriformis							ES			IT		SI								
39. lomatiaeformis																				
40. sareptana																				
41. volgensis																				
43. ichneumoniformis		GB	NO	SW	FI	PR	ES	FR	CO	IT	SA	SI	BL	LX	NL	DT	DK	CH	AU	PL
44. albanensis		GB					ES	FR	CO	IT	SA	SI		LX		DT		CH	AU	

Acronyms of the countries

IR	Ireland	PR	Portugal	SA	Sardinia	DT	Germany
GB	Great Britain	ES	Spain	SI	Sicily	DK	Denmark
NO	Norway	FR	France	BL	Belgium	CH	Switzerland
SW	Sweden	CO	Corsica	LX	Luxembourg	AU	Austria
FI	Finland	IT	Italy	NL	Netherlands	PL	Poland

				SL	HR	YU	BH	MK	AL	BG	GR	CR	TR		RU					tineiformis 1.
	SK	HG	RO	SL	HR	YU	BH	MK	AL	BG	GR		TR	UI	RU					brosiformis 2.
			RO					MK	AL	BG	GR		TR	UI	RU					myrmosaeformis 3.
											GR									hoplisiformis 4.
CZ	SK	HG	RO	SL	HR	YU	BH	MK	AL	BG	GR			UI	RU	BR	LT	LV	EE	hylaeiformis 5.
CZ															RU					bohemica 6.
CZ	SK	HG	RO	SL	HR	YU	BH	MK	AL	BG	GR	CR	TR	UI	RU	BR	LT	LV	EE	apiformis 7.
CZ	SK			SL										UI			LT			bembeciformis 8.
								MK		BG	GR				RU					pimplaeformis 9.
CZ	SK	HG	RO	SL	HR									UI	RU	BR	LT	LV	EE	melanocephala 10.
												CR								fenusaeformis 11.
CZ	SK	HG	RO	SL	HR	YU	BH	MK	AL	BG	GR	CR	TR	UI	RU	BR	LT	LV	EE	tabaniformis 12.
CZ	SK	HG		SL	HR		BH	MK		BG	GR			UI						insolita 13.
						YU	BH	MK		BG										diaphana 14.
CZ	SK		RO	SL	HR					BG				UI	RU	BR	LT	LV	EE	scoliaeformis 15.
		HG	RO		HR	YU	BH	MK		BG	GR			UI	RU		LT		EE	mesiaeformis 16.
CZ	SK	HG	RO	SL	HR	YU	BH	MK	AL					UI	RU	BR	LT	LV	EE	spheciformis 17.
CZ	SK	HG	RO	SL	HR	YU	BH	MK	AL	BG	GR		TR	UI	RU		LT			stomoxiformis 18.
CZ	SK	HG	RO	SL	HR	YU	BH	MK	AL	BG	GR		TR	UI	RU	BR	LT	LV	EE	culiciformis 19.
CZ	SK	HG	RO	SL	HR	YU	BH	MK	AL	BG	GR			UI	RU	BR	LT	LV	EE	formicaeformis 20.
															RU					polaris 21.
CZ	SK	HG	RO	SL											RU		LT	LV	EE	flaviventris 22.
CZ	SK	HG	RO	SL	HR	YU	BH	MK	AL	BG	GR			UI	RU					andrenaeformis 23.
CZ	SK														RU					soffneri 24.
														UI	RU					uralensis 25.
	SK	HG		SL	HR	YU	BH			BG										melliniformis 26.
															RU					martjanovi 27.
CZ	SK	HG	RO	SL	HR	YU	BH	MK	AL	BG	GR	CR	TR	UI	RU	BR	LT	LV	EE	myopaeformis 28.
CZ	SK	HG	RO	SL	HR	YU	BH	MK	AL	BG	GR		TR	UI	RU	BR	LT			vespiformis 29.
																				codeti 30.
																				theryi 31.
CZ	SK	HG	RO	SL	HR	YU	BH	MK	AL	BG	GR		TR	UI	RU	BR				conopiformis 32.
CZ	SK	HG	RO	SL	HR	YU	BH	MK	AL	BG	GR		TR	UI	RU	BR	LT	LV	EE	tipuliformis 33.
CZ	SK	HG	RO	SL	HR	YU	BH	MK	AL	BG	GR		TR		RU					spuleri 34.
											GR									geranii 35.
CZ	SK	HG	RO	SL	HR	YU	BH	MK	AL	BG	GR									loranthi 36.
CZ	SK		RO	SL	HR	YU	BH	MK	AL	BG	GR			UI						cephiformis 37.
																				hymenopteriformis 38.
											GR									lomatiaeformis 39.
															RU					sareptana 40.
															RU					volgensis 41.
CZ	SK	HG	RO	SL	HR	YU	BH	MK	AL	BG	GR	CR	TR	UI	RU	BR	LT		EE	ichneumoniformis 43.
CZ	SK	HG	RO	SL	HR	YU	BH	MK	AL	BG	GR	CR	TR		RU					albanensis 44.

CZ Czech Republic
SK Slovakia
HG Hungary
RO Romania
BG Bulgaria
SL Slovenia
HR Croatia
YU Yugoslavia
BH Bosnia & Herzegovina
MK Macedonia
AL Albania
GR Greece
CR Crete
TR European Turkey
UI Ukraine
RU Eur. Russia (Kazakhstan)
BR Belarus
LT Lithuania
LV Latvia
EE Estonia

45. pavicevici																				
46. fibigeri							ES	FR												
47. scopigera							ES	FR		IT								CH	AU	
49. iberica						PR	ES	FR	CO	IT										
50. blanka																				
51. fokidensis																				
52. megillaeformis								FR								DT			AU	PL
53. puella																				
54. sirphiformis						PR	ES	FR		IT										
55. sanguinolenta																				
56. flavida												SI								
57. himmighoffeni						PR	ES	FR		IT										
58. uroceriformis						PR	ES	FR	CO	IT	SA	SI						CH		
59. chrysidiformis		GB				PR	ES	FR	CO	IT	SA	SI	BL	LX		DT		CH	AU	
60. minianiformis																				
61. doryliformis						PR	ES			IT	SA	SI								
62. triannuliformis								FR		IT		SI				DT		CH	AU	PL
63. meriaeformis						PR	ES	FR	CO	IT	SA	SI								
64. hispanica						PR	ES	FR												
65. koschwitzi							ES													
66. muscaeformis	IR	GB		SW			ES	FR		IT						DT	DK	CH	AU	PL
67. kautzi							ES													
68. aistleitneri							ES													
69. leucomelaena						PR	ES	FR	CO	IT	SA	SI								
70. affinis						PR	ES	FR		IT		SI	BL	LX		DT		CH	AU	
71. umbrifera																				
72. cirgisa																				
73. mannii																				
74. lanipes																				
75. mysiniformis						PR	ES	FR		IT										
76. anatolica																				
77. chalciformis										IT									AU	
78. schmidtiiformis										IT										
79. anthrax							ES													
80. maurusia												SI								
81. alysoniformis																				
82. aerifrons						PR	ES	FR	CO	IT	SA	SI				DT		CH		
83. albiventris																				
84. osmiaeformis									CO	IT	SA	SI								
85. ramburi						PR	ES	FR												
86. doleriformis										IT									AU	
87. thracica										IT										
88. dumonti								FR		IT						DT		CH	AU	
89. oxybeliformis																				
90. annellata										IT						DT			AU	PL
91. staudingeri												SI								
92. proximata																				
93. masariformis										IT		SI							AU	
94. nigrifrons								FR	CO					LX		DT			AU	PL
95. bibioniformis							ES	FR		IT									AU	
96. anthraciformis									CO	IT	SA	SI								
97. palustris										IT									AU	
98. euceraeformis							ES	FR		IT									AU	
99. amygdaloidis																			AU	
100. crassicornis																			AU	
101. leucopsiformis							ES	FR		IT						DT		CH	AU	PL
102. guriensis																				
103. hungarica																			AU	PL
104. empiformis							ES	FR		IT			BL	LX	NL	DT		CH	AU	PL
105. tenthrediniformis							ES	FR					BL	LX		DT			AU	PL
106. astatiformis																			AU	
107. agdistiformis																				
Total per country	6	16	13	17	13	25	48	51	16	53	17	25	19	27	12	35	14	34	46	29

				SL	HR	YU	BH	MK	AL	BG	GR									pavicevici 45.
																				fibigeri 46.
CZ	SK	HG		SL	HR		BH	MK		BG	GR			UI	RU					scopigera 47.
																				iberica 49.
												CR								blanka 50.
											GR									fokidensis 51.
CZ	SK	HG	RO	SL	HR	YU	BH	MK	AL	BG	GR			UI	RU					megillaeformis 52.
	SK	HG	RO							BG	GR			UI	RU					puella 53.
																				sirphiformis 54.
										BG	GR									sanguinolenta 55.
																				flavida 56.
				SL	HR															himmighoffeni 57.
	SK	HG	RO	SL	HR	YU	BH	MK	AL		GR	CR	TR		RU					uroceriformis 58.
			RO	SL	HR															chrysidiformis 59.
			RO			YU		MK	AL	BG	GR	CR	TR	UI						minianiformis 60.
																				doryliformis 61.
CZ	SK	HG	RO	SL	HR	YU	BH	MK	AL	BG	GR	CR	TR	UI	RU		LT		EE	triannuliformis 62.
																				meriaeformis 63.
																				hispanica 64.
																				koschwitzi 65.
CZ	SK		RO					MK	AL	BG	GR									muscaeformis 66.
																				kautzi 67.
																				aistleitneri 68.
			RO	SL	HR	YU	BH	MK	AL	BG	GR	CR	TR	UI						leucomelaena 69.
CZ	SK	HG	RO	SL	HR	YU	BH	MK	AL		GR		TR	UI	RU					affinis 70.
											GR									umbrifera 71.
			RO											UI	RU					cirgisa 72.
										BG			TR							mannii 73.
										BG										lanipes 74.
																				mysiniformis 75.
		HG	RO			YU					GR									anatolica 76.
	SK	HG	RO		HR	YU	BH	MK	AL	BG	GR	CR	TR	UI	RU					chalciformis 77.
			RO		HR	YU	BH	MK	AL	BG	GR		TR	UI	RU					schmidtiiformis 78.
																				anthrax 79.
																				maurusia 80.
			RO			YU		MK	AL	BG	GR	CR	TR							alysoniformis 81.
					HR		BH		AL		GR	CR								aerifrons 82.
										BG	GR									albiventris 83.
																				osmiaeformis 84.
																				ramburi 85.
CZ	SK	HG	RO		HR	YU	BH	MK	AL	BG	GR		TR	UI	RU					doleriformis 86.
			RO			YU		MK		BG	GR		TR							thracica 87.
CZ	SK	HG	RO	SL	HR	YU	BH	MK		BG	GR			UI	RU					dumonti 88.
			RO			YU		MK		BG				UI	RU					oxybeliformis 89.
CZ	SK	HG	RO		HR	YU	BH	MK	AL	BG	GR		TR	UI	RU					annellata 90.
																				staudingeri 91.
						YU		MK	AL	BG	GR		TR							proximata 92.
CZ	SK	HG	RO	SL	HR	YU	BH	MK	AL	BG	GR	CR	TR	UI	RU					masariformis 93.
CZ	SK	HG	RO		HR	YU	BH	MK		BG	GR		TR	UI	RU					nigrifrons 94.
	SK	HG	RO	SL	HR	YU	BH	MK	AL	BG	GR	CR	TR	UI	RU					bibioniformis 95.
																				anthraciformis 96.
CZ	SK	HG	RO		HR	YU								UI	RU					palustris 97.
CZ	SK	HG	RO	SL	HR	YU	BH	MK	AL	BG				UI	RU					euceraeformis 98.
																				amygdaloidis 99.
CZ	SK	HG	RO		HR	YU				BG				UI	RU					crassicornis 100.
CZ	SK	HG	RO		HR	YU	BH	MK		BG				UI						leucopsiformis 101.
															RU					guriensis 102.
CZ	SK	HG	RO		HR	YU														hungarica 103.
CZ	SK	HG	RO	SL	HR	YU	BH	MK	AL	BG	GR			UI	RU	BR				empiformis 104.
CZ	SK	HG	RO	SL	HR	YU	BH	MK	AL	BG	GR				RU					tenthrediniformis 105.
CZ	SK	HG	RO	SL	HR	YU	BH	MK		BG	GR		TR	UI	RU					astatiformis 106.
															RU					agdistiformis 107.
42	47	44	53	41	50	50	44	50	40	55	55	17	31	45	54	14	17	11	14	Total per country

References

Aistleitner, E. & Aistleitner, U., 1997. In memoriam Dirk Hamborg: Neue und bemerkenswerte Nachweise von Glasflüglern aus Vorarlberg (Austria occ.) und dem Fürstentum Liechtenstein (Lepidoptera, Sesiidae).– *Entomofauna* **18**: 213-220.

Alphéraky, S. N., 1882. Lépidoptères du district de Kouldja et des montagnes environnantes.– *Hor. Soc. ent. Ross.* **17**: 15-103.

Anikin, V., Sachkov, S. A. & Zolotuhin, V. V., 2000. „Fauna lepidopterologica Volgo-Uralensis" 150 years later: changes and additions. Part 2. Bombyces and Sphinges (Insecta, Lepidoptera).– *Atalanta, Würzburg* **31**: 265-292.

Arheilger, T., 1992. Ein Nachweis von *Paranthrene novaki* Toševski für Griechenland (Lepidoptera, Sesiidae).– *Nachr. ent. Ver. Apollo (N. F.)* **13**: 249-251.

Arnscheid, W. R., 2000. Sesiidae. *In* Die Macrolepidopteren-Fauna Westliguriens (Riviera dei Fiori und Ligurische Alpen in Oberitalien)(Insecta, Lepidoptera).– *Neue ent. Nachr.* **47**: 58-59.

Auer, H., 1967. Aufstellung der von mir festgestellten Aegeriden (Sesien) in Oberösterreich, insbesondere im Raume von Linz, Mühlviertel u. Donauraum.– *Steyrer Entomologenrunde, Jahresabschlussber.* **9**: 38-39.

Bąkowski, M., 1995. *Chamaesphecia tenthrediniformis* (Denis & Schiffermüller, 1775)(Lepidoptera, Sesiidae) – a new species of clearwing moth to the fauna of Poland.– *Wiad. ent.* **14**: 169-172 (in Polish).

– 1996. The clearwing moths (Lepidoptera, Sesiidae) of xerothermic habitats of the Miechów – Sandomierz District.– *Wiad. ent.* **15**: 43-50 (in Polish).

– 1997. New records of *Synanthedon stomoxyformis* (Hübner, 1790)(Lepidoptera: Sesiidae) from Poland.– *Wiad. ent.* **16**: 121 (in Polish).

– 1998. A contribution to the knowledge of the clearwing moths of Turkey (Lepidoptera: Sesiidae).– *Phegea* **26**: 85-86.

Bąkowski, M. & Hołowiński, M., 1996. *Chamaesphecia hungarica* (Tomala, 1901), a species of the clearwing moth new to the Polish fauna (Lepidoptera, Sesiidae).– *Wiad. ent.* **15**: 51-54 (in Polish).

– 1997. Clearwing moths (Lepidoptera: Sesiidae) of the south-eastern part of Polesie Lubelskie region (E Poland).– *Wiad. ent.* **16**: 107-114 (in Polish).

Bąkowski, M. & Surmacki, A., 1995. A new record of *Synanthedon mesiaeformis* (Herrich-Schäffer, 1846) from Poland.– *Wiad. ent.* **14**: 60 (in Polish).

Bartel, M., 1906a. Eine neue Sesia-Art aus der Schweiz.– *Int. ent. Z. (Frankfurt)* **19**: 190-191.

– 1906b. Drei neue russische Sesia-Arten.– *Soc. ent.* **20**: 169-170.

– 1912. Familie Aegeriidae (Sesiidae).– *In* A. Seitz, *Die Gross-Schmetterlinge der Erde, I. Abt.: Die Gross-Schmetterlinge des palaearktischen Faunengebietes.* **2**: 375-416. Stuttgart.

Bartsch, D. & Bettag, E., 1997. Eine neue Art der Gattung *Bembecia* Hübner, 1819 aus Südwesteuropa: *Bembecia psoraleae* spec. nov. (Lepidoptera: Sesiidae).– *Nachr. ent. Ver. Apollo (N. F.)* **18**: 29-40.

Bartsch, D. & Pelz, V., 1997. Untersuchungen zur Biologie und Phänologie einer hochsubalpinen Population von *Synanthedon soffneri* Špatenka 1983 aus der Schweiz (Lepidoptera: Sesiidae).– *Mitt. ent. Ver. Stuttgart* **32**: 112-115.

Bellier, J. B. E., 1860. Observations sur la fauna entomologique de la Sicilia.– *Annls Soc. ent. Fr.* **8**: 677-713.

Bettag, E., 1991. Zur Biologie und Verbreitung einiger Glasflügler (Lep. Aegeriidae) in Rheinhessen-Pfalz.– *Pfälzer Heimat* **42**: 82-84.

Bettag, E. & Bläsius, R., 1999a. A propos de 4 Sésies récemment découvertes en France (Lepidoptera, Sesiidae).– *Rev. Assoc. roussillonn. Ent.* **8**: 62-68.

Bettag, E. & Bläsius, R., 1999b. Über den Status von *Dipsosphecia megillaeformis* var. *tunetana* (Lepidoptera: Sesiidae).– *Phegea* **27**: 93-101.

Bläsius, R., 1992. *Chamaesphecia aerifrons* Zeller, 1847 (Lepidoptera, Sesiidae) – erster sicherer Nachweis aus Mitteleuropa.– *Pfälzer Heimat* **43**: 129-134.

– 1993: Neues vom Eichenzweig-Glasflügler *Paranthrene insolita* Le Cerf, 1914 (= *Paranthrene novaki* Toševski, 1987)(Lep., Sesiidae).– *Melanargia* **5**: 37-45.

Bläsius, R. & Herrmann, R., 1992. *Synanthedon loranthi* (Králíček, 1966) auch an der Obermosel (Lep., Sesiidae).– *Melanargia* **4**: 35-36.

Blum, E., 1990. Drei weitere neue Glasflüglerarten in der Pfalz (Lepidoptera, Aegeriidae).– *Pfälzer Heimat* **41**: 184-189.

Blum, E. & Bläsius, R., 1991. *Bembecia albanensis* Rebel, 1918, eine "neue" Glasflüglerart in Rheinland-Pfalz (Lepidoptera, Aegeriidae).– *Pfälzer Heimat* **42**: 80-81.

Boisduval, J. B. A. D., [1828]. *Europaeorum Lepidopterorum Index methodicus, I.* 103 pp. Paris.

– 1840. *Genera et index methodicus Europaeorum Lepidopterorum.* 238 pp. Paris.

Borkhausen, M. B., 1789. *Naturgeschichte der europäischen Schmetterlinge nach systematischer Ordnung.* Vol. 2, 239 pp., 1 pl. Frankfurt.

Bournier, A. & Khial, B., 1968. *Dipsosphecia scopigera* Scop. la sésie du sainfoin.– *Annls Épiphyties* **19**: 235-260.

Bradley, J.D., Fletcher, D. S. & Whalley, P. E. S., 1972. Order XXIV: Lepidoptera. *In* G. S. Kloet & W. D. Hincks (eds): *A check list of British insects (2nd edn.).* 11(2), 153 pp. London.

Bradley, J. D. & Fletcher, D. S., 1974. Addenda & Corrigenda to the Lepidoptera part of Kloet & Hincks Check list of British Insects (Edn. 2), 1972.– *Entomologist's Gaz.* **25**: 219-223.

Brock, J. P., 1971. A contribution towards an understanding of the morphology and phylogeny of the ditrysian Lepidoptera.– *J. nat. Hist.* **5**: 29-102.

Buszko, J., 1973. [*Chamaesphecia triannuliformis* (Frr.)(Lep. Aeg.): a new clearwing moth species for Polish fauna].– *Przegl. zool.* **17**: 190-192 (in Polish).

Buszko, J. & Hołowiński, M., 1994. On the occurrence of *Aegeria mesiaeformis* (Herrich-Schäffer, 1845)(Lepidoptera, Sesiidae) in Poland.– *Wiad. ent.* **13**: 121-123 (in Polish).

Căpuşe, I. K., 1971. Zur Morphologie und Taxonomie einiger Typen der Aegeriidae (Lepidoptera) aus der R. Pungeler-Sammlung im Zoologischen Museum zu Berlin.– *Trav. Mus. Hist. nat. Gr. Antipa* **11**: 239-292.

– 1973. Zur Systematik und Morphologie der Typen der Sesiidae (Lepidoptera) in der R. Püngeler-Sammlung des Zoologischen Museums zu Berlin.– *Mitt. münch. ent. Ges.* **63**: 134-171.

Čila, P. & Špatenka, K., 1989. Faunistic records from Czechoslovakia.– *Acta ent. bohemoslov.* **86**: 78.

Clerck, C., 1759. *Icones Insectorum rariorum cum nominibus eorum trivialibus locisque e C. Linnaei Systema Naturae allegatis. Sectio prima.* 8 pp., 16 pls. Holmiae.

Contarini, E. & Fiumi, G., 1984. *Chamaesphecia palustris* Kautz, nuova specie per la fauna italiana ed osservazioni sulla biologia larvale (Lep. Aegeriidae).– *Lav. Soc. venez. Sci. nat.* **9**: 33-38.

Cungs, J., 1991. Beitrag zur Faunistik und Ökologie der Schmetterlinge im ehemaligen Erzabbaugebiet "Haardt" bei Düdelingen (Insecta, Lepidoptera).– *Trav. sc. Mus. natl Hist. nat. Luxemb.* **17**: 1-364, 14 pls.

– 1998. Beitrag zur Faunistik und Ökologie der Glasflügler (Lepidoptera, Sesiidae) im südlichen Erzbecken Luxemburgs.– *Bull. Soc. nat. luxemb.* **99**: 165-186.

Cungs, J. & Meyer, M., 1990: Beobachtungen von Glasflüglern im Jahre 1989 (Lepidoptera: Sesiidae).– *Paiperlek* **12**: 17-20.

Curtis, J., 1825. *British Entomology; being illustrations and descriptions of the genera of insects found in Great Britain and Ireland.* Vol. 2, 53 pp. London.

Curtis, W. P., 1957. *Sciapteron tabaniformis* Rott.– *Entomologist's Rec. J. Var.* **69**: 218-219.

Dalla Torre, K. W. & Strand, E., 1925. Aegeriidae. *In* K. W. Dalla Torre & E. Strand (eds): *Lepidopterorum catalogus,* 31, 202 pp. Berlin.

Dalman, J. W., 1816. Försök till Systematisk Uppställning af Sveriges Fjärillar.– *Kongl. Svenska Vetenskaps-Acad. Handl.* **37**: 48-101, 199-225.

Dehne, A., 1850. Beschreibung einer neuen *Setia* (*Sesia* Fabr.) mit Federfühlern, *Pennisetia anomala* m.– *Stettin ent. Ztg* **11**: 28-29.

[Denis, M. & Schiffermüller, I.], 1775. *Ankündung eines systematischen Werkes von den Schmetterlingen der Wienergegend.* 323 pp., 3 pls. Wien.

Doczkal, D. & Rennwald, E., 1992. Beobachtungen zur Lebensweise, Verbreitung und Gefährdung des "Kreuzdorn-Glasflüglers" *Synanthedon stomoxiformis* (Hübner, 1790) in Baden-Württemberg (Lepidoptera, Sesiidae).– *Atalanta, Würzburg* **23**: 259-274.

Duckworth, W. D. & Eichlin, T. D., 1974. Clearwing moths of Australia and New Zealand (Lepidoptera: Sesiidae).– *Smith. Contrib. Zool.* **180**: 1-45.

– 1977. A classification of the Sesiidae of America North of Mexico (Lepidoptera: Sesioidea).– *Occ. Pap. Ent.* **26**: 1-54.

Dutreix, C., 2000. L'entomofaune du Morvan. Contribution á l'étude des Sésies (Lepidoptera).– *Bull. Soc. hist. nat. Autun* **168(1998)**: 37-40.

Dutreix, C. & Morel, D., 2000a. *Paranthrene insolita* Le Cerf, espèce peu connue de France et nouvelle pour la Bourgogne (Lepidoptera Sesiidae).– *Bull. Soc. ent. Mulhouse* **2000**: 10-11.

Dutreix, C. & Morel, D., 2000b. Remarques sur *Synanthedon flaviventris* Stgr. en France, espèce á présence certaine dans le Morvan (Lepidoptera Sesiidae).– *Bull. Soc. ent. Mulhouse* **2000**: 18-20.

Embacher, G., 1994. Zwei neue Sesien-Arten für die Fauna Salzburgs (Lepidoptera, Sesiidae).– *NachrBl. bayer. Ent.* **43**: 46-47.

Emich von Emöke, G., 1872. Descriptions de Lépidoptères de Transcaucasie.– *Revue Mag. Zool.* **23**: 63-64.

Engelhard, H., 1974. Beiträge zur Verbreitung und Biologie der Aegeriidae (Lep.). 1. Folge: Beiträge I-III.– *Ent. Ber.* **1974**: 89-90.

– 1975. Beiträge zur Verbreitung und Biologie der Aegeriidae (Lep.). 2. Folge: Beitrag IV.– *Ent. Ber.*, **1975**: 27.

– 1978. Die Suche nach der Futterpflanze und den Praeimaginalstadien von *Pyropteron minianiforme* Frr., 1843 (Lep., Sesiidae).– *Ent. Ber.* **1978**: 10-12.

Esper, E. J. C., 1778-[1804]. *Die Schmetterlinge in Abbildungen nach der Natur mit Beschreibungen.* Vol. 2, 234 pp., 36 pls., Supplement 2(2), 52 pp., 11 pls. Erlangen.

Eversmann, E. F., 1841. *Fauna lepidopterologica Volgo-Uralensis exhibens lepidopterorum species.* 633 pp. Casani.

Fabricius, J. C., 1775. *Systema entomologiae, sistens insectorum classes, ordines, genera, species, adiectis synonymis, locis, descriptionibus, observationibus.* 832 pp. Flensburgi & Lipsiae.

– 1807. [Contribution] *In* J. C. Illiger: Die neueste Gattungs-Eintheilung der Schmetterlinge aus den linneischen Gattungen Papilio und Sphinx.– *Mag. Insektenk.* **6**: 277-295.

Failla-Tedaldi, L., 1890. Contribuzione alla fauna lepidotterologica della Sicilia: Descrizione di alcune nuove specie.– *Naturalista Sicil., Giorn. Sci. nat.* **10**: 25-31.

Fibiger, M. & Kristensen, N. P., 1974. The Sesiidae (Lepidoptera) of Fennoscandia and Denmark.– *Fauna ent. scand.* **2**: 1-91.

Fiumi, G. & Fabbri, R. A., 1996. *Chamaesphecia doleriformis* (Herrich-Schäffer, 1846) in Emilia-Romagna (Insecta, Lepidoptera, Sesiidae).– *Quad. Studi nat. Romagna* **5**: 41-47.

Freina, J. J. de, 1997. *Die Bombyces und Sphinges der Westpalaearktis (Insecta, Lepidoptera), 4. Sesioidea: Sesiidae.* 431 pp., 31 pls. München.

Freyer, C. F., 1833-1858. *Neuere Beiträge zur Schmetterlingskunde mit Abbildungen nach der Natur.* Vol. 2, 162 pp., 96 pls., vol. 4, 167 pp., 96 pls, vol. 5, 166 pp., 96 pls. Augsburg.

Fuchs, F., 1908. *Sesia spuleri* nov. spec. – *Int. ent. Z. (Guben)* **2**: 33.

Garrevoet, T. C., 1995. *Synansphecia affinis* (Staudinger, 1856): nieuw voor de Belgische fauna (Lepidoptera: Sesiidae).– *Phegea* **23**: 141-143.

Garrevoet, T. C. & Garrevoet, W. 2000. *Synanthedon flaviventris*, een nieuwe soort voor de Belgische fauna (Lepidoptera: Sesiidae).– *Phegea* **28**: 153-154.

Garrevoet, T. C. & Laštůvka, Z., 1998. *Chamaesphecia nigrifrons* new to the Czech Republic (Lepidoptera: Sesiidae).– *Phegea* **26**: 21-22.

Garrevoet, T. C. & Vanholder, B., 1996. *Synanthedon perigordensis* sp. n. (Lepidoptera: Sesiidae).– *Phegea* **24**: 141-148.

Gelin, H. & Lucas, D., 1913. Catalogue des Lépidoptères observés dans l' Ouest de la France, 1. Macrolépidoptères.– *Mem. Soc. hist. sci. Deux-Sèvres.* Niort, **1912**: 1-232.

Georges, P., 1984. Les Lépidoptères de la famille des Sesiidae (= Aegeriidae).– *Pares nationaux Ardenne & Gaume* **39**: 80-84.

Georgiev, G. & Beshkov, S., 2000. New and little known lepidopteran (Lepidoptera) phytophages on the poplars (*Populus* spp.) in Bulgaria.– *Anz. Schädlingsknd.* **73**: 1-4.

Gorbunov, O., 1991. Six new species of the clearwing moths from the Caucasus, USSR.– *Atalanta, Würzburg* **22**: 125-143.

– 1992a. [Revision of the types of the Sesiidae (Lepidoptera), preserved in the collection of the Zoological Museum of Kiev State University].– *Ent. Obozr.* **71**: 121-133 (in Russian).

– 1992b. [Lectotype designation of clearwing moths (Lepidoptera, Sesiidae) in collections of Zoological Institute of Russian Academy of Sciences and of Zoological Museum of Kiev University].– *Vest. Zool.* **3**: 69-71 (in Russian).

– 1994. A new species of the genus *Bembecia* Hübner, [1819] from the European part of Russia (Lepidoptera, Sesiidae).– *Atalanta, Würzburg* **25**: 563-566.

Gorbunov, O. & Tshistjakov, Yu. A., 1995. A review of the clearwing moths (Lepidoptera, Sesiidae) of the Russian Far East.– *Far East. Entomologist* **10**: 1-18.

Hamborg, D., 1991. Der Glasflügler *Paranthrene novaki* (Toševski, 1987), ein Neufund für Österreich (Lep., Sesiidae).– *Mitt. Abt. Zool. Landesmus. Joanneum* **44**: 35-42.

– 1993. Fünf für die Steiermark neue *Synanthedon*-Arten (Lepidoptera, Sesiidae).– *Entomofauna* **14**: 149-170.

– 1994a. Zur Lebensweise der Raupen sowie zur Variabilität der Imagines von *Synanthedon andrenaeformis* (Laspeyres, 1801) in der Steiermark (Lepidoptera, Sesiidae).– *Mitt. Abt. Zool. Landesmus. Joanneum* **48**: 19-36.

– 1994b. Weitere Glasflügler-Neufunde in der Steiermark (Lepidoptera, Sesiidae).– *Mitt. Abt. Zool. Landesmus. Joanneum* **48**: 37-40.

– 1994c. Zwei für Kärnten neu nachgewiesene Schmetterlingsarten sowie weitere Funde aus der Familie der Glasflügler (Lepidoptera, Sesiidae).– *Carinthia II*, **184/104**: 515-518.

Hampson, G. F., 1919. A classification of the Aegeriidae of the Oriental and Ethiopian Regions.– *Novit. zool.* **26**: 46-119.

Heath, J. & Emmet, A. M. (eds), 1985. *The moths and butterflies of Great Britain and Ireland*, Vol. 2, 460 pp., 16 pls. Colchester.

Heidemaa, M. & Kesküla, T., 1992. Estonian Sesiidae. Annotated checklist with distribution maps.– *Acta Mus. zool. Univ. Tartuensis* **5**: 1-32.

Heppner, J. B., 1982. Dates of selected Lepidoptera literature for the Western Hemisphere fauna.– *J. Lepid. Soc.* **36**: 87-111.

Heppner, J. B. & Duckworth, W. D., 1981. Classification of the superfamily Sesioidea (Lepidoptera: Ditrysia).– *Smithson. Contr. Zoll.* **314**: 1-144.

Herrich-Schäffer, G. A. W., 1843-1856. *Systematische Bearbeitung der Schmetterlinge von Europa, als Text, Revision und Supplement zu J. Hübner's Sammlung europäischer Schmetterlinge.* Vol. 2, 450 pp., 191 pls., Vol. 6, Nachtrag, 178 pp., 2 pls. Regensburg.

Herrmann, R. & Bläsius, R., 1991. *Chamaesphecia similis* Laštůvka 1983 an Mosel und Mittelrhein (Lep., Sesiidae).– *Melanargia* **3**: 101-103.

Hołowiński, M. & Miłkowski, M., 1999. *Synanthedon loranthi* (Králíček, 1966), a species of the clearwing moth new to the Polish fauna (Lepidoptera: Sesiidae).– *Wiad. ent.* **18**: 99-102 (in Polish).

Hübner, J., 1790. *Beiträge zur Geschichte der Schmetterlinge.* Vol. 2, 134 pp., 16 pls. Augsburg.

– 1796-[1825]. *Sammlung europäischer Schmetterlinge. Part 2, Sphinges*, 32 pp., 35 pls. Augsburg.

– 1816-[1826]. *Verzeichniss bekannter Schmettlinge* [sic]. 432 pp. Augsburg.

Ilinskij, A. I., 1962. [*A Key to forest pests*]. 391 pp., 80 pls. Moskva (in Russian).

Issekutz, L., 1950. *Chamaesphecia hungarica* Tomala: bona species (Lepidoptera).– *Folia ent. hung.* **3**: 49-55.

Joannis, J., 1908. Contribution a l'étude de lépidoptères du Morbihan.– *Annls Soc. ent. Fr.* **77**: 689-838.

– 1909. Description d'une nouvelle espèce française du genre *Sesia* (Lep., Sesiidae).– *Bull. Soc. ent. Fr.*, **1909**: 183-187.

Kallies, A., 1997a. Eine neue und bemerkenswerte *Synanthedon*-Art aus Griechenland (Lepidoptera: Sesiidae).– *Nachr. ent. Ver. Apollo (N. F.)* **18**: 59-66.

– 1997b. Synopsis der in der Bundesrepublik Deutschland nachgewiesenen Glasflügler-Arten (Lep., Sesiidae).– *Ent. Nachr. Ber.* **41**: 107-111.

– 1999. Revision of the south-western Palaearctic species of *Synansphecia* (Sesiidae).– *Nota lepid.* **22**: 82-114.

Kallies, A. & Hamborg, D., 1993. Wenig bekannte Sesiiden-Arten aus Deutschland mit Anmerkungen zur Biologie und Verbreitung (Lep., Sesiidae).– *Mitt. Thüringer ent. Verb.* **0**(Probeheft): 4-12.

Kallies, A., Petersen, M. & Riefenstahl, H. G., 1998. Drei neue Glasflüglerarten aus Anatolien (Lepidoptera, Sesiidae).– *Esperiana* **6**: 56-62.

Kallies, A. & Riefenstahl, H. G., 2000. A new species of *Bembecia* Hübner, [1819] from the Balearic Island of Mallorca (Lepidoptera: Sesiidae).– *Ent. Z.* **110**: 359-363.

Kallies, A. & Sobczyk, T., 1993. Zum Vorkommen von *Synansphecia triannuliformis* (Freyer, 1845) in Deutschland (Lepidoptera, Sesiidae).– *Ent. Nachr. Ber.* **37**: 133-136.

Karsholt, O. & Razowski, J. (eds), 1996. *The Lepidoptera of Europe. A distributional Checklist.* 380 pp. Stenstrup.

Kautz, H., 1927. Eine neue Sesiidae.– *Z. öst. EntVer.* **12**: 1-4.

Köhler, J., 1991. *Paranthrene novaki* Tosevski 1987 auch in Deutschland (Lepidoptera: Sesiidae).– *Ent. Z. Frankfurt* **101**: 273-278.

– 1992. Die Glasflügler (Lepidoptera: Sesiidae) im Hannoverschen Wendland (Ost-Niedersachsen). Biologische und ökologische Ergebnisse.– *Braunschw. naturkdl. Schr.* **4**: 101-141.

– 1996. Die Glasflügler (Lepidoptera: Sesiidae) im Hannoverschen Wendland (Ost-Niedersachsen): *Sesia bembeciformis* und *Synanthedon flaviventris*.– *Braunschw. naturkdl. Schr.* **5**: 55-70.

Králíček, M., 1966. Neue Glasflügler-Art der Gattung *Aegeria* F. aus Südmähren (Sesiidae, Lep.).– *Acta Mus. Moraviae, Sci. nat.* **51**: 231-236.

– 1969. Eine neue Glasflügler-Art der Gattung *Chamaesphecia* Spuler 1910 aus Südmähren (Sesiidae, Lep.).– *Acta Mus. Moraviae, Sci. nat.* **54**: 115-122.

– 1975a. Eine neue Glasflügler-Art aus der Gattung *Aegeria* Fabricius, 1807 (*Synanthedon* Hübner, 1819) aus der Slowakei (Lep., Sesiidae).– *Annot. Zool. Bot. (Bratislava)* **1975**: 1-9.

– 1975b. Zur Bionomie und Verbreitung einiger Glasflügler- Arten aus der Tschechoslowakei (Lepidoptera, Sesiidae).– *Acta ent. bohemoslov.* **72**: 115-120.

Králíček, M. & Povolný, D., 1974. *Pennisetia bohemica* sp. n. – a new species of clearwing moth (Lepidoptera, Sesiidae) from Czech Republic.– *Acta Mus. Moraviae, Sci. nat.* **59**: 165-182.

– 1977. Drei neue Arten und eine neue Untergattung der Tribus Aegeriini (Lepidoptera, Sesiidae) aus der Tschechoslowakei.– *Věst. čs. Spol. zool.* **41**: 81-104.

Kranjčev, R., 1979. *Synanthedon croaticus* sp. nov. (Lepid. Aegeridae).– *Acta ent. jugosl.* **14**(1978): 27-33.

Kristal, P. M., 1990. *Synanthedon loranthi* (Králíček 1966) auch in Deutschland (Lepidoptera, Sesiidae).– *Nachr. ent. Ver. Apollo (N. F.)* **11**: 61-74.

Kristensen, N. P. (ed.), 1999. Lepidoptera, Moths and Butterflies. Vol. 1: Evolution, Systematics, and Biogeography.– In *Handbook of Zoology, Vol. IV, Arthropoda: Insecta.* 491 pp., Berlin – New York.

Laspeyres, J. H., 1801. *Sesiae Europaeae iconibus et descriptionibus illustratae.* 32 pp. Berolini.

Laštůvka, A. & Laštůvka, Z., 1980. On the bionomics of four European species of Sesiidae (Lepidoptera).– *Acta ent. bohemoslov.* **77**: 424-425.

Laštůvka, Z., 1980. *Chamaesphecia crassicornis* Bartel, 1912 in der ČSSR (Lepidoptera, Sesiidae).– *Scripta Fac. Sci. nat. Univ. purk. brun.* **10**: 457-462.

– 1983a. Two new species of the genus *Chamaesphecia* Spul. (Sesiidae) from Central and South-east Europe.– *Acta Univ. Agric. (Brno), Fac. agron.* **31**(1-2): 199-214.

– 1983b. A contribution to the biology of clearwing moths (Lepidoptera, Sesiidae).– *Acta Univ. Agric. (Brno), Fac. agron.* **31**(1-2): 215-223.

– 1983c. Morphology and biology of clearwing moths *Synanthedon loranthi* (Kr.) and *Synanthedon cephiformis* (O.) (Lepidoptera, Sesiidae).– *Acta Univ. Agric. (Brno), Fac. agron.* **31**(3): 143-158.

– 1983d. On the ultrastructure of the ♀ genitalia of the genus *Chamaesphecia* Spuler (Lepidoptera, Sesiidae).– *Acta Univ. Agric. (Brno), Fac. agron.* **31**(4): 127-132.

– 1984. Generic and tribal positions of *Sesia palariformis* Lederer and *S. fenusaeformis* Lederer (Lepidoptera, Sesiidae).– *Acta ent. bohemosl.* **81**: 380-383.

– 1985. Praeimaginalstadien und Bionomie von *Zenodoxus brosiformis* (Hb.) (Lepidoptera, Sesiidae).– *Acta Univ. Agric. (Brno), Fac. agron.* **33**(1): 163-166.

– 1986. Interesting faunistic records of Lepidoptera from Czechoslovakia.– *Zpr. čs. Spol. ent. ČSAV* **22**: 2-8 (in Czech).

– [1987]. On the taxonomy of *Microsphecia tineiformis* (Esper) and *M. brosiformis* (Hübner)(Lepidoptera, Sesiidae).– *Acta Univ. Agric. (Brno), Fac. agron.* **33**(4)(1985): 183-190.

– 1988. A contribution to faunistics of clearwing moths in Czechoslovakia II (Lepidoptera, Sesiidae).– *Zpr. čs. Spol. ent. ČSAV* **24**: 93-98 (in Czech).

– [1989a]. Zur Taxonomie und Morphologie von *Synansphecia muscaeformis* (Esper) (Lepidoptera, Sesiidae).– *Acta Univ. Agric. (Brno), Fac. agron.* **34**(4)(1986): 177-180.

– 1989b. *Bembecia puella* sp. n. aus der Slowakei (Lepidoptera, Sesiidae).– *Scripta Fac. Sci. nat. Univ. purk. brun.* **19**: 85-92.

– [1990a]. Zur Taxonomie der Gattungen *Chamaesphecia* Spuler, *Synansphecia* Căpuşe und *Dipchasphecia* Căpuşe (Lepidoptera, Sesiidae).– *Acta Univ. Agric. (Brno), Fac. agron.* **36**(1)(1988): 93-103.

– [1990b]. Zur Taxonomie und Verbreitung der europäischen Arten der Gattung *Pyropteron* Newman (Lepidoptera, Sesiidae).– *Acta Univ. Agric. (Brno), Fac. agron.* **36**(1)(1988): 105-111.

– 1990c. Die Glasflügler Ungarns – Faunistik und Bionomie (Lepidoptera, Sesiidae).– *A Janus Pannonius Múzeum Évk. (Pécs)* **34**(1989): 39-46.

– [1990d]. Eine Übersicht der Futterpflanzen der europäischen Glasflügler (Lepidoptera, Sesiidae).– *Acta Univ. Agric. (Brno), Fac. agron.* **37**(1-2)(1989): 153-162.

– [1990e]. Zur Taxonomie von *Synansphecia triannuliformis* (Freyer, 1842) (Lepidoptera, Sesiidae).– *Acta Univ. Agric. (Brno), Fac. agron.* **37**(3-4)(1989): 128-131.

– 1990f. Der Katalog der europäischen Glasflügler (Lepidoptera, Sesiidae).– *Scripta – J. Fac. Sci. Masaryk Univ. Brno* **20**: 461-476.

– [1992a]. Zur Systematik der paläarktischen Gattungen der Tribus Synanthedonini. 1. Morphologie und Klassifikation (Lepidoptera, Sesiidae).– *Acta Univ. Agric. (Brno), Fac. agron.* **38**(1990)(3-4): 221-233.

– 1992b. Zur Systematik der paläarktischen Gattungen der Tribus Synanthedonini. 2. Phylogenese (Lepidoptera, Sesiidae).– *Acta Univ. Agric. (Brno), Fac. agron.* **38**(1990)(3-4): 235-243.

Laštůvka, Z., Bläsius, R., Bartsch, D., Bettag, E., Blum, E., Laštůvka, A., Lingehöle, A., Petersen, M., Riefenstahl, H. & Špatenka, K., 2000. Zur Kenntnis der Glasflügler Spaniens (Lepidoptera: Sesiidae).– *SHILAP Revta. lepid.* **28**: 227-237.

Laštůvka, Z. & Laštůvka, A., 1988. A contribution to the knowledge of clearwing moths (Lepidoptera, Sesiidae) in Hungary.– *Folia ent. hung.* **48**(1987): 97-104.

– 1994. *Bembecia fibigeri* sp. n. aus Spanien (Lepidoptera, Sesiidae).– *Nota lepid.* **16**: 233-239.

Laštůvka, Z., Malicky, H., Hüttinger, E., Rausch, H. & Ressl, F., 1990. Sesien-Funde aus Europa und dem Mediterrangebiet (Lepidoptera, Sesiidae).– *Z. Arb-Gem. öst. Ent.* **41**(1989): 105-110.

Laštůvka, Z. & Špatenka, K., 1984. Morphologie und Bionomie von *Chamaesphecia proximata* (Staudinger, 1891) und *Ch. lanipes* (Lederer, 1863)(Lepidoptera, Sesiidae).– *Acta Univ. Agric. (Brno), Fac. agron.* **32**(2): 111-119.

– 1995. Verzeichnis der französischen Glasflügler-Arten (Lepidoptera, Sesiidae).– *Alexanor* **18**: 475-482.

Le Cerf, F., 1911. Description d'une espèce nouvelle de *Sesia* (Lep., Aegeriidae).– *Bull. Soc. ent. Fr.* **1911**: 244-246.

– 1914a. Description d'une espèce nouvelle du genre *Zenodoxus* Gr. et Rob. (Lep. Aegeriidae).– *Bull. Soc. ent. Fr.* **1914**: 272-275.

– 1914b. Diagnoses sommaries d'espèces et variétés nouvelles d'Aegeriidae paléarctiques.– *Bull. Soc. ent. Fr.* **1914**: 421-424.

– 1916. Aegeriidae de Barbarie. Explication des Planches.– *Étud. Lépid. comp.* **11**: 11-17, pl. 316-322.

– 1917. Contributions a l'étude des Aegeriidae (I). Description et iconographie d'espèces et de formes nouvelles ou peu connues.– *Étud. Lépid. comp.* **14**: 137-388.

– 1920. Contributions a l'étude des Aegeriidae (II). Révision des Aegeriidae de Barbarie.– *Étud. Lépid. comp.* **17**: 181-583.

– 1922. Troisième contribution a l'étude des Aegeriidae: Descriptions d'espèces et variétés nouvelles.– *Étud. Lépid. comp.* **19**(2): 17-39.

Lederer, J., 1853. Versuch die europäischen Lepidopteren in möglichst natürliche Reihenfolge zu stellen.– *Verh. zool.-bot. Ver. Wien* **2**(1852): 14-54, 65-126.

– 1863. Verzeichnis der von Herrn Johann und Frau Ludmilla Haberhauer 1861 und 1862 bei Varna in Bulgarien und Sliwno in Rumelien gesammelten Lepidopteren.– *Wien. ent. Monatschr.* **7**: 17-27, 40-47.

Lepidopterologen-Arbeitsgruppe, 2000: *Schmetterlinge und ihre Lebensräume. Arten. Gefährdung. Schutz. Schweiz und angrenzende Gebiete. Band 3.* Schweizerischer Bund für Naturschutz, 914 pp.

Leraut, P. , 1985. *Chamaesphecia tengyraeformis* (Boisduval, 1840) comb. n. (= *monspeliensis* Staudinger, 1856) syn. n. (Lep. Sesiidae).– *Ent. gall.* **1**: 304-305.

Lewin, W., 1797. Observations respecting some rare British insects.– *Trans. Linn. Soc. London* **3**: 1-4.

Linnaeus, C., 1758. *Systema naturae per regna tria naturae, secundum classes, ordines, genera, species, cum characteribus, differentiis, synonymis, locis.* 10th ed., vol. 1, 824 pp. Holmiae.

– 1761. *Fauna svecica sistens animalia svecicae regnii mammalia, aves, amphibia, pisces, insecta, vermes.* 2nd ed., 578 pp. Holmiae.

Lipthay, B., 1961. Eine neue *Chamaesphecia*-Art (Lepidoptera: Aegeriidae).– *Acta zool. Acad. Sci. hung.* **7**: 213-218.

Löfqvist, E., 1922. Eine neue Aegeriide (Lepid.) aus Finnland.– *Notul. ent.* **2**: 82-84.

Loos, K. & Bittermann, J., 1995. *Synanthedon flaviventris* (Staudinger, 1883) eine für die bayerische Fauna neue Glasflüglerart (Lepidoptera: Sesiidae).– *Beitr. bayer. Entomofaunistik* **1**: 175-178.

Lucas, P. H., 1849. Insectes.– *In* Histoire naturelle des Animaux Articules. *Exploration scientifique de l'Algérie.* Zoologie. Vol. 2, part 3, 527 pp. Paris.

MacKay, M. R., 1968. The North American Aegeriidae (Lepidoptera): A revision based on late-instar larvae.– *Mem. ent. Soc. Can.* **58**: 1-112.

Malicky, H., 1968. Richtigstellung zur Bionomie und Systematik von *Chamaesphecia stelidiformis amygdaloidis* Schleppnik.– *NachrBl. bayer. Ent.* **17**: 96-99.

Mann, J., 1859. Verzeichniss der im Jahre 1858 in Sicilien gesammelten Schmetterlinge.– *Wien. ent. Monatschr.* **3**: 78-106, 161-178.

– 1864. Nachtrag zur Schmetterling-Fauna von Brussa.– *Wien. ent. Monatschr.* **8**: 173-190.

Marek, J., 1962. On the occurrence of some clearwing moth species in Czech Republic and Slovakia (Lep., Sesiidae).– *Čas. čsl. Spol. ent.* **59**: 281-284 (in Czech).

Meyrick, E., 1928. *A revised handbook of British Lepidoptera.* 914 pp. London.

Minet, J., 1991: Tentative reconstruction of the ditrysian phylogeny (Lepidoptera: Glossata).– *Ent. scand.* **22**: 69-95.

Moraal, L. G., 1989. Artificial rearing of the poplar clearwing moth, *Paranthrene tabaniformis*.– *Ent. exp. appl.* **52**: 173-178.

Naumann, C. M., 1971. Untersuchungen zur Systematik und Phylogenese der holarktischen Sesiiden (Insecta, Lepidoptera).– *Bonn. zool. Monogr.* **1**: 1-190.

Naumann, C. M. & Schroeder, D., 1980. Ein weiteres Zwillingsarten-Paar mitteleuropäischer Sesiiden: *Chamaesphecia tenthrediniformis* ([Denis & Schiffermüller], 1775) und *Chamaesphecia empiformis* (Esper, 1783)(Lepidoptera, Sesiidae).– *Z. ArbGem. öst. Ent.* **32**: 29-46.

Newman, E., 1832. Monographia Aegeriarum Angliae.– *Ent. Mag.* **1**: 66-84.

Niculescu, E. V., 1964. Les Aegeriidae: Systematique et phylogenie.– *Linn. belg.* **3**: 34-45.

Oberthür, C., 1872. Catalogue raisonné des Lépidoptères rapportés par M. Théophile Deyrolle de son exploration scientifique en Asie Mineure.– *Rev. Mag. Zool.* **23**: 480-488.

– 1881. Lépidoptères d'Algérie.– *Étud. Ent.* **6**: 41-96.

Ochsenheimer, F., 1808. *Die Schmetterlinge von Europa.* Vol. 2, 256 pp. Leipzig.

– 1816. *Die Schmetterlinge von Europa.* Vol. 4, 226 pp. Leipzig.

Olivier, A., 2000. Christian Friedrich Freyer's „Neuere Beiträge zur Schmetterlingskunde mit Abbildungen nach der Natur": an analysis, with new data on its publication dates (Insecta, Lepidoptera).– *Beitr. Ent.* **50**: 407-486.

Pazsiczky, S., 1941. *Synanthedon mesiaeformis* HS. elöfordulása Somogyban.– *Folia ent. hung.* **6**: 36-37 (in Hungarian).

Petersen, M. & Bartsch, D., 1998. Ein Beitrag zur Sesienfauna Griechenlands (Lepidoptera, Sesiidae).– *Esperiana* **6**: 50-55.

Petersen, M. & Ernst, M., 1991. Zur Verbreitung von *Paranthrene novaki* Toševski 1987 im südlichen Hessen (Lepidoptera, Sesiidae).– *Nachr. ent. Ver. Apollo (N.F.)* **12**: 197-202.

Pinker, R., 1954. Eine interessante Sesie aus Mazedonien.– *Z. Wien. ent. Ges.* **39**: 182-185.

Popescu-Gorj, A., 1962. Noi date pentru cunoasterea lepidopterelor Aegeriidae din fauna R.P.R.– *Commun. Acad. Rep. pop. rom.* **12**: 859-864.

Popescu-Gorj, A. & Căpuşe, I. K., 1965. *Chamaesphecia deltaica* sp. n. (Lepidoptera-Aegeriidae).– *Bull. Annls Soc. R. ent. belg.* **101**: 341-344.

– 1966. [A description of a new species of clearwing moth, *Chamaesphecia djakonovi*, sp. n. (Lepidoptera, Aegeriidae) from the Crimea].– *Ent. Obozr.* **45**: 862-864.

Popescu-Gorj, A., Niculescu, E. V. & Alexinschi, A., 1958. Lepidoptera, Familia Aegeriidae.– *In* Insecta. *Fauna R.P.R.*, 11(1). 195 pp. Bucharest.

Priesner, E., 1993. Pheromontest an einer südbayerischen Population von *Synanthedon soffneri* Špatenka, 1983 (Lepidoptera, Sesiidae).– *NachrBl. bayer. Ent.* **42**: 97-107.

Priesner, E., Ryrholm, N. & Dobler, G., 1989: Der Glasflügler *Synanthedon polaris* (Stgr.) in den Schweizer Hochalpen, nachgewiesen mit Sexualpheromon.– *NachrBl. bayer. Ent.* **38**: 89-97.

Priesner, E. & Špatenka, K., 1990. Pheromonfänge zum Verbreitungsbild von *Pennisetia bohemica* Králíček & Povolný, 1974 (Lepidoptera: Sesiidae) in Mitteleuropa.– *Mitt. schweiz. ent. Ges.* **63**: 87-98.

Prola, C. & Beer, S., 1991. Le Sesiidae della fauna italiana (Lepidoptera).– *Mem. Soc. ent. ital., Genova* **70**: 279-312.

Prola, C. & Beer, S., 1995. I feromoni in lepidotterologie e per la conoscenza delle Sesiidae italiane.– *Mem. Soc. ent. ital.* **73**: 231-271.

Pühringer, F., 1996. Erstnachweis von *Chamaesphecia tenthrediniformis* ([Denis & Schiffermüller] 1775)(Eselswolfsmilchglasflügler) in Oberösterreich (Lepidoptera, Sesiidae).– *Beitr. Naturk. Oberösterreich* **4**: 143-151.

– 1997. Glasflüglernachweise in Österreich (Lepidoptera, Sesiidae).– *Mitt. ent. Arb-Gem. Salzkammergut* **2**: 1-172.

– 1998. Zwei weitere für Oberösterreich neue Glasflügler (Lepidoptera, Sesiidae).– *Beitr. Naturk. Oberösterreich* **6**: 313-318.

Pühringer, F., Ortner, S. & Pöll, N., 1998. Interessante Glasflüglernachweise aus dem Salzkammergut mit zwei für das Bundesland Salzburg neuen Arten und Anmerkungen zur Biologie (Lepidoptera, Sesiidae).– *Beitr. Naturk. Oberösterreich* **6**: 133-138.

Pühringer, F. & Pöll, N., 1999. Das bisher unbekannte Männchen von *Synansphecia kautzi* (Reisser, 1930)(Lepidoptera, Sesiidae).– *Z. ArbGem. öst. Ent.* **51**: 1-7.

Pühringer, F. & Pöll, N., 2001a. Zur Biologie von *Synansphecia kautzi* (Reisser, 1930)(Lepidoptera, Sesiidae).– *Z. ArbGem. öst. Ent.* in prep.

Pühringer, F. & Pöll, N., 2001b. Zur Verbreitung und Biologie von *Synansphecia aistleitneri* Špatenka, 1992 (Lepidoptera, Sesiidae).– *Z. ArbGem. öst. Ent.* in prep.

Pühringer, F., Ryrholm, N. & Dobler, G., 1999. *Synanthedon polaris* (Staudinger 1877), der Polarglasflügler (Lepidoptera, Sesiidae) auch in Südtirol !– *Linzer biol. Beitr.* **31**: 533-539.

Pühringer, F. & Scheuchenpflug, A., 1995. Erstnachweis von *Synanthedon soffneri* Špatenka 1983 (Heckenkirschenglasflügler) in Oberösterreich (Lepidoptera, Sesiidae).– *Z. ArbGem. öst. Ent.* **47**: 65-69.

Ragusa, E., 1922. Le Aegeriidae (Sesiidae) della Sicilia.– *Boll. Lab. Zool. gen. agrar. Portici* **16**: 211-220.

Rambur, J. P., 1832. Catalogue des lépidoptères de l'ile de Corse, avec la description et la figure des espèces inédites.– *Annls Soc. ent. Fr.* **1**: 245-295.

– 1858-[1866]. *Catalogue systematique des lépidoptères de l'Andalusie.* 412 pp. Paris.

Rämisch, F. & Sobczyk, Th., 1998. Aktuelle Verbreitung und Biologie des Glasflüglers *Synanthedon flaviventris* (Staudinger, 1883) in Brandenburg und im angrenzenden Sachsen (Lep., Sesiidae).– *Ent. Nachr. Ber.* **42**: 37-40.

Rebel, H., 1910. Bericht der Sektion für Lepidopterologie.– *Verh. zool.-bot. Ges. Wien* **60**: (4)-(6).

– 1916. Die Lepidopterenfauna Kretas.– *Ann. naturhist. Hofmus. Wien* **30**: 67-172.

– 1917. Eine Lepidopterenausbeute aus dem Amanusgebirge (Alman Dagh).– *Sber. Akad. Wiss., mathem.-naturwiss. Klasse Wien, ser. 1.* **26**: 243-282.

– 1918. Lepidopteren aus Mittelalbanien.– *Z. öst. EntVer.* **3**: 75-77.

– 1927. Beitrag zur Lepidopterenfauna der Insel Cypern.– *Verh. zool.-bot. Ges. Wien* **77**: (58)-(63).

Reisser, H., 1930. Eine neue europäische Sesiide.– *Z. öst. EntVer.* **15**: 101-104.

Rossi, P., 1792-1794. *Mantissa insectorum, exhibens species nuper in Etruria collectas, adjectis faunae Etruscae illustrationibus ad emendationibus.* 148 + 154 pp., 8 pls. Pisa.

Rottemburg, S. A. von, 1775. Anmerkungen zu den hufnagelschen Tabellen der Schmetterlinge.– *Naturforscher* **7**: 105-112.

Rungs, C. E. E., 1972. Lépidoptères nouveaux du Maroc et de la Mauritanie.– *Bull. Mus. natn. Hist. nat.(3), (Zoologie)* **46**: 669-692.

Saglioco, J.- L. & Coupland, J. B., 1995. Biology and host specificity of *Chamaesphecia mysiniformis* (Lepidoptera: Sesiidae), a potential biological control agent of *Marrubium vulgare* (Lamiaceae) in Australia.– *Biocontr. Sci. Technol.* **5**: 509-515.

Saramo, Y., 1973. *Synanthedon mesiaeformis* H. S. (Lep., Aegeriidae) at Kotka and in the eastern archipelago of the Gulf of Finland.– *Annls ent. Fenn.* **39**: 24-28.

Schantz, M., 1959. Studien über *Synanthedon polaris* Stgr. (Lep.).– *Notulae ent.* **39**: 33-42.

Schawerda, K., 1922. Zwölfter Nachtrag zur Lepidopterenfauna Bosniens und der Herzegowina.– *Verh. zool.-bot. Ges. Wien* **71**(1921): 145-170.

Scheuringer, E., 1991. *Paranthrene novaki* Toševski, 1987, eine für Bayern neue Sesie (Lepidoptera, Sesiidae).– *Nachrbl. bayer. Ent.* **40**: 84-86.

Schintlmeister, A. & Rämisch, F., 1986. Die Sesiidae der Dresdner Gegend (Lepidoptera).– *Ent. Nachr. Ber.* **30**: 65-68.

Schleppnik, A., 1933. *Chamaesphecia stelidiformis* Frr. f. n. *amygdaloidis*.– *Z. öst. EntVer.* **18**: 24-25.

– 1936. VI. Versammlungen der Sektion für Lepidopterologie.– *Verh. zool.-bot. Ges. Wien* **85**(1935): 129.

Schnaider, J., [1939]. *Paranthrene polonica* sp. n.– *Polskie Pismo ent.* **16/17**: 140-143.

Schnaider, J., Schnaider, J. & Schnaider, Z., 1961. Przezierniki – Aegeriidae. *Klucze do oznaczania owadów Polski.* Vol. 37, 42 pp. Warszawa (in Polish).

Schwarz, R., 1953: *Motýli* [*Lepidoptera*] *3.* 158 pp., 48 pls. Praha (in Czech).

Schwarz, R. & Tolman, V., 1961. *Bembecia pectinata* Stgr. et ses premiers états (Lepidoptera, Sesiidae).– *Acta Soc. ent. Čechosl.* **58**: 284-287.

Schwingenschuss, L., 1938. Sechster Beitrag zur Lepidopterenfauna Inner-Anatoliens.– *Ent. Rdsch.* **55**: 141-147, 157-164, 173-177, 181-184, 199-202, 223-226, 299-300, 337-340, 411-412, 454-457, 532, 700.

Scopoli, J. A., 1763. *Entomologica carniolica, exhibens insecta carniolia indigena et distributa in ordines, genera, species, varietates, methodo Linneana.* 422 pp., 43 pls. Vindobonae.

– 1777. *Introductio ad Historiam naturalem, sistens Genera Lapidum, Plantarum et Naturalium.* 506 pp. Pragae.

Sheljuzhko, L., 1918. Diagnoses lepidopterorum novorum sibirae.– *Neue Beitr. syst. Insektenk.* **1**: 104.

– 1924. Zwei neue palaearktische Aegeriiden-Arten.– *Dt. ent. Z. Iris* **38**: 183-185.

Silbernagel, A., 1943. Die ostasiatische Art *Bembecia pectinata* Stgr. hat auch in Europa das Heimatrecht.– *Z. wien. ent. Ges.* **28**: 145-148.

Sobczyk, Th., 1995. Kommentiertes Verzeichnis der Glasflügler (Lep., Sesiidae) des Freistaates Sachsen.– *Mitt. sächs. Ent.* **31**: 22-26.

– 1996. *Synanthedon loranthi* (Králíček, 1966) in Ostdeutschland (Lep., Sesiidae).– *Ent. Nachr. Ber.* **40**: 49-51.

– 1997. *Synanthedon flaviventris* (Staudinger, 1883) in Sachsen (Lepidoptera, Sesiidae).– *Abh. Ber. Naturk-Mus. Görlitz* **69**: 17-25.

Sobcyzk, Th. & Rämisch, F., 1997. Zur Faunistik und Ökologie der Schmetterlinge der Mark Brandenburg. VIII. Ausbreitung und ökologische Ansprüche von *Synansphecia triannuliformis* (Freyer, 1845) in der Mark Brandenburg und den angrenzenden Bundesländern (Lep., Sesiidae).– *Ent. Nachr. Ber.* **41**: 33-37.

Špatenka, K., 1983. *Synanthedon soffneri* sp. n. (Lepidoptera, Sesiidae) aus der Tschechoslowakei.– *Acta ent. bohemoslov.* **80**: 297-303.

– 1987. Fünf neue paläarktische Glasflügler (Lepidoptera, Sesiidae).– *Z. ArbGem. öst. Ent.* **39**: 12-26.

– 1992a. Weitere neue paläarktische Sesiiden.– *Alexanor* **17**: 427-446.

– 1992b. Contribution à la stabilisation de la taxinomie des Sésiides paléarctiques (Lepidoptera, Sesiidae).– *Alexanor* **17**: 479-503.

– 1997. Neue Glasflügler-Arten und Unterarten aus Europa und der Türkei (Sesiidae, Lepidoptera).– *Bonn. zool. Beitr.* **47**(1996): 43-57.

– 2001. Neue paläarktische Glasflügler-Arten (Lepidoptera: Sesiidae).– *Ent. Z.* in press.

Špatenka, K. & Gorbunov, O., 1992. Vier neue paläarktische Glasflügler (Sesiidae, Lepidoptera).– *Entomofauna* **13**: 377-393.

Špatenka, K., Gorbunov, O., Laštůvka, Z., Toševski, I. & Arita, Y. 1996. Die Futterpflanzen der paläarktischen Glasflügler (Lepidoptera: Sesiidae).– *Nachr. ent. Ver. Apollo (N. F.)* **17**: 1-20.

Špatenka, K., Gorbunov, O., Laštůvka, Z., Toševski, I. & Arita, Y. 1999. Sesiidae – Clearwing moths.– *In* C. M. Naumann (ed.), *Handbook of Palaearctic Macrolepidoptera.* Vol. 1, 569 pp., 57 pls., Wallingford.

Špatenka, K. & Laštůvka, Z., 1983. *Bembecia uroceriformis* (Treitschke, 1834) und *Chamaesphecia sevenari* (Lipthay, 1961) (Lepidoptera, Sesiidae) in der ČSSR.– *Acta Univ. Agric. (Brno), Fac. agron.* **31**(3): 159-165.

– 1988. Typen der Glasflügler aus der Staudinger- und Püngeler Sammlung im Zoologischen Museum Berlin (Lepidoptera, Sesiidae).– *Dt. ent. Z. (N. F.)* **35**: 331-339.

– 1990: Zur Taxonomie von *Bembecia scopigera* (Scopoli, 1763), *B. ichneumoniformis* ([Denis & Schiffermüller], 1775) und *B. albanensis* (Rebel, 1918) (Lepidoptera, Sesiidae).– *Entomofauna* **11**: 109-121.

– 1997. Zur Verbreitung und Variabilität von *Paranthrene insolita* Le Cerf, 1914 (Lepidoptera, Sesiidae).– *Nachr. ent. Ver. Apollo (N. F.)* **18**: 13-21.

Špatenka, K., Laštůvka, Z., Gorbunov, O., Toševski, I. & Arita, Y., 1993. Die Systematik und Synonymie der paläarktischen Glasflügler-Arten (Lepidoptera, Sesiidae).– *Nachr. ent. Ver. Apollo (N.F.)* **14**: 81-114.

Špatenka, K. & Tesař, F., 1980. Sesiidae of southern Czech Republic.– *Acta Sci. nat. Mus. Bohem. merid., České Budějovice* **20**: 83-90 (in Czech, German summary).

Spuler, A., 1910. *Die Schmetterlinge Europas*. Vol. 2, 524 pp., pls 56-91. Stuttgart.

Stadie, D., 1995. Lebesweise und Verbreitung des Kreuzdornglasflüglers *Synanthedon stomoxiformis* (Hübner, 1790) in Thüringen und Sachsen-Anhalt (Lep., Sesiidae).– *Ent. Nachr. Ber.* **39**: 219-223.

– 1997. Erster sicherer Nachweis von *Chamaesphecia tenthrediniformis* ([Denis & Schiff.], 1775) in Sachsen-Anhalt (Lep., Sesiidae).– *Ent. Nachr. Ber.* **41**: 37-38.

– 1998. *Chamaesphecia dumonti* (Le Cerf, 1922) – ein neuer Glasflügler für die Fauna Thüringens und Sachsen-Anhalt (Lep., Sesiidae).– *Ent. Nachr. Ber.* **42**: 167-169.

Staudinger, O., 1854. *De Sesiis agri berolinensis*. 66 pp., 1 table, 2 pls. Berlin.

– 1856. Beitrag zur Feststellung der bisher bekannten Sesien-Arten Europas und des angrenzenden Asien's.– *Stettin ent. Ztg* **17**: 193-224, 257-288, 323-338.

– 1866. Drei neue Sesien und Berichtigung über einige ältere Arten.– *Stettin ent. Ztg* **27**: 50-55.

– 1870. Beitrag zur Lepidopteren Fauna Griechenlands.– *Hor. Soc. ent. Ross.* **7**: 3-304.

– 1877. Neue Lepidopteren des europäischen Faunengebietes aus meiner Sammlung.– *Stettin ent. Ztg* **38**: 175-208.

– 1883. Einige neue Lepidopteren Europa's.– *Stettin ent. Ztg* **44**: 177-186.

– 1891. Neue Arten und Varietäten von Lepidopteren des paläarktischen Faunengebietes.– *Dt. ent. Z. Iris* **4**: 224-339.

– 1894. Neue Lepidopteren-Arten und Varietäten aus dem paläarktischen Faunengebietes.– *Dt. ent. Z. Iris* **7**: 241-296.

– 1895. Neue paläarktischen Lepidopteren.– *Dt. ent. Z. Iris* **8**: 288-366.

Steffny, H., 1985. Zur Biologie und Mimikry der Sesien unter besonderer Berücksichtigung der Ökologie und Verbreitung des Goldwespenglasflüglers in Rheinland-Pfalz (*Bembecia chrysidiformis* Esper, 1782, Sesiidae, Lepidoptera).– *Dendrocopus* **12**: 118-129.

– 1990. Ein Beitrag zur Faunistik und Ökologie der Glasflügler Südbadens (Lep., Sesiidae).– *Melanargia* **2**: 32-57.

Stephens, J. F., 1828. Haustellata.– *Illustrations of British Entomology* **1**: 57-152, pls 10-12. London.

Sterzl, O., 1967. Prodromus der Lepidopterenfauna von Niederösterreich.– *Verh. zool.-bot. Ges. Wien* **107**: 75-193.

Tomala, F., 1901. *Sesia empiformis* Esp. var. *hungarica* n. var.– *Rovart. Lapok* **8**: 47-50 (in Hungarian).

– 1913. Adatok a *Synanthedon flaviventris* Stgr. és a *Paranthrene tabaniformis* Rott. var. *rhingiaeformis* Hbn. életmódjának ismeretéhez és azok magyar honossága.– *Rovart. lapok* **20**: 196-197 (in Hungarian).

Torstenius, S. & Lindmark, H., 2000. *Synanthedon andrenaeformis* (Laspeyres 1801), Lepidoptera: Sesiidae, a clearwing moth new to Sweden.– *Ent. Tidskr.* **121**: 21-22 (in Swedish).

Toševski, I., 1986a. Morphology and bionomics of *Chamaesphecia schmidtiiformis* (Freyer, 1836) and *Chamaesphecia doleriformis* (Herrich-Schäffer, 1846) and bionomics of *Bembecia* (*Synansphecia*) *leucomelaena* (Zeller, 1847)(Lepidoptera, Sesiidae).– *Fragm. balc. Mus. maced. Sci. nat.* **12**: 179-189.

– 1986b. A new species of clear-wing moth from the Balkan Peninsula (Lepidoptera, Sesiidae).– *Fragm. balc. Mus. maced. Sci. nat.* **12**: 191-196.

– 1987. A supplement to the present knowledge of the genus *Paranthrene* Hübner, 1819 on the territory of Yugoslavia (Lepidoptera, Sesiidae).– *Acta Mus. Maced. Sci. nat.* **18**: 177-195.

– 1989. A new species of the genus *Bembecia* Hübner, 1819 from Macedonia (Lepidoptera, Sesiidae).– *Fragm. balc. Mus. maced. Sci. nat.* **14**: 81-89.

– 1991. A new species of clearwing moths from Greece: *Bembecia fokidensis* spec. nov. (Lepidoptera, Sesiidae).– *Atalanta, Würzburg* **22**: 169-172.

– 1992. *Bembecia pavicevici* Toševski, 1989 – bionomics and female description (Lepidoptera, Sesiidae).– *Zaštita Bilja* **43**: 293-298.

Treitschke, F., 1834. *Die Schmetterlinge von Europa.* Vol. 10, suppl., part 1, 210 pp. Leipzig.

Troukens, W., 1979. Onze inheemse Sesiidae of Wespvlinders.– *Atalanta, Würzburg* **7**: 33-43.

Tuttin, T. G., Heywood, V. H., Burges, N. A., Moore, D. M., Valentine, D. H., Walters, S. M. & Webb, D. A. (eds), 1964-1980. *Flora Europaea.* 5 vols. Cambridge.

Walker, F., 1856. *List of the specimens of lepidopterous insects in the collection of the British Museum.* Vol. 8, 271 pp. London.

– [1865]. *List of the specimens of lepidopterous insects in the collection of the British Museum.* Vol. 31, 321 pp. London.

Weitzel, M. 1993. Funde von *Chamaesphecia nigrifrons* (Le Cerf, 1911), *Synanthedon spuleri* (Fuchs, 1908) und *Bembecia albanensis* (Rebel, 1918) im Trierer Raum (Lep., Sesiidae).– *Melanargia* **5**: 13-15.

– 1994. Einige Nachweise von *Chamaesphecia tenthrediniformis* (Den. & Schiff.) im Moselgebiet (Insecta, Sesiidae).– *Dendrocopus* **21**: 190-191.

Westwood, J. O., 1840. *Introduction to the modern classification of Insects.* Vol. 2, 587 pp. *Synopsis of the genera of British Insects.* 158 pp. London.

Wichra, J., 1966. *Chamaesphecia colpiformis* (Staudinger)(Lep., Sesiidae) in Czechoslovakia.– *Zpr. čs. Spol. ent. ČSAV* **2**: 16-17 (in Czech).

Zeller, Ph. C., 1847a. Verzeichniss der vom Professor Dr. Loew in der Türkei und in Asien gesammelten Lepidopteren.– *Isis, Leipzig* **40**: 3-39.

– 1847b. Bemerkungen über die auf einer Reise nach Italien und Sicilien beobachteten Schmetterlingsarten.– *Isis, Leipzig* **40**: 121-914.

Zuccherelli, G., 1969. Un insetto dannoso al kaki. Osservazioni su "*Synanthedon tipuliformis* Clerck" e sui danni in Emilia-Romagna al *Diospyros kaki* L.– *Frutticoltura* **31**: 235-238.

Zukowsky, B., 1915. Aphoristische Skizze über die bisher bekannt gewordenen Futterpflanzen der paläarktischen Aegeriidae.– *Int. ent. Z. (Guben)* **9**: 77-79.

– 1929. Neue europäische Aegeriidae.– *Int. ent. Z. (Guben)* **23**: 20-22.

– 1935. Neue paläarktische Aegeriidae.– *Int. ent. Z. (Guben)* **29**: 39-41.

Index to Subfamily, Tribe and Genus Names

The numbers refer to page numbers. Synonyms are in *italics*.

Index to Species Names

The numbers refer to the serial number of a species. Synonyms, subspecies and forms are in *italics*.

Index to Hostplants

The numbers refer to the serial number of a species.

Printed in the United States
by Baker & Taylor Publisher Services

Printed in the United States
by Baker & Taylor Publisher Services